ADVANCES IN MECHATRONICS

Edited by **Horacio Martínez-Alfaro**

Advances in Mechatronics
Edited by Horacio Martínez-Alfaro

CBS, Edition 2016

Published by InTech
Janeza Trdine 9, 51000 Rijeka, Croatia

Publishing Process Manager Mia Devic

Technical Editor Teodora Smiljanic

Cover Designer InTech Design Team

Image Copyright Tonis Pan, 2010. Used under license from Shutterstock.com

Additional hard copies can be obtained from orders@intechopen.com

Advances in Mechatronics, Edited by Horacio Martínez-Alfaro
p. cm.
ISBN 978-953-307-373-6

Contents

Preface

The community of researchers claiming the relevance of their work to the field of mechatronics is growing faster and faster, despite the fact that the term itself has been in the scientific community for more than 40 years. Numerous books have been published specializing in any one of the well known areas that comprised it: mechanical engineering, electronic control and systems, but attempts to bring them together as a synergistic integrated areas are scarce. Yet some common application areas clearly appear since then.

The goal of this book is to collect state-of-the-art contributions that discuss recent developments that show more more synergistic integration among the areas. The book is divided in three sections with out and specific special order. The first section is about Automatic Control and Artificial Intelligence with five chapters, the second section is Robotics and Vision with six chapters, and the third section is Other Applications and Theory with two chapters.

The first chapter on Automatic Control and Artificial Intelligence by Wuwei Chen, Hansong Xiao, Liqiang Liu, Jean W. Zu, and HuiHui Zhou is some theory and experiments of integrated control vehicle dynamics. The second chapter by Wahidah Mansor, Saifulrizal Ab Rani, and Nurfatehah Wahi is about integrating neural signal and embedded system for controlling a small motor. Ismaila B. Tijani, Akmeliawati Rini, and Jimoh E. Salami Momoh in the third chapter shows an artificial intelligent based friction modelling and compensation for motion control system. The fourth chapter by Corneliu Cristescu, Petrin Drumea, Dragos Ion Guta, and Catalin Dumitrescu is about a mechatronic systems for kinetic energy recovery at the braking of motor vehicles. The fifth chapter and last of this section by Chin-Yin Chen, I-Ming Chen, and Chi-Cheng Cheng is about integrated mechatronic design for servo-mechanical systems.

For the Robotics and Vision section, the first chapter is on the design of underactuated finger mechanisms for robotic hands by Pierluigi Rea. The following chapter by Akhtar Khurshid deals with robotic grasping and fine manipulation using soft fingertip. In the next chapter, Chiharu Ishii talks about recognition of finger motions for myoelectric prosthetic hand via surface EMG. Yanfei Liu and Carlos Pomalaza-Ráez in the following chapter talks about self-landmarking for robotics applications. The next chapter is about robotic waveguide by free space optics by Koichi Yoshida, Kuniaki Tanaka, and Takeshi Tsujimura. And the last chapter for this section by Kun Mo and Zhoupin Yin is about surface reconstruction of defective point clouds based on dual off-set gradient functions.

For the Other Applications and Theory section, the first chapter by Angela Elia, Cinzia Di Franco, Adeel Afzal, Nicola Cioffi and Luisa Torsi is about advanced NOx sensors for mechatronic applications. The last chapter but not the least by Ioan G.Pop and Vistrian Măties is about a transdisciplinary approach of the mechatronics in the knowledge based society.

I do hope you will find the book interesting and thought provoking. Enjoy!

Horacio Martínez-Alfaro
Mechatronics and Automation Department,
Tecnológico de Monterrey, Monterrey,
México
July 2011

Part 1

Automatic Control and Artificial Intelligence

Integrated Control of Vehicle System Dynamics: Theory and Experiment

Wuwei Chen[1], Hansong Xiao[2], Liqiang Liu[1],
Jean W. Zu[2] and HuiHui Zhou[1]
[1]Hefei University of Technology,
[2]University of Toronto,
P. R. China
Canada

1. Introduction

Modern motor vehicles are increasingly using active chassis control systems to replace traditional mechanical systems in order to improve vehicle handling, stability, and comfort. These chassis control systems can be classified into the three categories, according to their motion control of vehicle dynamics in the three directions, i.e. vertical, lateral, and longitudinal directions: 1) suspension, e.g. active suspension system (ASS) and active body control (ABC); 2) steering, e.g. electric power steering system (EPS) and active front steering (AFS), and active four-wheel steering control (4WS); 3) traction/braking, e.g. anti-lock brake system (ABS), electronic stability program (ESP), and traction control (TRC). These control systems are generally designed by different suppliers with different technologies and components to accomplish certain control objectives or functionalities. Especially when equipped into vehicles, the control systems often operate independently and thus result in a parallel vehicle control architecture. Two major problems arise in such a parallel vehicle control architecture. First, system complexity in physical meaning comes out to be a prominent challenge to overcome since the amount of both hardware and software increases dramatically. Second, interactions and performance conflicts among the control systems occur inevitably because the vehicle motions in vertical, lateral, and longitudinal directions are coupled in nature. To overcome the problems, an approach called integrated vehicle dynamics control was proposed around the 1990s (Fruechte et al., 1989). Integrated vehicle dynamics control system is an advanced system that coordinates all the chassis control systems and components to improve the overall vehicle performance including safety, comfort, and economy.

Integrated vehicle dynamics control has been an important research topic in the area of vehicle dynamics and control over the past two decades. Comprehensive reviews on this research area may refer to (Gordon et al., 2003; Yu et al., 2008). The aim of integrated vehicle control is to improve the overall vehicle performance through creating synergies in the use of sensor information, hardware, and control strategies. A number of control techniques have been designed to achieve the goal of functional integration of the chassis control systems. These control techniques can be classified into two categories, as suggested by (Gordon et al., 2003): 1) multivariable control; and 2) hierarchical control. Most control techniques used in the previous studies fall into the first category. Examples include nonlinear predictive control (Falcone et al., 2007), random sub-optimal control (Chen et al., 2006), robust H_∞ (Hirano et al., 1993), sliding mode (Li et al., 2008), and artificial neural networks (Nwagboso et al., 2002), etc. In contrast, hierarchical control has not yet been applied extensively to integrated vehicle control system. It is indicated by the relatively small volume of research publications (Gordon et al., 2003; Gordon, 1996; Rodic and Vukobratovie, 2000; Karbalaei et al., 2007; He et al., 2006; Chang and Gordon, 2007; Trächtler, 2004). In the studies, there are two types of hierarchical control architecture: two-layer architecture (Gordon et al., 2003; Gordon, 1996; Rodic and Vukobratovie, 2000; Karbalaei et al., 2007; He et al., 2006) and three-layer architecture (Chang and Gordon, 2007; Trächtler, 2004). For instance in (Chang and Gordon, 2007), a three-layer model-based hierarchical control structure was proposed to achieve modular design of the control systems: an upper layer for reference vehicle motions, an intermediate layer for actuator apportionment, and a lower layer for stand-alone actuator control.

In the review of the past studies on integrated vehicle dynamics control, we address the following two aspects in this study. First, hierarchical control has been identified as the more effective control technique compared to multivariable control. In addition to improving the overall vehicle performance including safety, comfort, and economy, application of hierarchical control brings a number of benefits, among which: 1) facilitating the modular design of chassis control systems; 2) mastering complexity by masking the details of the individual chassis control system at the lower layer; 3) favoring scalability; and 4) speeding up development processes and reducing costs by sharing hardware (e.g. sensors). Second, most of the research activities on this area were focused solely on simulation investigations. There have been very few attempts to conduct experimental study to verify the effectiveness of those proposed integrated vehicle control systems. However, the experimental verification is an essential stage in developing those integrated vehicle control systems in order to transfer them from R&D activities to series production.

In this chapter, a comprehensive and intensive study on integrated vehicle dynamics control is performed. The study consists of three investigations: First, a multivariable control technique called stochastic sub-optimal control is applied to integrated control of electric power steering system (EPS) and active suspension system (ASS). A simulation investigation is performed and comparisons are made to demonstrate the advantages of the proposed integrated control system over the parallel control system. Second, a two-layer hierarchical control architecture is proposed for integrated control of active suspension system (ASS) and electronic stability program (ESP). The upper layer controller is designed to coordinate the interactions between the ASS and the ESP. A simulation investigation is conducted to demonstrate the effectiveness of the proposed hierarchical control system in improving vehicle overall performance over the non-integrated control system. Finally, a hardware-in-the-loop (HIL) experimental investigation is performed to verify the simulation results.

2. System model

In this study, two types of vehicle dynamic model are established: a non-linear vehicle dynamic model developed for simulating the vehicle dynamics, and a linear 2-DOF reference model used for designing controllers and calculating the desired responses to driver's steering input.

2.1 Vehicle dynamic model

A vehicle dynamic model is established and the three typical vehicle rotational motions, including yaw motion, pitch motion, and roll motion, are considered. They are illustrated in Fig. 1(a), Fig. 1(b), and Fig. 1(c), respectively. In the figures, we denote the front-right wheel, front-left wheel, rear-right wheel, and rear-left wheel as wheel 1, 2, 3, and 4, respectively. The equations of motion can be derived as:

For yaw motion of sprung mass shown in Fig. 1(a)

$$I_z \dot{\omega}_z - I_{xz} \ddot{\phi} = a(F_{y1} + F_{y2}) - b(F_{y3} + F_{y4}) \tag{1}$$

And the equations of motion in the longitudinal direction and the lateral direction can be written as

$$m(\dot{v}_x - v_y \omega_z) - m_s h \dot{\omega}_z \phi = F_{x1} + F_{x2} + F_{x3} + F_{x4} - f_r mg \tag{2}$$

$$m(\dot{v}_y + v_x \omega_z) + m_s h \ddot{\phi} = F_{y1} + F_{y2} + F_{y3} + F_{y4} \tag{3}$$

For pitch motion of sprung mass shown in Fig. 1(b)

$$I_y \ddot{\theta} = b(F_{z3} + F_{z4}) - a(F_{z1} + F_{z2}) \tag{4}$$

And for roll motion of sprung mass shown in Fig. 1(c)

$$I_x \ddot{\phi} + m_s(\dot{v}_y + \dot{v}_x \omega_z)h - I_{xz} \dot{\omega}_z = m_s g h \phi + (F_{z2} + F_{z3} - F_{z1} - F_{z4})d \tag{5}$$

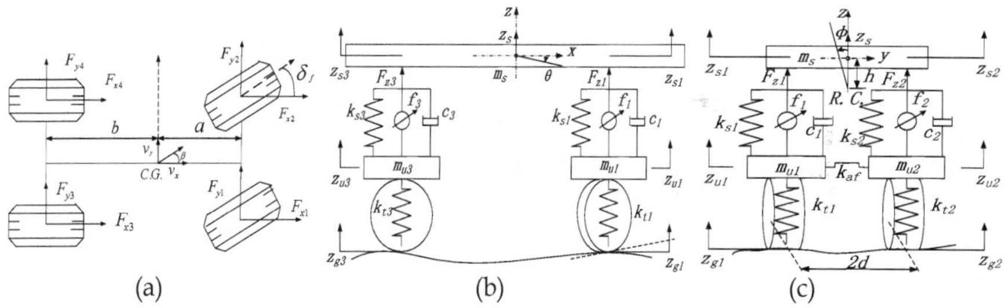

Fig. 1. Three typical vehicle rotational motions: (a) yaw motion; (b) pitch motion; (c) roll motion.

We also have the equations for the vertical motions of sprung mass and unsprung mass

$$m_s \ddot{z}_s = F_{z1} + F_{z2} + F_{z3} + F_{z4} \tag{6}$$

$$m_{ui} \ddot{z}_{ui} = k_{ti}(z_{gi} - z_{ui}) - F_{zi} \qquad (i=1,2,3,4) \tag{7}$$

where

$$F_{z1} = k_{s1}(z_{u1} - z_{s1}) + c_1(\dot{z}_{u1} - \dot{z}_{s1}) - \frac{k_{af}}{2d}[\phi - \frac{(z_{u2} - z_{u1})}{2d}] + f_1 \tag{8}$$

$$F_{z2} = k_{s2}(z_{u2} - z_{s2}) + c_2(\dot{z}_{u2} - \dot{z}_{s2}) + \frac{k_{af}}{2d}[\phi - \frac{(z_{u2} - z_{u1})}{2d}] + f_2 \tag{9}$$

$$F_{z3} = k_{s3}(z_{u3} - z_{s3}) + c_3(\dot{z}_{u3} - \dot{z}_{s3}) + \frac{k_{ar}}{2d}[\phi - \frac{(z_{u3} - z_{u4})}{2d}] + f_3 \tag{10}$$

$$F_{z4} = k_{s4}(z_{u4} - z_{s4}) + c_4(\dot{z}_{u4} - \dot{z}_{s4}) - \frac{k_{ar}}{2d}[\phi - \frac{(z_{u3} - z_{u4})}{2d}] + f_4 \tag{11}$$

When the pitch angle of sprung mass θ and the roll angle of sprung mass ϕ are small, the following approximation can be reached

$$z_{s1} = z_s - a\theta - d\phi \tag{12}$$

$$z_{s2} = z_s - a\theta + d\phi \tag{13}$$

$$z_{s3} = z_s + b\theta + d\phi \tag{14}$$

$$z_{s4} = z_s + b\theta - d\phi \tag{15}$$

Considering the rotational dynamics of the wheel of the vehicle shown in Fig. 2, the equation of motion is derived as

$$I_w \dot{\omega}_i = -F_{xwi}R_w + T_i \qquad (i = 1,...4) \tag{16}$$

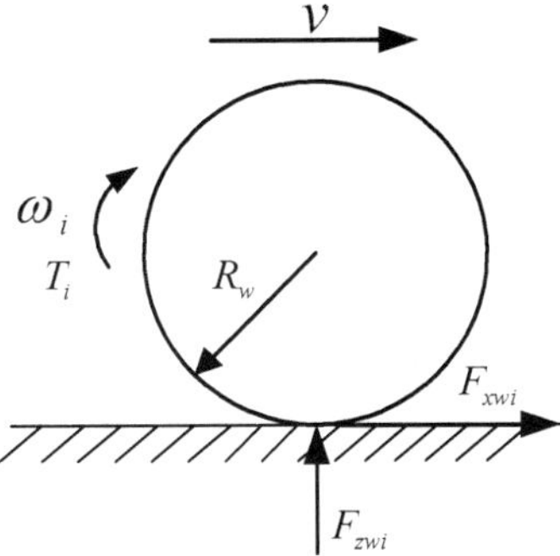

Fig. 2 Wheel dynamic model.

It is noted that the longitudinal and lateral forces acting on the i-th wheel, F_{xi} and F_{yi}, have the following relationships with the tyre forces along the wheel axes, F_{xwi} and F_{ywi}, because of the steering angle of the i-th wheel δ_i,

$$\begin{bmatrix} F_{xi} \\ F_{yi} \end{bmatrix} = \begin{bmatrix} \cos\delta_i & \sin\delta_i \\ \sin\delta_i & -\cos\delta_i \end{bmatrix} \begin{bmatrix} F_{xwi} \\ F_{ywi} \end{bmatrix} \qquad (i = 1,...,4) \tag{17}$$

For simplicity, the steering angles are assumed as: $\delta_1 = \delta_2 = \delta_f$, and $\delta_3 = \delta_4 = \delta_r$.

It is worthy to mention that: 1) for the above-mentioned first investigation, both the ASS controller and EPS controller are designed respectively. Eq. 4 through Eq. 15 are used to develop the ASS controller, while the other equations are employed to design the EPS controller; 2) for the second investigation, the same set of equations, i.e. Eq. 4 through Eq. 15, is used to design the ASS controller. While for the ESP controller, the yaw motion of sprung mass described in Eq. 1 is replaced by the following equations of motion.

For yaw motion of sprung mass

$$I_z \dot{\omega}_z - I_{xz} \ddot{\phi} = a(F_{y1} + F_{y2}) - b(F_{y3} + F_{y4}) + M_{zc} \tag{18}$$

where M_{zc} is the corrective yaw moment generated by the ESP controller, which is given as

$$M_{zc} = d(F_{x1} + F_{x3} - F_{x2} - F_{x4}) \tag{19}$$

2.2 EPS model

The major components of a rack-pinion EPS as shown in Fig. 3 consist of a torque sensor, a control unit (ECU), a motor, and a gear assist mechanism. The torque sensor measures the torque from the steering wheel and sends a signal to the ECU. The ECU also receives steering position signal from a position sensor and the vehicle speed signal. These signals are processed in the ECU and an assist command is generated. The command is in turn given to the motor, which provides the torque to the gear assist mechanism. The torque is amplified by the gear mechanism and the amplified torque is applied to the steering column, which is connected to the rack-pinion mechanism.

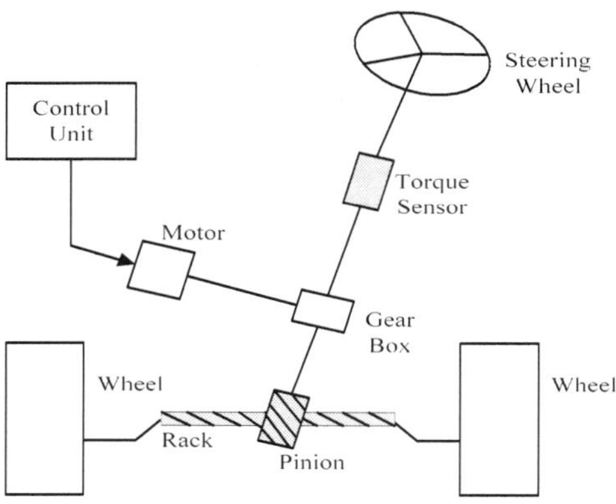

Fig. 3. EPS system.

The following governing equations for the pinion can be obtained by applying force analysis to the pinion

$$I_p \ddot{\delta}_1 = T_m + T_c - T_r - c_e \dot{\delta}_1 \tag{20}$$

where T_c is the torque applied on the steering wheel, which can be calculated by

$$T_c = k_s(\theta_h - \delta_1) \tag{21}$$

Let the speed reduction ratio of the rack-pinion mechanism be N_2, we have

$$\delta_1 = N_2\delta_f \tag{22}$$

2.3 Tyre model

The Pacejka nonlinear tyre model (Bakker et al., 1987; Pacejka, 2002) is used to determine the dynamic forces of each tyre i. The inputs of the tyre model include the vertical tyre force, tyre sideslip angle and tyre slip ratio; and the outputs include the longitudinal tyre force F_{xwi}, lateral tyre force F_{ywi} and self-aligning torque T_{zwi}. The Pacejka's magic formula is presented as

$$F_{xwi} = -(\sigma_x / \sigma)F_{x0} \tag{23}$$

$$F_{ywi} = -(\sigma_y / \sigma)F_{y0} \tag{24}$$

$$T_{zwi} = D_z \sin\left[C_z \tan^{-1}(B_z\phi_z)\right] \tag{25}$$

where T_{zwi} is the aligning torque acting on the tyre; and

$$F_{x0} = D_x \sin\left[C_x \tan^{-1}(B_x\phi_x)\right] \tag{26}$$

$$F_{y0} = D_y \sin\left[C_y \tan^{-1}(B_y\phi_y)\right] \tag{27}$$

$$\sigma = \sqrt{\sigma_x{}^2 + \sigma_y{}^2} \,, \ \sigma_x = -\lambda / (1+\lambda), \ \sigma_y = -\tan\alpha / (1+\lambda) \tag{28}$$

where the coefficients depend on the tyre characteristics and road conditions, the physical definitions of these coefficients can be found in the references (Bakker et al., 1987; Pacejka, 2002).

2.4 Road excitation model

A filtered white noise signal (Yu and Crolla, 1998) is selected as the road excitation to the vehicle, which can be expressed as

$$\dot{z}_{gi} = -2\pi f_0 z_{gi} + 2\pi w_i\sqrt{G_0 v} \qquad (i = 1,...,4) \tag{29}$$

2.5 2-DOF vehicle rreference model

A 2-DOF linear bicycle model is used as the vehicle reference model to generate the desired vehicle states in this study since the 2-DOF model reflects the desired relationship between the driver's steer input and the vehicle yaw rate. This model is employed for both the upper layer controller design and the ESP controller design later in the paper. The equations of motion are expressed as follows by assuming a small sideslip angle and a constant forward speed.

$$m(\dot{v}_y + v_x\omega_z) = C_f(\beta + \frac{a\omega_z}{v_x} - \delta_f) + C_r(\beta - \frac{b\omega_z}{v_x})$$ (30)

$$I_z\dot{\omega}_z = aC_f(\beta + \frac{a\omega_z}{v_x} - \delta_f) - bC_r(\beta - \frac{b\omega_z}{v_x}) + M_{zc}$$ (31)

3. Investigation 1: Multivariable control

As mentioned earlier in the chapter, the first investigation addresses the coupling effects between dynamics of the steering system and the suspension system. With this in mind, a full-car dynamic model that integrates EPS and ASS is established. Then based on the integrated model, a multivariable control method called stochastic sub-optimal control strategy based on output feedback is applied to coordinate the control of both EPS and ASS.

3.1 State space formulation

For further analysis, it is convenient to formulate the full car dynamic model in state space form by combining the dynamic models for the sub-systems that we developed earlier in Section 2. Firstly, the state variables are defined as

$$X = \begin{bmatrix} \dot{\delta} & \delta & \beta & \omega_z & \dot{\theta} & \theta & \dot{z}_{u1} & \dot{z}_{u2} & \dot{z}_{u3} & \dot{z}_{u4} z_{u1} & z_{u2} & z_{u3} & z_{u4} & \dot{\phi} & \phi & \dot{z}_s & z_s & z_{g1} & z_{g2} & z_{g3} & z_{g4} \end{bmatrix}^T$$ (32)

and the output variables are chosen as

$$Y = \begin{bmatrix} \delta & T_c & \beta & \omega_z & \dot{\phi} & \dot{z}_s & \dot{\theta} & z_{u1}-z_{s1} & z_{u2}-z_{s2} & z_{u3}-z_{s3} & z_{u4}-z_{s4} & k_{t1}(z_{g1}-z_{u1}) & k_{t2}(z_{g2}-z_{u2}) & k_{t3}(z_{g3}-z_{u3}) & k_{t4}(z_{g4}-z_{u4}) \end{bmatrix}^T$$ (33)

where $z_{ui} - z_{si}$ represents the suspension dynamic deflection at wheel i, and $k_{ti}(z_{gi} - z_{ui})$ represents the tyre dynamic load at wheel i. Therefore the state equation and output equation can be written as

$$\begin{cases} \dot{X}(t) = AX(t) + B_1U(t) + B_2U_2(t) + B_3W(t) \\ Y(t) = CX(t) \end{cases}$$ (34)

where $U(t)$ is the control input vector, and $U(t) = [T_m(t) \quad f_1(t) \quad f_2(t) \quad f_3(t) \quad f_4(t)]^T$; $U_2(t)$ is the steering input vector, and $U_2(t) = [\theta_h(t)]^T$; $W(t)$ is the Gaussian white noise disturbance input vector, and $W(t) = [w_1(t) \quad w_2(t) \quad w_3(t) \quad w_4(t)]^T$.

3.2 Integrated controller design

The stochastic sub-optimal control strategy based on output feedback is applied to design the integrated controller. This control strategy monitors the vehicle states and adjusts or tunes the control forces for the ASS and the assist torque for the EPS by using the measured outputs. The major advantage of the algorithm is that the critical parameters suggested by the original dynamic system are automatically adjusted by the sub-optimal feedback law. This overcomes the disadvantage resulted from that some of the state variables are immeasurable in practice. To apply the control strategy, we first propose the objective function (or performance indices) for the integrated control system defined in Eq. 34.

Since it is a full-car dynamic model that integrates EPS and ASS, the multiple vehicle performance indices must be considered, which include maneuverability, handling stability, ride comfort, and safety. These performance indices can be measured by the following physical terms: the torque applied on the steering wheel T_c, the yaw rate of the full car ω_z, the pitch angle of sprung mass θ, the roll angle of sprung mass ϕ, the vertical acceleration of sprung mass $\ddot{z}_s$, the suspension dynamic deflection $z_s - z_u$, and the tyre dynamic load $k_t(z_u - z_g)$. In addition, we also take into account the consumed control energy, which is represented by the assist torque T_m and the control force of the active suspension f_i. Therefore, the integrated performance index is defined as

$$
\begin{aligned}
J = E\Bigg\{ \int_0^\infty [& q_1\left(T_c - T_0\right)^2 + q_2\omega_z^2 + q_3\dot{\phi}^2 + q_4\ddot{z}_s^2 + q_5\dot{\theta}^2 + q_6\left(z_{u1} - z_{s1}\right)^2 + \\
& q_7\left(z_{u2} - z_{s2}\right)^2 + q_8\left(z_{u3} - z_{s3}\right)^2 + q_9\left(z_{u4} - z_{s4}\right)^2 + q_{10}\left(k_{t1}\left(z_{g1} - z_{u1}\right)\right)^2 + \\
& q_{11}\left(k_{t2}\left(z_{g2} - z_{u2}\right)\right)^2 + q_{12}\left(k_{t3}\left(z_{g3} - z_{u3}\right)\right)^2 + q_{13}\left(k_{t4}\left(z_{g4} - z_{u4}\right)\right)^2 + r_m T_m^2 + \\
& r_1 f_1^2 + r_2 f_2^2 + r_3 f_3^2 + r_4 f_4^2]dt \Bigg\}
\end{aligned}
\tag{35}
$$

where $q_1,\cdots,q_{13}$, r_m, $r_1\cdots,r_4$ are the weighting coefficients. We rewrite Eq. 35 in matrix form

$$
\begin{aligned}
J &= E\left\{ \int_0^\infty \left[Y^T Q_0 Y + U^T R U\right]dt \right\} = E\left\{ \int_0^\infty \left[X^T\left(C^T Q_0 C\right)X + U^T R U\right]dt \right\} \\
&= E\left\{ \int_0^\infty \left[X^T Q X + U^T R U\right]dt \right\}
\end{aligned}
\tag{36}
$$

where $Q = C^T Q_0 C$; $Q_0 = diag\{q_1, q_2, \cdots, q_{13}\}$; $R = diag\{r_1, r_2, r_3, r_4, r_m\}$.
To minimize the above performance index, the sub-optimal feedback control law is developed as follows.
The control matrix U can be expressed by

$$
U = -KY \tag{37}
$$

where K is the output feedback gain matrix, which can be derived through the following procedure.
Step 1. We first can derive the state feedback gain matrix F^* using optimal control method:

$$
F^* = R^{-1}B^T P \tag{38}
$$

where the matrix B is calculated as $B = AA_1^{-1}B_1$; and the matrix P is the solution of the following *Riccati* equation:

$$
PA + A^T P - PBR^{-1}B^T P + Q = 0 \tag{39}
$$

Step 2. Since there is no inverse matrix for the non-square (or rectangular) matrix C, the output feedback gain matrix K cannot be directly obtained through the equation $KC = F^*$. In this case, the norm-minimizing method is used to find the approximate solution of K (Gu et al., 1997). First, the following objective function is constructed

$$H = \left\| F - F^* \right\| = \sqrt{\sum_{i=1}^{22}\sum_{j=1}^{22}\left(F_{ij}^* - F_{ij} \right)^2} \tag{40}$$

and then we can find F by minimizing the objective function H

$$F = F^* C^T \left(C C^T \right)^{-1} C \tag{41}$$

we also have

$$F = KC \tag{42}$$

Thus K is derived by combining Eq. 41 and Eq. 42

$$K = F^* C^T \left(C C^T \right)^{-1} \tag{43}$$

and the control matrix U becomes

$$U = -KY = -F^* C^T \left(C C^T \right)^{-1} Y \tag{44}$$

3.3 Simulations and discussions

The integrated control system is analyzed using Matlab/Simulink. We assume that the vehicle travels at a constant speed v_x = 20m/s, and is subject to a steering input from steering wheel. The steering input is set as a step signal with amplitude of 120°.

The road excitation shown in Fig. 4 is assumed to be independent for each wheel and the power of the white noise for each wheel equals 20dB. The assumption of independent road excitation for each wheel has practical significance because in real road conditions, the road excitations on the four wheels of the vehicle are different and independent. It must be noted that this assumption on the road excitation is different from the assumption commonly made in other studies. The commonly made assumption states that the rear wheels follow the front wheels on the same track and hence the excitations at the rear wheels are just the same as the front wheels except for a time lag. Such a simplification is not applied in this simulation. The values of the vehicle physical parameters used in the simulation are listed in Table 1.

The parameter setting for the weighting coefficient matrices Q_0 and R defined in Eq. 36 plays an important role in the simulation performance. After tuning these weighting coefficients, we choose the following parameter setting when a satisfactory system performance is achieved: $q_1 = 10$, $q_2 = 10^6$, $q_3 = 5.0 \times 10^5$, $q_4 = q_5 = 2 \times 10^6$, $q_6 = q_7 = \cdots = q_{13} = 10^3$, $r_m = 0.1$, and $r_1 = r_2 = r_3 = r_4 = 1$.

It must be noted that different levels of importance are assigned to the different performance indices with such a parameter setting for the weighting coefficients. For example, the vertical acceleration of sprung mass is considered to be more important than the suspension dynamic deflection. In order to study comprehensively the characteristics of

N_2	20	c_3/c_4	1760/ 1760 (N·s/m)
k_s	90 (N·m/ rad)	k_t	138000 (N/m)
I_p	0.06 (kg·m^2)	h	0.505 (m)
c_e	0.3 (N·s·m/rad)	d	0.64 (m)
M	1030 (kg)	a	0.968 (m)
m_s	810 (kg)	b	1.392 (m)
m_{u1}/m_{u2}	26.5/ 26.5 (kg)	I_x	300 (kg·m^2)
m_{u3}/m_{u4}	24.4/ 24.4 (kg)	I_y	1058.4 (kg·m^2)
k_{s1}/k_{s2}	20600/ 20600 (N/m)	I_z	1087.8 (kg·m^2)
k_{s3}/k_{s4}	15200/ 15200 (N/m)	f_0	0.01 (Hz)
k_{af}/k_{ar}	6695/ 6695 (N·m/ rad)	G_0	5.0×10^{-6} (m^3/cycle)
c_1/c_2	1570/ 1570 (N·s/m)	v_x	20m/s

Table 1. Vehicle Physical Parameters.

the integrated control system, the integrated control system is compared to two other systems. One is the system without control, i.e. the passive mechanical system. While the other is the system that only has ASS (denoted as ASS-only) or EPS (denoted as EPS-only). For each of the two control systems, the sub-optimal control strategy is applied and the identical parameter setting for the weighting coefficient matrices Q_0 and R is selected.

It can be observed from the simulation results that all the performance indices are improved for the integrated control system, compared to those for the passive system, and those for ASS-only or EPS-only. For brevity, only the performance indices with higher lever of importance are selected to illustrate in Fig. 5 through Fig. 8. The following discussions are made:

1. As shown in Fig. 5, the roll angle for the integrated control system is reduced significantly compared to that for the ASS-only system and the passive system. A quantitative analysis of the results shows that the peak value of the roll angle for the integrated control system is decreased by 37.6%, compared to that for the ASS-only system, and 55.3% for the passive system. Moreover, the roll angle for the integrated control is damped quickly and thus less oscillation is observed for the integrated control system, compared to the other two systems. Therefore the results indicate that the anti-roll ability of the vehicle is greatly enhanced and thus a better handling stability is achieved through the application of the integrated control system.

2. It is presented clearly in Fig. 6 that the overshoot of the yaw rate for the integrated control system is decreased compared to that for the EPS-only system and the passive system. Furthermore, the yaw rate for the integrated control system and the EPS-only system becomes stable more quickly than the passive system after the overshoot. However, there is no significant time difference for the integrated control system and the EPS-only system to stabilize the yaw rate after the overshoot. The results demonstrate that the application of the integrated control system contributes a better lateral stability to the vehicle, compared to the EPS-only system and the passive system.

3. A quantitative analysis is performed for the vertical acceleration of sprung mass as shown in Fig. 7. The obtained R.M.S. (Root-Mean-Square) value of the vertical acceleration of sprung mass for the integrated control system is reduced by 23.1%,

compared to that for the ASS-only system, and 35.5% for the passive system. The results show that the vehicle equipped with the integrated control system has a better ride comfort than that with the ASS-only system and the passive system. In addition, the dynamic deflection of the front suspension as shown in Fig. 8 also suggests similar results.

In summary, the integrated control system improves the overall vehicle performance including handling, lateral stability, and ride comfort, compared to either the EPS-only system or the ASS-only system, and the passive system.

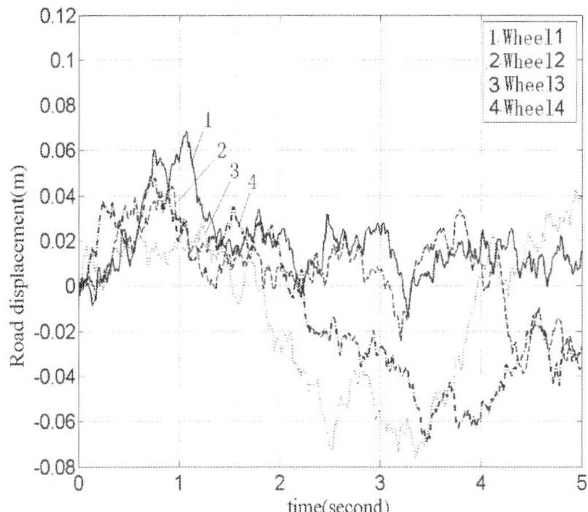

Fig. 4. Road Input.

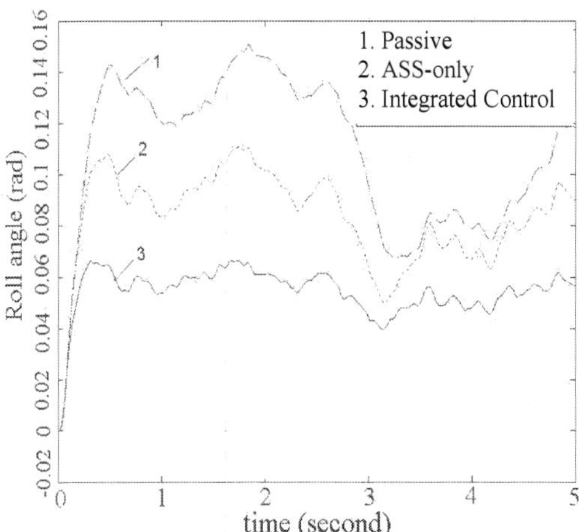

Fig. 5. Roll angle.

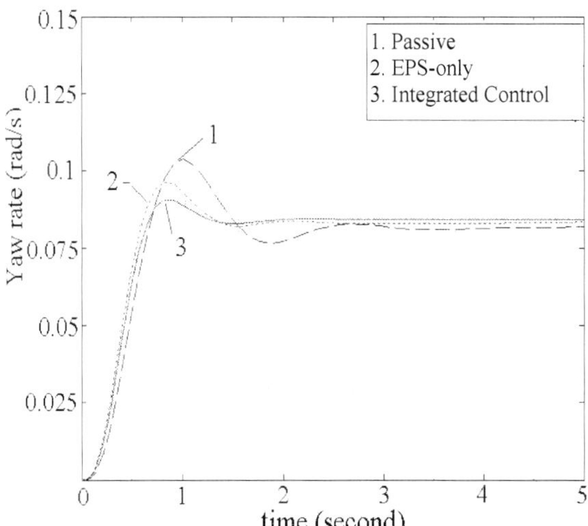

Fig. 6. Yaw rate.

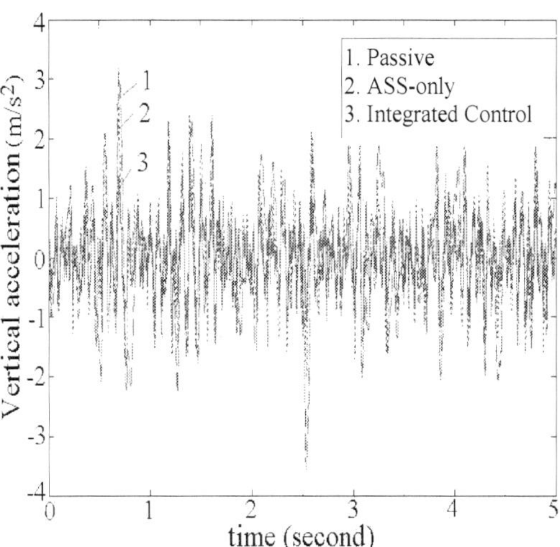

Fig. 7. Vertical acceleration of sprung mass.

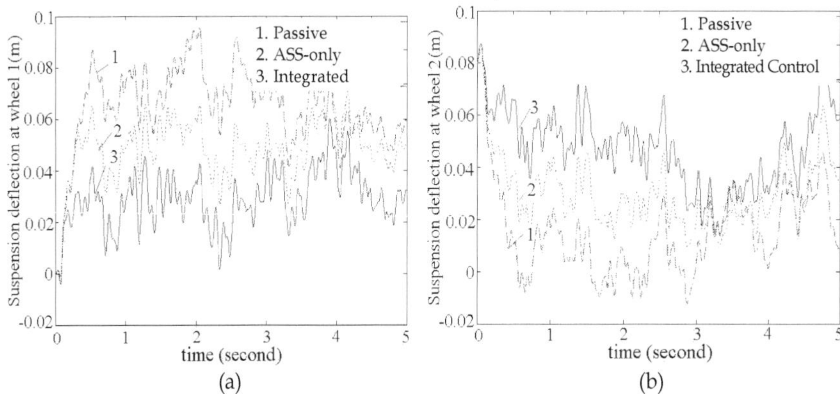

Fig. 8. Front suspension deflection: (a) at wheel 1; (b) at wheel 2.

In this investigation, a full-car dynamic model has been established through integrating electrical power steering system (EPS) with active suspension system (ASS) in order to address the coupling effects between the dynamics of the steering system and the suspension system. Thereafter, a multivariable control approach called stochastic sub-optimal control strategy based on output feedback has been applied to coordinate the control of both the EPS and ASS. Simulation results show that the integrated control system is effective in fulfilling the integrated control of the EPS and the ASS. This is demonstrated by the significant improvement on the overall vehicle performance including handling, lateral stability, and ride comfort, compared to either the EPS-only system or the ASS-only system, and the passive system. However, the development of the integrated vehicle control system requires fully understanding the vehicle dynamics in both the global level and system or subsystem level. Thus the development task for the integrated vehicle control system becomes very difficulty when the number of control systems increases. Furthermore, a whole new design is required for the integrated vehicle control system including both control logic and hardware, when a new control system, e.g. anti-lock brake system (ABS), is equipped with.

4. Investigation 2: Hierarchical control

In the above investigation, we demonstrated the effectiveness of one of the integrated control approaches called multivariable control on coordinating the control of the ASS and the EPS. While the second investigation moves up a step further on developing the integrated control approach. To this end, a hierarchical control architecture is proposed for integrated control of active suspension system (ASS) and electronic stability program (ESP). The advantages of the hierarchical control architecture are demonstrated through the following design practice of the integrated control system.

4.1 Hierarchical controller design

The architecture of the proposed hierarchical control system is shown in Fig. 9. The control system consists of two layers. The upper layer controller monitors the driver's intentions and the current vehicle states including the steering angle of the front wheel δ_f, the sideslip angle β, the yaw rate ω_z and the lateral acceleration a_y, etc. Based on these input signals, the upper layer controller computes the corrective yaw moment M_{zc} in order to track the desired vehicle motions. Thereafter, the upper layer controller generates the distributed torques M_{ESP} and M_{ASS} to the two lower layer controllers, i.e., the ESP and the ASS, respectively, according to a rule-based control strategy. Moreover, the distributed torques M_{ESP} and M_{ASS} are converted into the corresponding control commands for the two individual lower layer controllers. Finally, the ESP and the ASS execute respectively their local control objectives to control the vehicle dynamics. The upper layer controller and the two lower layer controllers are designed as follows.

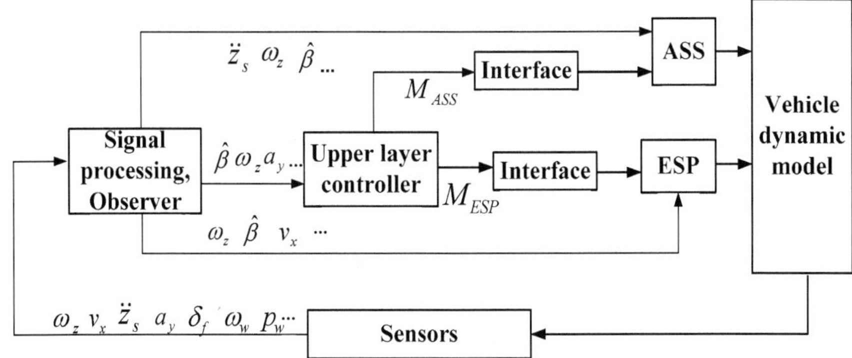

Fig. 9. Block diagram of the hierarchical control system.

4.2 Upper layer controller design

It is known that both the applications of the ESP and the ASS are able to develop corrective yaw moments (either directly or indirectly). To coordinate the interactions between the ASS and the ESP, a simple rule-based control strategy is proposed to design the upper layer controller. The aim of the proposed control rule is to distribute the corrective yaw moment appropriately between the two lower layer controllers. The control rule is described as follows.

First, the corrective yaw moment M_{zc} is calculated by using the 2-DOF vehicle reference model defined in Section 2.5, based on the measured and estimated vehicle input signals.

Second, the braking/traction torque M_d and the pitch torque M_p are computed by using the following equations

$$M_d = c_p \cdot p_w - 0.5 M_{zc} + I_w \cdot \dot{\omega}_w \tag{45}$$

$$M_p = \frac{k_\alpha tan\alpha}{c_\lambda \lambda_w} M_d \tag{46}$$

where Eq. (45) is derived by considering the dynamics of one of the front wheels. It should be noted that although a front wheel drive vehicle is assumed, the main conclusions of this study can be easily extended to vehicles with other driveline configurations; In general, the brake torque at each wheel is a function of the brake pressure p_w at that wheel, and c_p is an equivalent braking coefficient of the braking system, which is determined by using the equation $c_p = A_w \mu_b R_b$; The number "0.5" represents that the corrective yaw moment is evenly shared by the two front wheels.

Finally, the distributed torques M_{ESP} and M_{ASS} are generated by using a linear combination of the braking/traction torque M_d and the pitch torque M_p, which is given as

$$\begin{cases} M_{ESP} = n_1 M_d + (1 - n_1) M_p \\ M_{ASS} = n_2 M_p + (1 - n_2) M_d \end{cases} \tag{47}$$

where n_1 and n_2 are the weighting coefficients, and $1 > n_1 > 0.5$, $1 > n_2 > 0.5$. Therefore, through tuning the weighting coefficients n_1 and n_2, the upper layer controller is able to coordinate the two lower layer controllers and determine to what extent the two lower layer controllers to be controlled.

4.3 Lower layer controller design
4.3.1 ASS controller design

The LQG control method is used to control the active suspension system. The state variables are defined as $X = [z_s \ \dot{z}_s \ z_{u1} \ z_{u2} \ z_{u3} \ z_{u4} \ \dot{z}_{u1} \ \dot{z}_{u2} \ \dot{z}_{u3} \ \dot{z}_{u4} \ \theta \ \phi \ \dot{\theta} \ \dot{\phi}]^T$; and the output variables are chosen as $Y = [\ddot{z}_s \ z_{u1} \ z_{u2} \ z_{u3} \ z_{u4} \ \theta \ \varphi]^T$. Therefore, based on Eq. 4 through Eq. 16, together with the road excitation model presented in Section 2.4, the state equation and the output equation can be written as

$$\begin{cases} \dot{X} = AX + BU \\ Y = CX + DU \end{cases} \tag{48}$$

where $U = [U_1 \ U_2]^T$ is the control input vector. $U_1 = [f_1 \ f_2 \ f_3 \ f_4]^T$ is the control force vector, and $U_2 = [z_{g1} \ z_{g2} \ z_{g3} \ z_{g4}]^T$ is the road excitation vector. The multiple vehicle performance indices are considered to evaluate the vehicle handling stability, ride comfort, and safety. These performance indices can be measured by the following physical terms: vertical displacement of each wheel z_{u1}, z_{u2}, z_{u3}, z_{u4}; the suspension dynamic deflections $(z_{s1} - z_{u1})$, $(z_{s2} - z_{u2})$, $(z_{s3} - z_{u3})$, $(z_{s4} - z_{u4})$; the vertical acceleration of sprung mass $\ddot{z}_s$; the pitch angular acceleration $\ddot{\theta}$; the roll angular acceleration $\ddot{\varphi}$; and the control forces of the active suspension f_1, f_2, f_3, f_4. Therefore, the combined performance index is defined as

$$\begin{aligned} J = \lim_{T \to \infty} \frac{1}{T} \int_0^T [&q_1 z_{u1}^2 + q_2 z_{u2}^2 + q_3 z_{u3}^2 + q_4 z_{u4}^2 + q_5 (z_{s1} - z_{u1})^2 \\ &+ q_6 (z_{s2} - z_{u2})^2 + q_7 (z_{s3} - z_{u3})^2 + q_8 (z_{s4} - z_{u4})^2 + q_9 \ddot{\theta}^2 \\ &+ q_{10} \ddot{\varphi}^2 + q_{11} \ddot{z}_s^2 + r_1 f_1^2 + r_2 f_2^2 + r_3 f_3^2 + r_4 f_4^2] dt \end{aligned} \tag{49}$$

where $q_1,...,q_{11}$, and $r_1,...,r_4$ are the weighting coefficients. The above equation can be rewritten as the following matrix form

$$J = \lim_{T \to \infty} \frac{1}{T} \int_0^T (X^T QX + U^T RU + 2X^T NU)dt \tag{50}$$

where Q, R, N are the weighting matrices.

The state feedback gain matrix K is derived using the optimal control method, and it is the solution of the following *Riccati* equation

$$KA + A^T K + Q - KB_1 R^{-1} B_1{}^T K + B_2 U_2 B_2{}^T = 0 \tag{51}$$

4.3.2 ESP controller design

In this study, an adaptive fuzzy logic (AFL) method is applied to the design of the ESP controller. Fuzzy logic controller (FLC) has been identified as an attractive control method in vehicle dynamics control (Boada et al., 2005). This method has advantages when the following situations are encountered: 1) there is no explicit mathematical model that describes how control outputs functionally depend on control inputs; 2) there are experts who are able to incorporate their knowledge into the control decision-making process. However, traditional FLC with a fixed parameter setting cannot adapt to changes in the vehicle operating conditions or in the environment. Therefore, an adaptive mechanism must be introduced to adjust the controller parameters in order to achieve a satisfactory vehicle performance in a wide range of changing conditions.

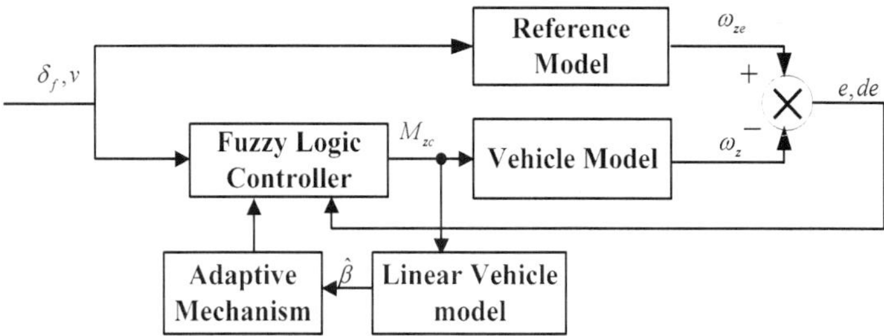

Fig. 10. Block diagram of the adaptive fuzzy logic controller for ESP.

As shown in Fig. 10, the AFL controller consists of a FLC and an adaptive mechanism. To design the AFL controller, the yaw rate and the sideslip angle of the vehicle are selected as the control objectives. The yaw rate can be measured by a gyroscope, but the sideslip angle cannot be directly measured and thus has to be estimated by an observer. The observer is designed by using the 2-DOF vehicle model described in Section 2.4. The linearized state space equation of the 2-DOF vehicle model is derived as follows, with the assumptions of a constant forward speed and a small sideslip angle.

$$\begin{cases} \dot{X} = A_E \cdot X + B_E \cdot U \\ Y = C_E \cdot X + D_E \cdot U \end{cases} \tag{52}$$

where

$$X = \begin{bmatrix} \beta \\ \omega_z \end{bmatrix}, \ U = \begin{bmatrix} \delta_f \\ M_{zc} \end{bmatrix}, \ A_E = \begin{bmatrix} -\dfrac{C_f+C_r}{mv} & -1-\dfrac{aC_f-bC_r}{mv^2} \\ -\dfrac{aC_f-bC_r}{I_z} & -\dfrac{a^2C_f+b^2C_r}{I_z v} \end{bmatrix}, \ B_E = \begin{bmatrix} \dfrac{C_f}{mv} & 0 \\ \dfrac{aC_f}{I_z} & \dfrac{1}{I_z} \end{bmatrix}, \ C_E = \begin{bmatrix} 1 & 0 \\ 0 & 1 \end{bmatrix},$$

$$D_E = \begin{bmatrix} 0 & 0 \\ 0 & 0 \end{bmatrix}.$$

The aim of the AFL is to track both the desired yaw rate and the desired sideslip angle. The desired yaw rate is calculated as

$$\omega_{ze} = \frac{v_x \cdot \delta_f}{L \cdot (1 + S \cdot v_x^2)} \tag{53}$$

where L is the wheel base; S is the stability factor of the vehicle, and $S = m(b/C_f - a/C_r)/L^2$. As shown in Fig. 10, the FLC has two input variables, the tracking error of the yaw rate e and the difference of the error de. They are defined as, at the kth sampling time

$$e(k) = \omega_z(k) - \omega_{ze}(k) \tag{54}$$

$$de(k) = e(k) - e(k-1) \tag{55}$$

The output variable of the FLC is defined as the corrective yaw moment M_{zc}. To determine the fuzzy controller output for the given error and its difference, the decision matrix of the linguistic control rules is designed and presented in Table 2. These rules are determined based on expert knowledge and a large number of simulation results performed in the study. In designing the FLC, the scaling factors k_e and k_{de} have great effects on the performance of the controller. Therefore the adaptive mechanism is applied to adjust the parameters in order to achieve a satisfactory control performance when there are changes in the vehicle operating conditions or in the environment. The adaptive law is given as

$$\beta(k_e) = \beta_0 + k_e \int_0^t (\frac{a_y}{v} - \omega_z)dt \tag{56}$$

$$\dot{\beta}(k_{de}) = -\omega_z + k_{de}\frac{1}{v}(a_y cos\beta - a_x sin\beta) \tag{57}$$

where $\beta_0 = 0$. Full details of the derivation of the above equations are given in the Appendix.

4.4 Simulations and discussions

In order to evaluate the performance of the developed hierarchical control system, a simulation investigation is performed. The performance and dynamic behaviors of the hierarchical control system are analyzed using Matlab/Simulink. We assume that the vehicle travels at a constant speed $v = 90$ km/h. Two driving conditions are performed: 1) step steering input; and 2) double lane change. For the first case, the vehicle is subject to a

e \\ de	PB	PM	PS	O	NS	NM	NB
PB	NB	NB	NB	NB	NM	O	O
PM	NB	NB	NB	NB	NM	O	O
PS	NM	NM	NM	NM	O	PS	PS
PO	NM	NM	NS	O	PS	PM	PM
NO	NM	NM	NS	O	PS	PM	PM
NS	NS	NS	O	PM	PM	PM	PM
NM	O	O	PM	PB	PB	PB	PB
NB	O	O	PM	PB	PB	PB	PB

Table 2. Fuzzy rule bases for ESP control.

steering input from the steering wheel and the steering input is set as a step signal with amplitude of 120°. The road excitation is assumed to be independent for the four wheels.

After tuning the parameter setting for the hierarchical control system, we select the weighting parameters for the ASS: $r_1 = r_2 = r_3 = r_4 = 1$, $q_1 = q_2 = q_3 = q_4 = 10^3$, $q_5 = q_6 = q_7 = q_8 = 10^4$, $q_9 = 2 \times 10^3$, $q_{10} = 10^5$, and $q_{11} = 10^6$. Moreover, the weighting parameters for the upper layer controller are selected as: $n_1 = 0.80$ and $n_2 = 0.85$.

The simulation results for the multiple performance indices are shown in Fig. 11 and Fig. 12 (For brevity, only some representative performance indices are presented here).

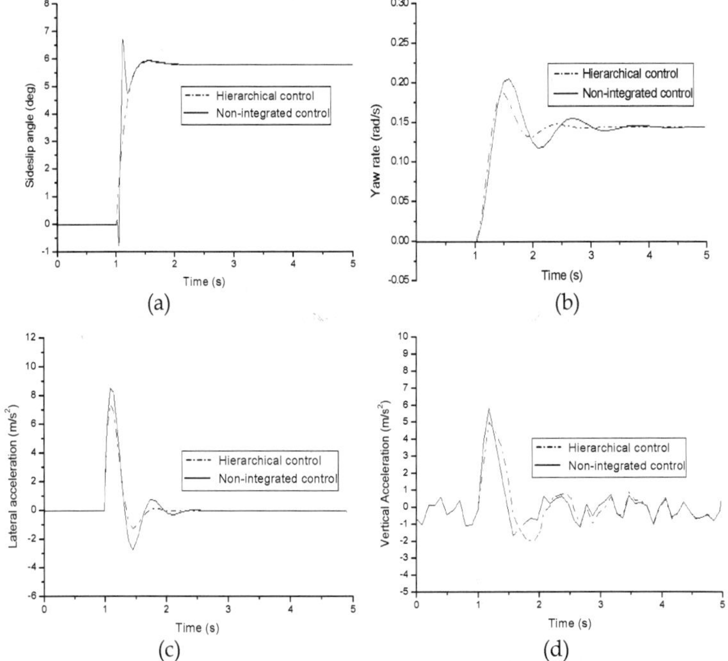

Fig. 11. Comparison of responses for the manoeuvre of step steering input: (a) sideslip angle; (b) yaw rate; (c) lateral acceleration; (d) vertical acceleration.

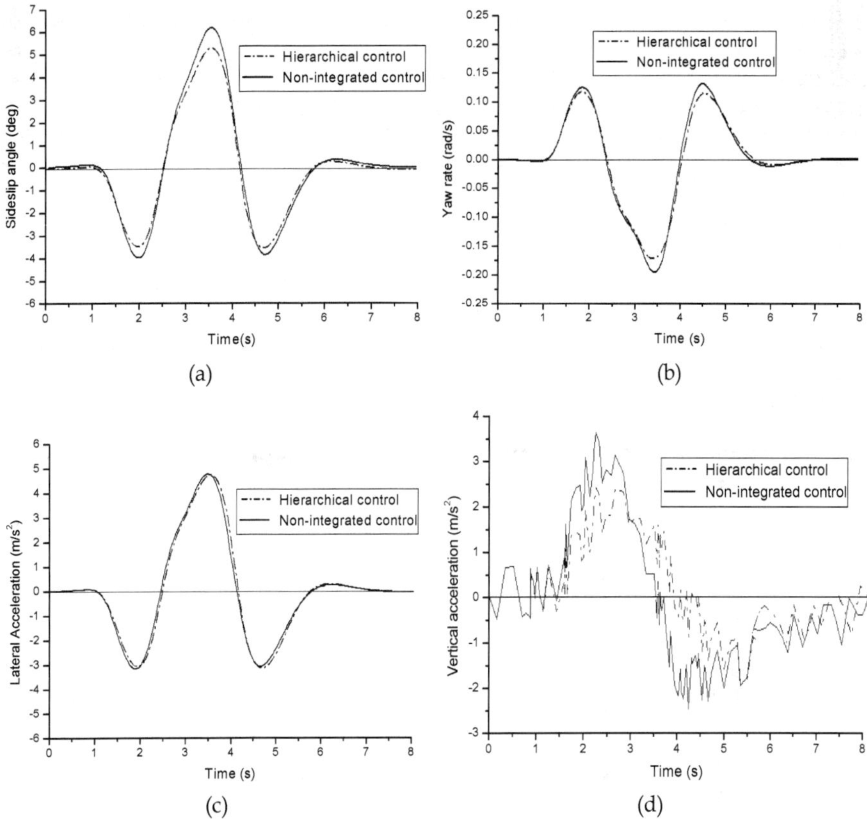

Fig. 12. Comparison of responses for the manoeuvre of double lane change: (a) sideslip angle; (b) yaw rate; (c) lateral acceleration; (d) vertical acceleration.

For comparisons, the simulation investigation for non-integrated control is also performed. In the case, we simply eliminate the upper layer controller. The following discussions are made:

1. For the manoeuvre of step steering input, it can be seen that the peak value of the sideslip angle for hierarchical control, as shown in Fig. 11(a), is reduced by 11.6% compared to that for non-integrated control. Moreover, the sideslip angle for hierarchical control is damped quickly and thus has less oscillation than that for non-integrated control. Similar patterns can be observed for the yaw rate and the lateral acceleration illustrated in Fig. 11(b) and Fig. 11(c), respectively. The results indicate that the vehicle lateral stability is improved by the proposed hierarchical control system in comparison with the non-integrated control system. In addition, the vertical acceleration of sprung mass, one of ride comfort indices, is presented in Fig. 11(d). It can be observed that the peak value of the performance index is decreased by 13.8% for hierarchical control, compared to that for non-integrated control.

2. For the manoeuvre of double lane change, it is observed that the peak value of the sideslip angle for hierarchical control is reduced by 15.3% compared to that for non-integrated control, as shown in Fig. 12(a). Moreover, for the peak value of the yaw rate shown in Fig. 12(b), the percentage of decrease is 7.9. However, as shown in Fig. 12(c), there is no significant difference on the lateral acceleration between the two control cases. While for the vertical acceleration of sprung mass shown in Fig. 12(d), it can be seen clearly that the peak value of this performance index for hierarchical control is reduced significantly by 30.5%, compared to that for non-integrated control. In addition, a quantitative analysis of the vertical acceleration shows that the R.M.S. (Root-Mean-Square) value of the vertical acceleration for hierarchical control is reduced by 21.9% compared to that for non-integrated control.

In summary, the application of the hierarchical control system improves the overall vehicle performance including the ride comfort and the lateral stability under the critical driving conditions. The results show that the hierarchical control system is able to coordinate the interactions between the ASS and the ESP and thus expand the functionalities of the two individual control systems.

5. Investigation 3: Experiment

To verify the effectiveness of the proposed hierarchical control architecture, an experimental study is performed. A physical configuration of the two-layer hierarchical control architecture is illustrated in Fig. 13. The upper layer controller determines the corrective yaw moment to track the desired vehicle motions by using the signals from the CAN-bus, e.g. driver's intentions, environment information, and current vehicle dynamic states. Thereafter, the upper layer controller generates the distributed torques to the two lower layer controllers, i.e., the ESP and the ASS, respectively, according to a rule-based control strategy. Moreover, the distributed torques are converted into the corresponding control commands for the two individual actuators to regulate or track respectively the vehicle dynamic states.

Development and test of complex control systems often benefit from a technique called hardware-in-the-loop (HIL) simulation. The advantages of this technique over real plant tests include: greater flexibility and higher safety in the test scenarios, shorter development time and reduced cost, and measurable/reproducible criteria for system and subsystem evaluation. With those in mind, the HIL simulation is applied to verify the effectiveness of the proposed hierarchical control system. Fig. 14 shows the developed hardware-in-the-loop test platform for the hierarchical control system. The client computer (PXI-8196 by National Instruments Inc.) collects the signals measured by the sensors, which include the pressure of each brake wheel cylinder, the pressure of brake master cylinder, and the vertical acceleration of sprung mass at each suspension, etc. These signals are in turn provided to the host computer (PC) through CAN-bus. Based on these input signals, the host computer computes the vehicle states and the desired vehicle motions, such as the desired yaw rate. Thereafter, the host computer generates control commands to the client computer. Through the hardware interface circuits, the client computer in turn sends the control commands to the corresponding actuators.

The experimental setup is shown in Fig. 15. A test vehicle was equipped with the developed control units for the upper layer controller, ESP controller and ASS controller. The test vehicle was running on a road simulator, which is mounted on the test ground as shown in the figure. Therefore the road excitation signal can be generated through the road simulator. Again, the two same driving conditions as those used in the simulation investigation were performed, i.e., the manoeuvre of step steering input and the manoeuvre of double lane change. Two cases were tested in the experiment, one is "with hierarchical control", and the other is "non-integrated control". For both testing cases, numerous vehicle tests were performed to validate the developed control units. The measured dynamic responses of the vehicle performance indices are illustrated in Fig. 16 for the manoeuvre of step steering input and Fig. 17 for the manoeuvre of double lane change, respectively.

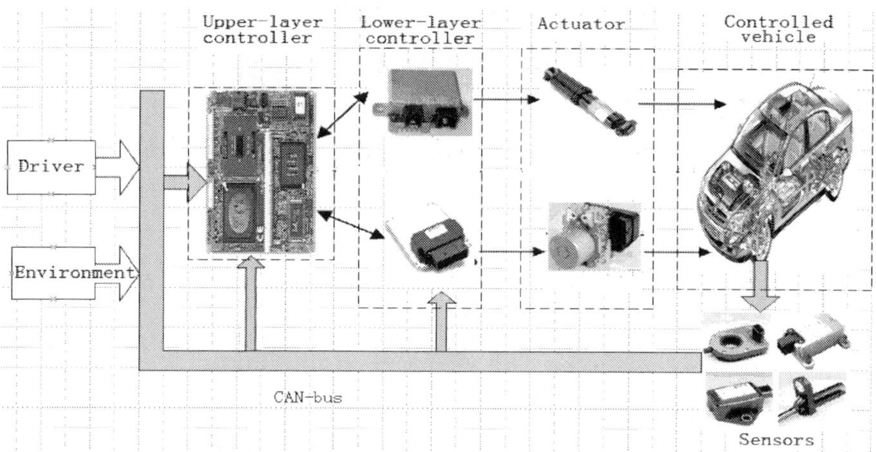

Fig. 13. Physical configuration of the hierarchical control architecture.

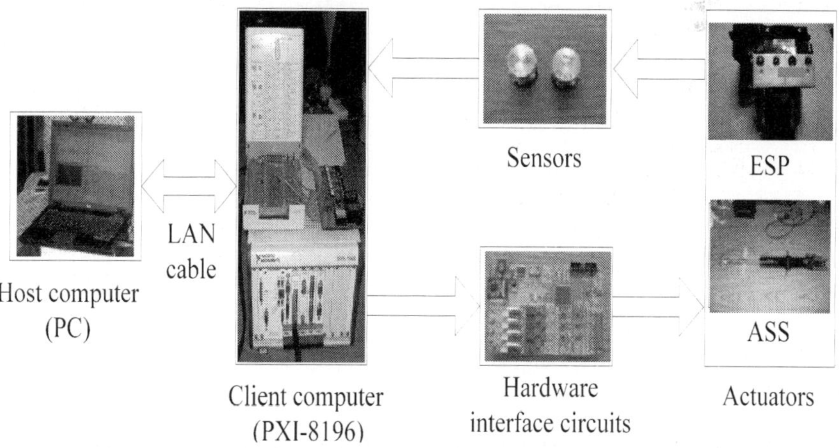

Fig. 14. HIL experimental configuration.

Fig. 15. Experimental setup.

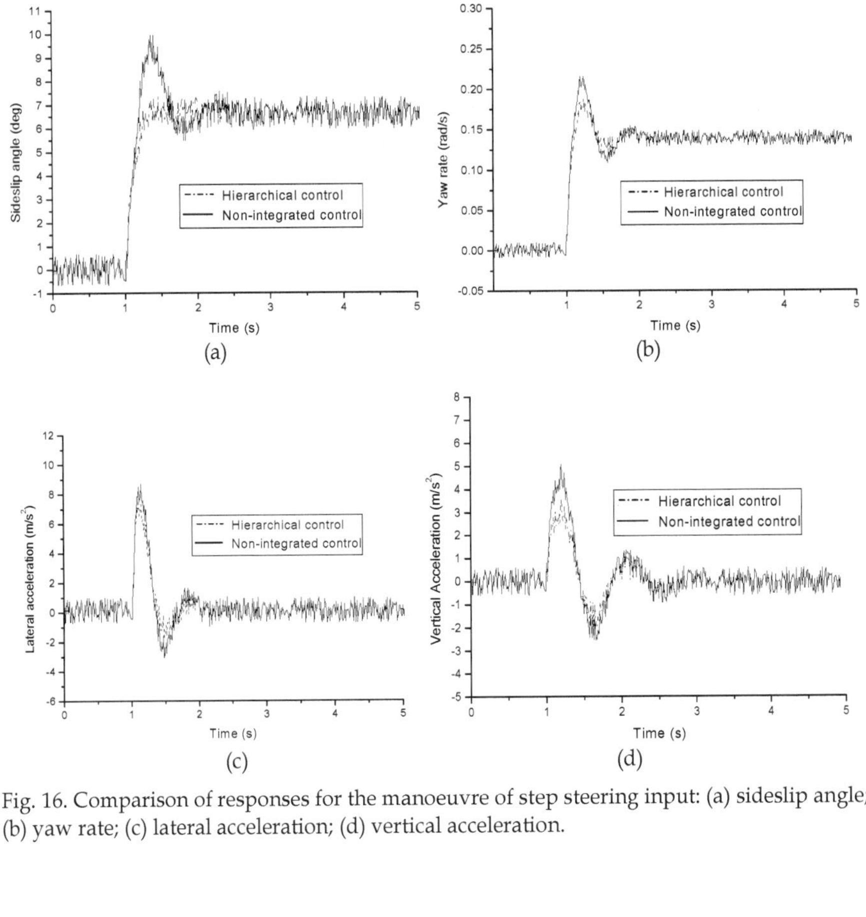

Fig. 16. Comparison of responses for the manoeuvre of step steering input: (a) sideslip angle; (b) yaw rate; (c) lateral acceleration; (d) vertical acceleration.

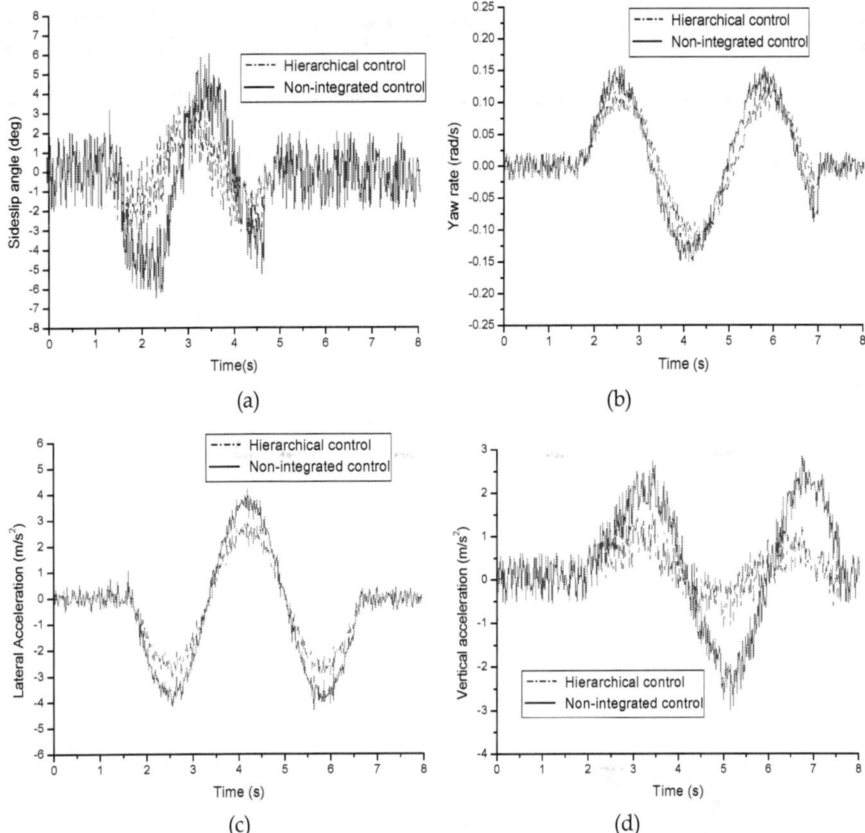

Fig. 17. Comparison of responses for the manoeuvre of double lane change: (a) sideslip angle; (b) yaw rate; (c) lateral acceleration; (d) vertical acceleration.

The following discussions are made by comparing the corresponding performance indices for hierarchical control and non-integrated control:

1. For the manoeuvre of step steering input, it is shown clearly in Fig. 16(a) that the peak value of the sideslip angle for hierarchical control is reduced by 25.1%, compared to that for non-integrated control. The similar phenomena can be observed in Fig. 16(b) for the yaw rate and Fig. 16(c) for the lateral acceleration, except that the percentages of decrease for the two performance indices are slightly smaller than that for the sideslip angle. In addition, as shown in Fig. 16(d), the peak value of the vertical acceleration of sprung mass is decreased greatly by 30.1% for hierarchical control, compared to that for non-integrated control. The results indicate that both the lateral stability and the ride comfort are improved by the proposed hierarchical control system in comparison with the non-integrated control system.

2. For the manoeuvre of double lane change, it is observed in Fig. 17(a) through Fig. 17(c) that the peak values of the sideslip angle, the yaw rate, and the lateral acceleration have

certain amount of decrease for hierarchical control, compared to those for non-integrated control. Moreover, a smaller R.M.S. value can be observed for those performance indices even without calculation. Finally, as presented in Fig. 17(d), the peak value of the vertical acceleration of sprung mass for hierarchical control is reduced significantly by 59.2%, compared to that for non-integrated control. A quantitative analysis of the vertical acceleration shows that the R.M.S. value of the vertical acceleration for hierarchical control is reduced by 47.9% compared to that for non-integrated control.

3. The experimental results have good agreement with the simulation results on demonstrating the vehicle performance improvements by the proposed hierarchical control system.

In summary, the experimental results demonstrate that the proposed hierarchical control system is able to improve both the lateral stability and the ride comfort, in comparison with the non-integrated control system. The experimental results verify the effectiveness of the hierarchical control system.

In the second and third investigations, integrated control and coordination of active suspension system (ASS) and electronic stability program (ESP) have been studied by using hierarchical control strategy. A two-layer hierarchical control architecture has been proposed to achieve the goal of function integration for the two chassis control systems. The upper layer controller has been designed to coordinate the interactions between the ASS and the ESP. A rule-based control method has been used to design the upper layer controller. In addition, the two lower layer controllers including the ASS and the ESP, have been designed independently to achieve their local control objectives. The LQG control strategy and the adaptive fuzzy logic control method have been used to design the ASS and the ESP, respectively. Both a simulation investigation and a hardware-in-the-loop experimental study have been performed. Simulation results demonstrate that the proposed hierarchical control system is able to improve the multiple vehicle performance indices including both the ride comfort and the lateral stability. Moreover, the experimental results verify the effectiveness of the design of the hierarchical control system.

6. Conclusions

In this chapter, integrated control and coordination of vehicle system dynamics have been studied comprehensively and intensively through theoretical developments and experimental verifications. The study consists of three investigations. The first investigation has been focused on coordinating the interactions and function conflicts between the steering system and the suspension system by using a multivariable control approach called stochastic sub-optimal control strategy. Simulation results show that the integrated control system is effective in improving the overall vehicle performance including handling, lateral stability, and ride comfort, compared to either the EPS-only system or the ASS-only system, and the passive system. Moreover, a more advanced integrated control approach called hierarchical control method has been applied to coordinate control of the ASS and the ESP. The design flexibility of the hierarchical control method has been demonstrated through the design practice of the two-layer control system. The upper layer controller has been designed to coordinate specifically the interactions between the ASS and the ESP. While the two lower layer controllers including the ASS and the ESP, have been designed independently to achieve their local control objectives. The application of the hierarchical control method to upper layer controller design has been focused on function coordination of the two lower layer control systems and thus few modifications are required for the two subsystems, in contrast to the multivariable control approach. Finally, both a simulation investigation and a hardware-in-the-loop experimental study have been performed. Simulation and experimental results demonstrate that the proposed hierarchical control system is able to improve the multiple vehicle performance indices including both the ride comfort and the lateral stability, compared to the non-integrated control system.

7. Acknowledgement

This research was sponsored in part by the Natural Science Foundation of China under Grant No. 51075112, and the Royal Society of UK under Grant No. 16558.

8. Nomenclature

a, b: horizontal distance between the C.G. of the vehicle and the front, rear axle;
A, B: state matrix, input matrix;
A_w: brake area of the wheel;
c_e : equivalent damping coefficient reflected to the pinion axis;
c_i: damping coefficient of the suspension at wheel i;
c_p: equivalent braking coefficient of the braking system;
c_λ : lateral stiffness of the tyre;
C: output matrix;
C_f, C_r: corning stiffnesses of the front tyre and the rear tyre, respectively;
d: half of the wheel track;
de: difference of the yaw rate tracking error;
D: feedforward matrix;
e: yaw rate tracking error;
f_0: low cut-off frequency;
$f_1 \sim f_4$: control force of each active suspension controller;
f_r: rolling resistance coefficient;
$F_{x1} \sim F_{x4}$ and $F_{y1} \sim F_{y4}$: longitudinal and lateral forces of the four wheels, respectively;
$F_{z1} \sim F_{z4}$: total force of the suspension acting on the sprung mass;
G_0 : road roughness coefficient;
h: vertical distance between the C.G. of sprung mass and the roll center;
I_p: equivalent moment of inertia of multiple parts reflected to the pinion axis. The multiple parts include the motor, the gear assist mechanism, and the pinion;
I_w: wheel moment of inertia about its spin axis;
I_x, I_y, I_z: roll moment of inertia, pitch moment of inertia, and yaw moment of inertia of sprung mass;
I_{xz}: product of inertia of sprung mass about the roll and yaw axes;
J: performance index;
k_{af}, k_{ar}: stiffness of the anti-roll bars for the front, rear suspension;
k_e, k_{de}: scaling factor;
k_s : torsional stiffness of the torque sensor;
k_{si}: stiffness of the suspension at wheel i;
k_{ti}: stiffness of tyre at wheel i;

3

k_α : cornering stiffness of the tyre;

K: state feedback gain matrix;

L: wheel base;

m, m_s, m_{ui} : mass of the vehicle, sprung mass, and unsprung mass at wheel i;

M_{ASS}, M_{ESP}: distributed torques for the ASS and the ESP, respectively;

M_d, M_p: braking/traction torque and pitch torque;

M_{ZC}: corrective yaw moment generated by the ESP controller;

n_1, n_2: weighting coefficient;

N: weighting matrix;

N_2: speed reduction ratio of the rack-pinion mechanism;

p_w : pressure of the brake wheel cylinder;

$q_1,..., q_{11}, r_1,..., r_4$: weighting coefficient;

Q, R: weighting matrix;

R_b: brake radius;

R_w: tyre rolling radius;

S: vehicle stability factor;

T_0: ideal steering torque applied on the steering wheel;

T_c: torque applied on the steering wheel;

T_i: wheel torque at wheel i;

T_m: assist torque applied on the steering column;

T_r: aligning torque transferred from tyres to the pinion;

T_{zwi}: aligning torque acting on the tyre i;

U, U_1, U_2: control input vector, control force vector, and road excitation vector, respectively;.

v, v_x, and v_y: vehicle speed, vehicle speed in the longitudinal direction and the lateral direction, respectively;

w_i : zero-mean Gaussian white noise with intensity of 1;

X, Y: state vector, output vector;

z_{gi}: road excitation;

z_s: vertical displacement of sprung mass;

z_{ui}: vertical displacement of unsprung mass;

α : sideslip angle of the tyre;

β : sideslip angle of the vehicle at the C.G.;

δ_1 : rotation angle of the pinion;

δ_f , δ_r : steering angles of the front, rear wheels;

δ_i : steering angle of wheel i;

ϕ : roll angle of sprung mass;

λ_w : pneumatic trail of the tyre;

μ_b : brake friction coefficient;

θ : pitch angle of sprung mass;

θ_h : rotation angle of the steering wheel;

ω_i : angular velocity of wheel i;

ω_z , ω_{ze} : yaw rate of the vehicle, desired yaw rate of the vehicle;

9. Appendix

The acceleration of the vehicle can be expressed by

$$a = \left(\dot{v}_x - v_y \omega_z\right)i + \left(\dot{v}_y + v_x \omega_z\right)j \tag{a1}$$

where $v_x = v\cos\beta$ and $v_y = v\sin\beta$; the above equation can be derived as the following equation by assuming the vehicle speed v is constant

$$a = -v\left(\dot{\beta} + \omega_z\right)\sin\beta\, i + v\left(\dot{\beta} + \omega_z\right)\cos\beta\, j \tag{a2}$$

Therefore

$$a_x = -v\left(\dot{\beta} + \omega_z\right)\sin\beta \tag{a3}$$

$$\text{and} \quad a_y = v\left(\dot{\beta} + \omega_z\right)\cos\beta \tag{a4}$$

and hence

$$a_x \cos\beta = -a_y \sin\beta \tag{a5}$$

Combining Eq. (a3) and (a5), the following equation can be easily derived

$$\dot{\beta} = -\omega_z + \frac{1}{v}(a_y \cos\beta - a_x \sin\beta) \tag{a6}$$

When β is small, the following equation can be easily derived from Eq. (a4)

$$\dot{\beta} = \frac{a_y}{v} - \omega_z \tag{a7}$$

10. References

Bakker, E, Nyborg, L, and Pacejka, H.B. (1987), Tyre Modeling for Use in Vehicle Dynamics Studies, *SAE Technical Paper 870421*, pp. 2190-2198.

Boada, B.L., Boada, M.J.L., and Diaz, V. (2005), Fuzzy-logic Applied to Yaw Moment Control for Vehicle Stability, *Vehicle System Dynamics*, Vol. 43, pp. 753-770.

Chang, S., and Gordon, T.J. (2007), Model-based Predictive Control of Vehicle Dynamics, *International Journal of Vehicle Autonomous Systems*, Vol. 5(1-2), pp. 3-27.

Chen, W.W., Xiao, H.S., Liu, L.Q., and Zu, J.W. (2006), Integrated Control of Automotive Electrical Power Steering and Active Suspension Systems Based on Random Sub-optimal Control, *International Journal of Vehicle Design*, Vol. 42(3/4), pp. 370-391.

Falcone, P., Borrelli, F., Asgari, J., Tseng, H. E., and Hrovat, D. (2007), Predictive Active Steering Control for Autonomous Vehicle Systems, *IEEE Transactions on Control Systems Technology*, Vol. 15(3), pp. 566-580.

Fruechte, R.D., Karmel, A.M., Rillings, J.H., Schilke, N.A., Boustany, N.M., and Repa, B.S. (1989), Integrated Vehicle Control, *Proceedings of the 39th IEEE Vehicular Technology Conference*, Vol. 2, pp. 868-877.

Gordon, T.J. (1996), An Integrated Strategy for the Control of a Full Vehicle Active Suspension System, *Vehicle System Dynamics*. Vol. 25, pp. 229-242.

Gordon, T.J., Howell, M., and Brandao, F. (2003), Integrated Control Methodologies for Road Vehicles, *Vehicle System Dynamics*, Vol. 40(1-3), pp. 157-190.

Gu, Z.Q., Ma, K.G., and Chen, W.D. (1997), *Active Control of Vibration* (in Chinese), China National Defense Industry Press, Beijing, China.

He, J.J., Crolla, D.A., Levesley, M.C., and Manning, W.J. (2006), Coordination of Active Steering, Driveline, and Braking for Integrated Vehicle Dynamics Control, *Proceedings of Institution of Mechanical Engineers - Part D: Journal of Automobile Engineering*, Vol. 220, pp. 1401-1421.

Hirano, Y., Harada, H., Ono, E., and Takanami, K. (1993), Development of An Integrated System of 4WS and 4WD by H Infinity Control, *SAE Technical Paper 930267*, pp. 79-86.

Karbalaei, R., Ghaffari, A., Kazemi, R., and Tabatabaei, S.H. (2007), A New Intelligent Strategy to Integrated Control of AFS/DYC Based on Fuzzy Logic, *International Journal of Mathematical, Physical and Engineering Sciences*, Vol. 1(1), pp. 47-52.

Li, D.F., Du S.Q., and Yu, F. (2008), Integrated Vehicle Chassis Control Based on Direct Yaw Moment, Active Steering, and Active Stabiliser, *Vehicle System Dynamics*, Vol. 46(1), pp. 341-351.

Nwagboso, C. O., Ouyang, X., and Morgan, C. (2002), Development of Neural Network Control of Steer-by-wire System for Intelligent Vehicles, *International Journal of Heavy Vehicle Systems*, Vol. 9(1), pp. 1-26.

Pacejka, H.B. (2002), *Tyre and Vehicle Dynamics*, Butterworth-Heinemann, Boston.

Rodic, A.D., and Vukobratovie, M.K. (2000), Design of an Integrated Active Control System for Road Vehicles Operating with Automated Highway Systems, *International Journal Computer Application Technology*, Vol. 13, pp. 78-92.

Trächtler, A. (2004), Integrated Vehicle Dynamics Control Using Active Brake, Steering, and Suspension Systems, *International Journal of Vehicle Design*, Vol. 36(1), pp. 1-12.

Yu F., and Crolla D.A. (1998), An Optimal Self-tuning Controller for An Active Suspension, *Vehicle System Dynamic*, Vol. 29, pp. 51-65.

Yu, F., Li, D.F., Crolla, D.A. (2008), Integrated Vehicle Dynamic Control – State-of-the Art Review, *Proceedings of IEEE Vehicle Power and Propulsion Conference (VPPC)*, September 3-5, Harbin, China, pp. 1-6.

2

Integrating Neural Signal and Embedded System for Controlling Small Motor

Wahidah Mansor, Mohd Shaifulrizal Abd Rani and Nurfatehah Wahy
Universiti Teknologi Mara
Malaysia

1. Introduction

Nowadays, controlling electronic devices without the use of hands is essential to provide a communication interface for disable persons to have control over their environment and to enable multi-tasking operation for normal person. Various methods of controlling electronic devices without the use of hands have been investigated by researchers, for examples sip-and-puff, electro-oculogram (EOG signals), light emitter and others [Ding et al., 2005, Kumar et al., 2002; Breau et al., 2004]. In our previous study, EOG signal was found to be suitable for activating a television using a specific protocol [Harun et al., 2009], however, it could not be used when a person is not facing the system. Thus, a method that is more flexible has to be investigated.

The use of neural signals to directly control a machine via a brain computer interface (BCI) has been studied since 1960s. Using an appropriate electrode placement and digital signal processing technique, useful information can be extracted from neural signals [Holzner et al, 2009; Jian et al., 2010; Gupta et al., 1996.] One of the events that can be detected from this signal is eye blink. It can be used as a mechanism to activate and control a machine which can help disable people to do their everyday routines.

Most BCI systems employs a computer to process neural signals and perform control. Since portable system offers benefits such as flexibility, mobility and convenience to use, it is more preferred than a fixed system. An embedded system can be designed to provide portability feature. To include this feature, a microcontroller is required to control its operation and provide a communication link between human and machine.

This chapter discusses how neural signal and embedded system can be combined together to activate a fan connected to a motor. It covers the introduction to neural signal, neural signal processing, embedded system and EEG based fan system hardware and software.

2. Neural signal

Neural signal or commonly known as electroencephalogram (EEG) is the representation of electrical activity of the brain. The overall excitation of the brain determines the amplitude and patterns of the signal. The excitation depends on the activity of the reticular activating system in the brain stem. The pattern changes markedly between states of sleep and wakefulness. The EEG signal is divided into five frequency bands; beta, alpha, theta, delta and gamma. Beta frequency is in the range of 12 Hz – 22 Hz and occurs when the person is awaken and in the state of alertness. Alpha is in the range of 8 – 13 Hz and is present when a person is awaken and relaxed with eyes close. Theta exists in the frequency range of 4 – 8 Hz when a person is sleepy, already sleep and in the sleep transition. The slowest wave is delta which is in the frequency range of 0.5 to 4 Hz and is associated with deep asleep. And finally gamma (22 – 30 Hz) consists of low amplitude & high frequency waves resulting from attention or sensory stimulation. Figure 1 shows the normal EEG signal of a relaxed patient. The signal consists of beta waves which lie in the frequency range of 13 to 22 Hz. Figure 2 shows EEG signal with eye closure and eye opening. The negative amplitude shows the eyelids closure and positive value shows the opening of eyelids.

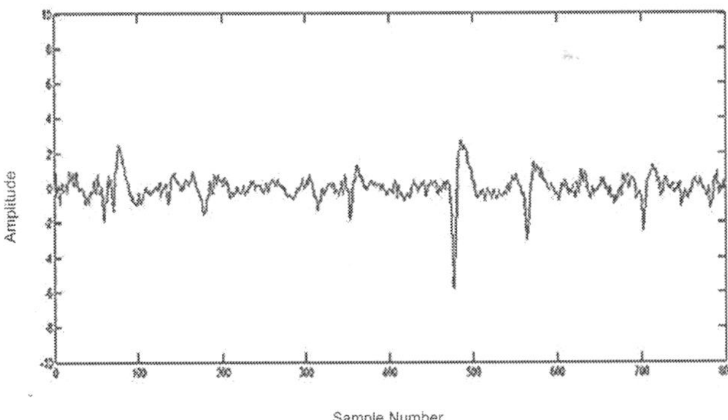

Fig. 1. Normal EEG signal when a person is relax.

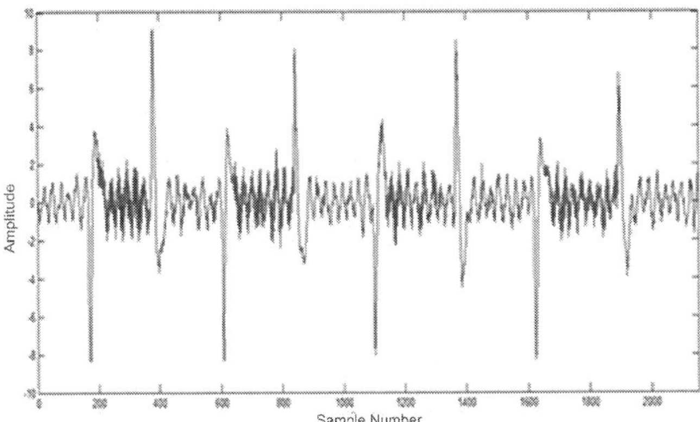

Fig. 2. Normal EEG signal with eye closure and eye opening.

The recorded EEG signals always contain artifacts which impede the analysis of the signals. The artifacts include muscle signals, heart signals, eye movements, power line interference, eye blinks and others. Artifacts in EEG signals typically are characterized by high amplitudes. Eye blinking artifacts always present in EEG signals since it is difficult to make the subject open his/her eye for a long time. In some cases, eye blinking artifacts may be useful and are required as a parameter for activating a system.

In EEG signals, eye blinks occur as peaks with relatively strong voltages. Eye blinks can be classified as short blinks if the duration of blink is less than 200ms or long blinks if it is greater or equal to 200ms [Bulling et al, 2006]. The amplitude of the peaks varies between different subjects. They are often located by setting a threshold in EEG and classified for all activity exceeding the threshold value.

Eye blinks can be classified into three types: reflexive, spontaneous and voluntary. The eye blink reflexive is the simplest response and does not require the involvement of cortical structures. Spontaneous eye blinks are those with no external stimuli specified and they are associated with the psycho-physiological state of the person . The amplitude of spontaneous eye blink is in the range of -4 to 3 V with duration of less than 400 ms and frequency of below 5 Hz. The EEG signal obtained when the eyes moved to the right and left is shown in Figure 3. This signal contains a lot of artifacts caused by spontaneous eye blinking and eyelid movements as the eyeball moved. The signal obtained from these eye movements are not suitable for activating a system as the occurrence of eye movements is difficult to detect. Figure 4 shows EEG with eye movements upward and downwards. This signal consists of noise which covers the required information to be extracted.

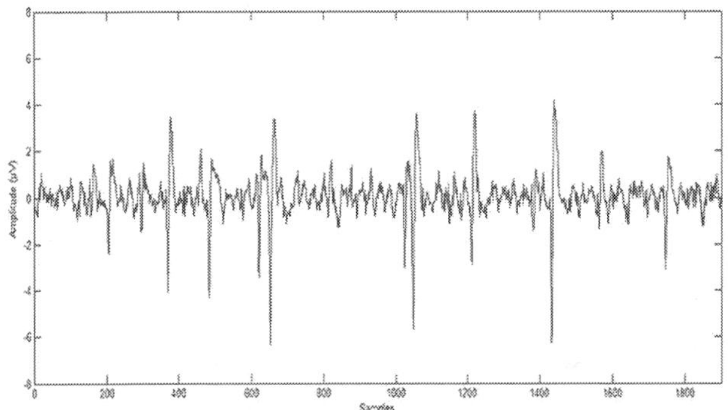

Fig. 3. EEG signal obtained when the eyes are moved to the right and left. [Abd Rani et al., 2009]

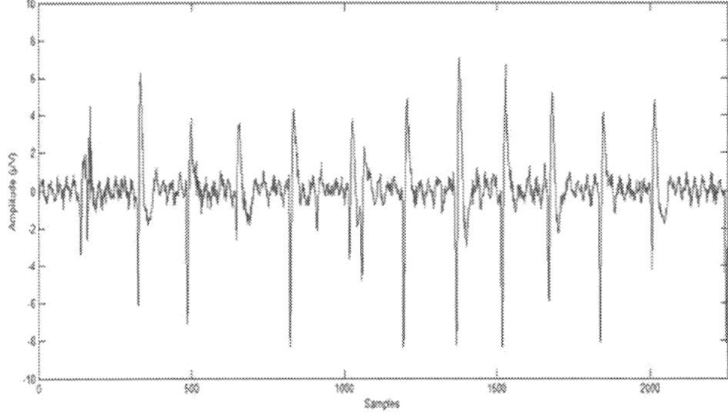

Fig. 4. EEG signal with eye movement upwards and downwards.

Voluntary eye blinking which is intentional blinking due to predetermined condition, involves multiple areas of the cerebral cortex as well as basal ganglion, brain stem and cerebella structures. Figure 5 shows the EEG signal with voluntary eye blinks. This EEG signal has larger amplitude and longer duration (400 -500 ms) compared to that obtained from spontaneous eye blink. This signal has been filtered which remove the signals above 5 Hz leaving only very clear eye blinking signals. Other artifacts such as 50 Hz power line interference and noise have also been removed using analogue filtering provided by the EEG instrument.

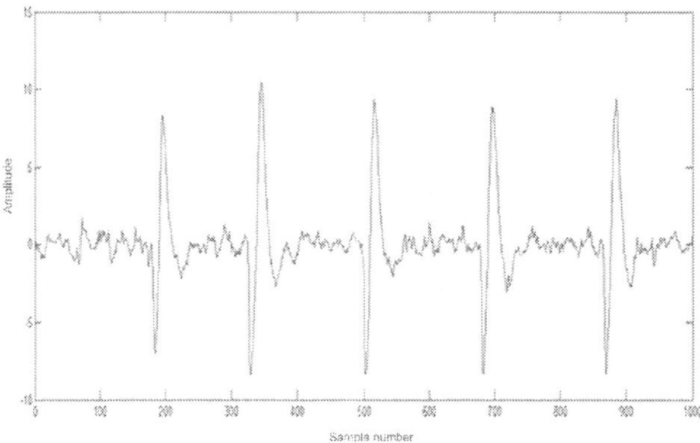

Fig. 5. EEG signal for voluntary eye blinking condition.

The suitable event for activating a system is three continuous eye blinks (with a duration of 1.5 to 2.5 seconds between eye closure and opening as they are not present when the subject is in relax condition [Abd Rani et al., 2009]. The duration between the first cycle of eye opening and closure and the second cycle should be 3 to 4.5 seconds.

3. Neural signal processing

Basically, there are two ways of acquiring the EEG signals from the subjects; invasive and non-invasive techniques. In the invasive technique, electrodes are implanted in the subject's brain and located on the brain surface whereas the non-invasive technique uses electrodes that are placed on the scalp. In most cases, non-invasive technique is more preferable than the invasive technique since it is harmless and easy to use. The standard electrode placement for the non-invasive technique is called International 10-20 system where 10% and 20% of a measured distance starting from craniometric reference points such as nasion, inion, left and right pre-auricular points are used to locate the EEG electrodes. The placement of electrodes for 10-20 system is shown in Figure 6. In this arrangement, a reference and ground electrodes are placed either on the ear lobe or mastoid.

As mentioned previously, the recorded EEG signals contain artefacts which have to be removed in order to obtain good morphological signals. Once a clean EEG signal is obtained, the second stage is to amplify the signal. EEG signal amplitude obtained from the scalp is very small, range up to 100mV which is difficult to see without amplification. The signal also has low frequency. It is necessary to analyse the signal to examine the characteristics of the signal and to ensure the noise has been removed. The signal can be analysed using Fast Fourier Transform (FFT), time-frequency analysis or time scale analysis. The FFT only gives frequency information of the signal, thus, time-frequency analysis or spectrogram is normally used to view the frequency at each time point.

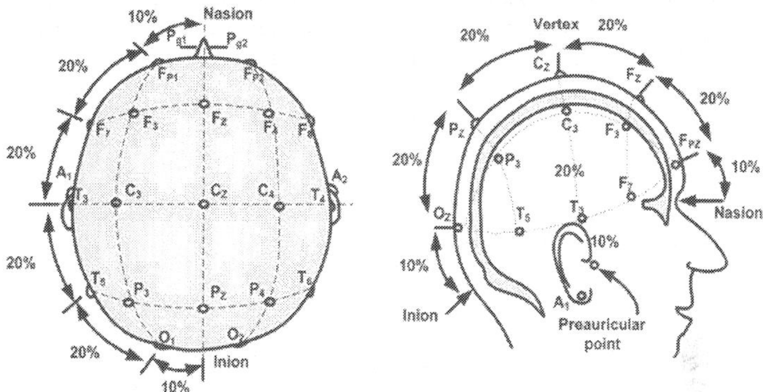

Fig. 6. The International 10-20 System of Electrodes Placement. (Redrawn from http://www.bem.fi/book/13/13.htm#03) [Norani et al., 2010]

The next stage is extracting the underlying information in the signal. Depending on the purpose of the study, this stage can be feature extraction or event detection as shown in Figure 7. If the EEG signal is to be used for activating equipment, a simple and an easy way is to detect an event from the signal, for example eye blinks and use the output which in the form of pulses to activate the equipment. Classification process is necessary if specific features are needed to perform the activation. This stage is also called translation process where the pattern classified is translated into suitable signal to activate equipment.

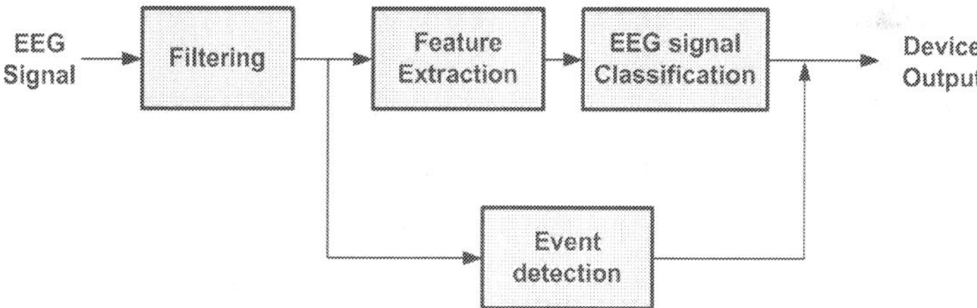

Fig. 7. EEG signal processing.

4. Embedded system

A computer system that is embedded in an electronic device to perform specific functions is called embedded system. It forms part of the system and controls one device or many devices. The main controller in this system is either a microcontroller or digital signal processing. A microcontroller is a small computer on a single integrated circuit which is designed to control devices. It consists of CPU, memory, oscillator, watchdog and input output units on the same chip. The microcontroller is available in wide range from 4 to 64 bits.

A PIC microcontroller is commonly used in embedded system due to its simplicity and ease of use. It offers several advantages such as design time saving, space saving and no compatibility problems. However, it has limited memory size and input/output capabilities. Figure 8 shows the block diagram of internal architecture of PIC16F877 microcontroller. The PIC microcontroller is built around Harvard architecture where two memories; one for program and the other one for data are separated. Separate buses are used for program and data memories. This eliminates jumping of program code into data or vice versa. PIC microcontroller uses RAM memory or known as file registers to store data during execution and a working register called W register to perform arithmetic and logic functions. User program is stored in the flash program memory and a status register is used to indicate the status of microcontroller through flag bits such as carry, zero, digit carry flags and others. PIC16F877 microcontroller has three 8 bits parallel input/ouput ports, 8 channels of analog inputs and serial outputs.

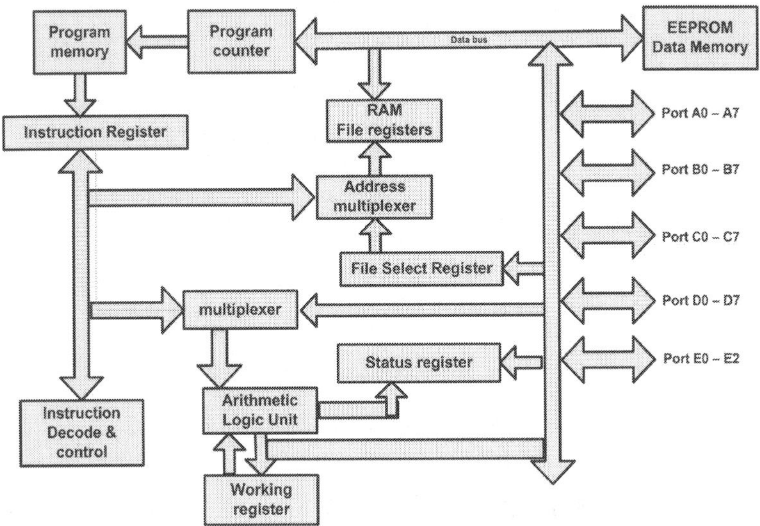

Fig. 8. Internal architecture of PIC16F877A microcontroller.

5. EEG based fan system incorporating microcontroller

A block diagram of EEG based fan system incorporating microcontroller is shown in Figure 9. It consists of EEG acquisition system, a microcontroller and a motor circuit. Three electrodes are connected to the EEG acquisition system and located on the subject's scalp at frontal, occipital and ear lobe. The EEG acquisition system is responsible for recording the EEG signals and passing the signals to the microcontroller system. The recorded EEG signal consists of voluntary and spontaneous eye blinks. Thus, to activate the motor, four seconds eye blinks in EEG signal is used. The functions of the microcontroller are to process the EEG signal, detect four-second eye blinks and use the detection results to control the movements of motor that is connected to a fan. Here, PIC16F877 is used as it can read analogue signal directly without the need of external analog to digital converter circuit. Three eye blinks within duration of four seconds are used since it is the best technique to activate a system [Abd Rani et al, 2009].

There are a few ways of connecting a motor to the microcontroller. If a dc motor is used, a circuit shown in Figure 10 can be implemented. This is a simple circuit which requires 5V supply to operate. A relay can be used to activate the motor if it is connected to 240V ac supply. Figure 11 shows the connection of the microcontroller to the devices on the motor circuit that comprises a transistor, a diode, a relay and a motor.

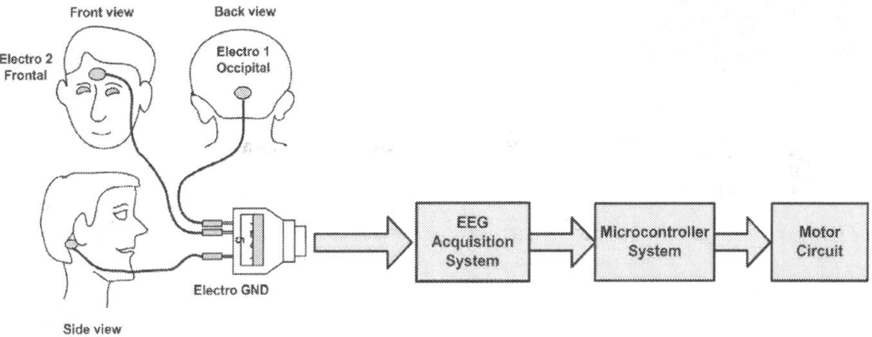

Fig. 9. Block diagram of EEG based fan system.

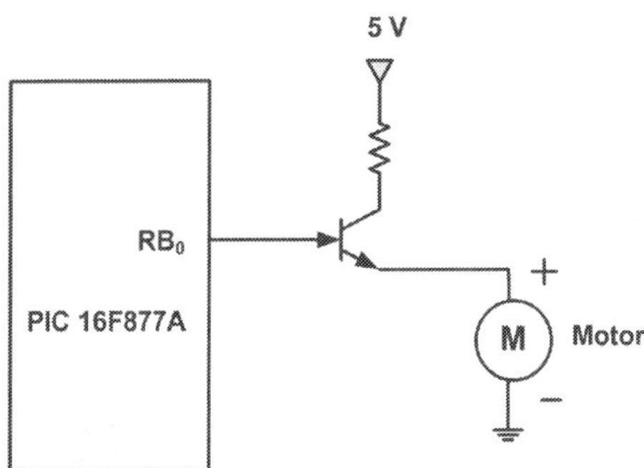

Fig. 10. A simple connection of a dc motor to PIC16F877A.

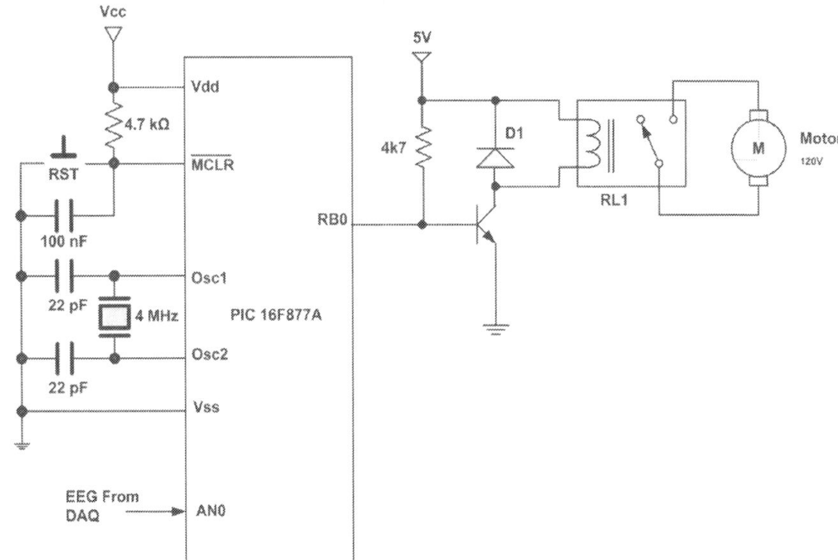

Fig. 11. Connection of a motor to PIC16F877A for EEG based fan system.

5.1 Controlling software for the EEG based fan system

The PIC microcontroller cannot work without software. A controlling program is required to read EEG signals from a subject, detect three time eye blinks and activate the motor. A program written in C language or PIC assembly language can be used to perform the detection and control operation. The process of detecting three time eye blink and activating the ac motor is shown in Figure 12. Initially, the program examines whether the EEG signal amplitude exceeds the maximum threshold voltage, is below minimum threshold voltage or lies between the threshold voltages. When three eye blinks within duration of four seconds is detected, the program sends logic 1 to the output of the microcontroller to drive the relay to activate the motor.

5.2 Examining system functionality

The functionality of the system can be examined in two stages. In the first stage, the performance of the system in detecting four second eye blink is evaluated. Here, the recorded EEG signals that are stored in excel file are used. A digital acquisition card is used as a communication medium between the computer and microcontroller system. The function of DAQ card is to transfer the recorded EEG signals from the computer to the PIC microcontroller. A program written in Visual Basic is used to send the EEG signal from the computer to the microcontroller. To view the EEG signal received at the output of DAQ, an oscilloscope is placed at one of the DAQ analogue channels. The transmitted EEG signal displayed on the computer screen is compared with the signal observed on the oscilloscope. The eye blinks are detected using software written in C language which is programmed on the microcontroller. In order to view the signal send to the motor, the output of the microcontroller is connected to the oscilloscope.

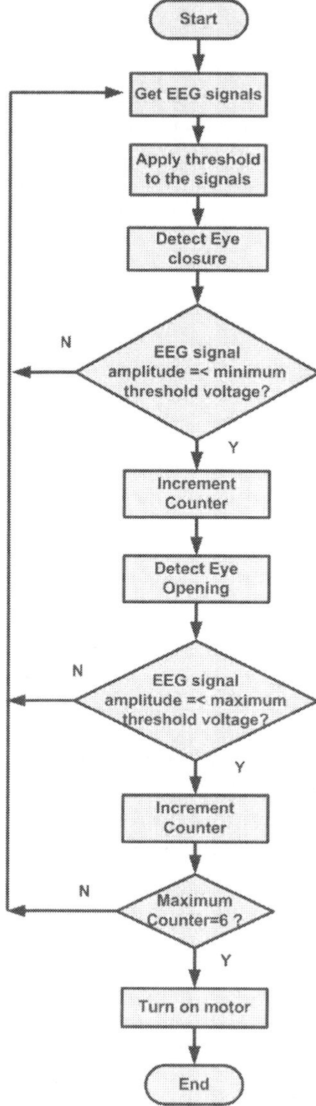

Fig. 12. Process of activating a motor using eye blink detected from EEG.

In the second stage, the functionality of the motor circuit is tested using a simple routine shown in Figure 13. A switch is connected to the input of the microcontroller to initiate the testing. When the switch is turned on, the routine activates the relay that is connected to the motor. Once it is confirmed that the eye blink detection module and motor activation routines are working successfully, these routines can be combined together.

```
//Program written for PIC programming to run the a motor connected to ac supply.
#include "16F877a.h"
#byte    PORTB=0x06
#byte    TRISB=0x86
#byte    TRISA=0x85
#byte    PORTA=0x05
#use delay(clock=400000)


//main function
void main()
{
        TRISA = 0x01;     //set PORTA to input
        PORTA = 0x00;     //set RA0-RA7 low
        TRISB = 0x00;     //set PORTB to output

  while(1)                        // Loop always
      {
    while(input(PIN_A0))                    // Read status of a switch
            {
                    delay_ms(20);    // Delay for 20 ms
                    PORTB = 0x01;    // Turn on the fan connected to a motor
            }
            PORTB = 0x00;            // turn off the motor if the switch
                                     // is off

      }
}
```

Fig. 13. A simple routine to test the functionality of the motor.

6. Results and discussions

The EEG signal containing eye blinks observed at the output of DAQ is shown in Fig. 14. This signal contains three eye blinks and the length of the signal is 4 seconds. The eyelid closure and opening can be observed clearly through the signal negative and positive amplitudes.

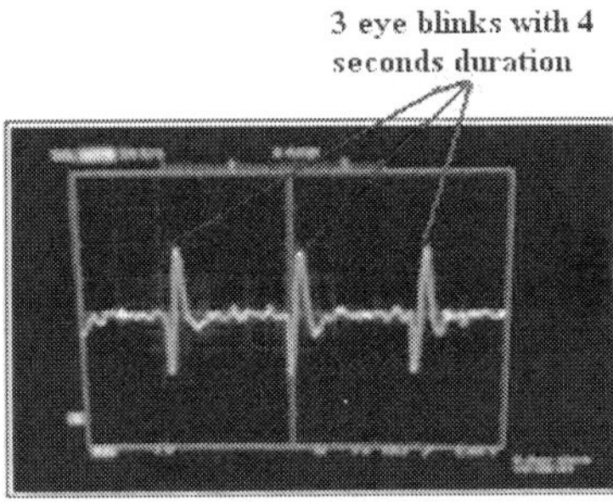

Fig. 14. EEG signal observed at the output of DAQ. [Wahy et al, 2010]

Figure 15 shows the EEG signal with voluntary eye blinks and the pulse obtained when four second eye blinks is detected. This pulse is observed at the Port B of the PIC16F877A microcontroller. The motor starts moving once the transistor connected to port B is switched on.

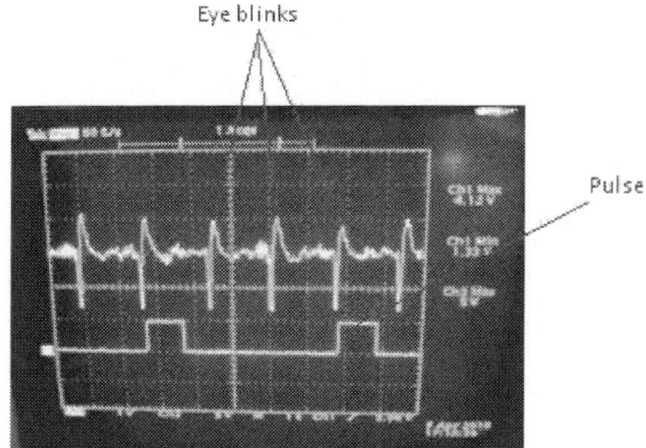

Fig. 15. Pulses generated at the output of PIC microcontroller when three eye blinks are detected. [Wahy et al, 2010]

Figure 16 shows the EEG signal when the subject is in relax condition. This EEG signal contains spontaneous eye blinks which are not detected by the PIC microcontroller. The amplitude of spontaneous eye blinks is below the threshold value which causes the PIC ignores them and no pulse is generated at the output.

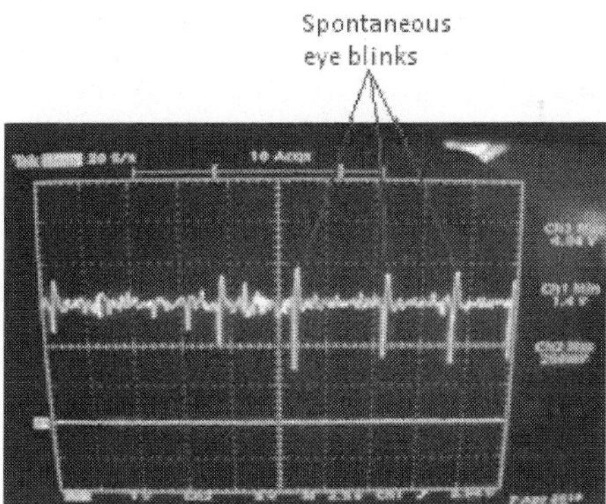

Fig. 16. EEG with spontaneous eye blinks observed at the output of PIC microcontroller. [Wahy et al, 2010]

7. Conclusion

A system that can activate a fan using EEG signal detected by a microcontroller has been described in this paper. The results showed that eye blinks can be detected successfully using PIC16F877A. With a program running on PIC16F877 microcontroller, a simple motor can be activated using neural signal. This application is suitable for people who cannot move their hands or the whole body to control a fan. Using this system, users can control a fan easily without any conventional remote controller. This system is useful for elderly people and disable persons as well as able-bodied people.

For future work, wireless electrodes should be employed in this system. The purpose is to make the users to feel comfortable with no wires hanging on their head. With wireless connection, the microcontroller module can be located at a distance from the user which provides more freedom for normal person to move around. However, this system requires intelligent software to eliminate interference and prevent false detection.

8. Acknowledgment

The authors would like to thank Universiti Teknologi MARA, Shah Alam, Malaysia for providing facilities to carry out this research project.

9. References

Abd Rani, M. S. ; Mansor, W.; Detection Of Eye Blinks From EEG Signals For Home Lighting System Activation, *Proc of International Symposium on Mechatronics and Its Applications*, pp 1- 4, ISBN 978-1-4244-3480-0, Sharjah, UAE, March 24-26, 2009

Bulling, A; Ward, J. A.; Gellersen, H; Troster, G.; Eye Movement Analysis for Activity Recognition Using Electroculography, *IEEE Transaction on Pattern Analysis and Machine Intelligence*, Vol. (33), No. 4, (April 2011), pp. 2006, ISSN 0162-8828

Breau, F.; Marsden, B.; McCluskey, J.; Ellwood, R. J.; Lewis, J.; Light Activated Position Sensing Array for Persons with Disabilities, *Proceedings of IEEE on Bioengineering*, ISBN 0-7803-8285-4 , pp 204-205, April 17-18, 2004

Ding, Q.; Tong, K., Li, G.; Development of an EOG (Electro-Oculography) Based Human-Computer Interface, *Proceedings of IEEE EMBS*, Shanghai, China, ISBN 1-4244-0032-5, pp. 6829-6831, Aug 30 – Sept 3, 2006

Harun, H. and Mansor, W.; EOG Signal Detection for Home Appliances Activation", *IEEE Colloquium on Signal Processing and Its Applications, Kuala Lumpur, Malaysia*, ISBN 978-1-4244-4151-8, pp 195-197, March 6-8, 2009

Kumar, D. and Poole, E.; Classification of EOG for Human Computer Interface, *Proceedings of IEEE EMBS/BMES, USA*, ISBN 0-7803-7612-9, pp 64–67, 2002

Norani, N.A.M.; Mansor, W.; Khuan, L.Y.; A review of Signal Processing in brain computer interface system, *Proceedings of IEEE on Biomedical Engineering and Sciences*, ISBN 978-1-4244-7599-5, pp 443 – 449, Nov 30 – Dec 2, 2010.

Wahy, N.; Mansor, W.;EEG Based Home-lighting System, *Proceedings of IEEE on Computer Applications and Industrial Electronics*, ISBN 978-1-4244-9054-7, pp 379-381, Dec 5-8, 2010.

3

Artificial Intelligent Based Friction Modelling and Compensation in Motion Control System

Tijani Ismaila B., Rini Akmeliawati and Momoh Jimoh E. Salami
Intelligent Mechatronics Systems Research Unit,
Department of Mechatronics Engineering,
International Islamic University Malaysia
Kuala Lumpur,
Malaysia

1. Introduction

The interest in the study of friction in control engineering has been driven by the need for precise motion control in most of industrial applications such as machine tools, robot systems, semiconductor manufacturing systems and Mechatronics systems. Friction has been experimentally shown to be a major factor in performance degradation in various control tasks. Among the prominent effects of friction in motion control are: steady state error to a reference command, slow response, periodic process of sticking and sliding (stick-slip) motion, as well as periodic oscillations about a reference point known as hunting when an integral control is employed in the control scheme. Table 1 shows the effects and type of friction as highlighted by Armstrong et. al. (1994). It is observed that, each of task is dominated by at least one friction effect ranging from stiction, or/and kinetic to negative friction (Stribeck). Hence, the need for accurate compensation of friction has become important in high precision motion control. Several techniques to alleviate the effects of friction have been reported in the literature (Dupont and Armstrong, 1993; Wahyudi, 2003; Tjahjowidodo, 2004; Canudas, et.al., 1986).

One of the successful methods is the well-known model-based friction compensation (Armstrong et al., 1994; Canudas de Wit et al., 1995 and Wen-Fang, 2007). In this method, the effect of the friction is cancelled by applying additional control signal which generates a torque/force. The generated torque/force has the same value (or approximately the same) with the friction torque/force but in opposite direction. This method requires a precise modeling of the characteristics of the friction to provide a good performance. Hence, in the context of model-based friction compensation, identification of the friction is one of the important issues to achieve high performance motion control.

However, as discussed in the literatures, several types of friction models have been identified (Armstrong et al., 1994; Canudas et. al., 1995; Makkar et. al., 2005) and classified as static or dynamic friction models. Among the static models are Coulomb friction model, Tustin model, Leuven model, Karnop model, Lorentzian model. Meanwhile Dahl model, Lugre model, Seven parameters model, and the most recent Generalized Maxwell-Slip (GMS) model, are among the dynamic friction models (Tjahjowidodo, 2004). The static friction model is simple and easy in the identification process, however using such model

Tasks	Friction Effects	Dominant Friction
Regulator (pointing/position control)	Steady-state error, hunting	Stiction
Tracking with velocity reversal	Standstill, and lost of motion	Stiction
Tracking at low velocity	Stick-slip	Stribeck friction, stiction
Tracking at high velocity	Large tracking error	Viscous behavior of lubricant

Table 1. Control tasks and associated friction effects

for friction compensation usually lead to poor performance especially at very low velocity control.

On the other hand, the accuracy of the dynamic friction model is anchored on the dependency of friction on immeasurable internal states such as velocity and position. Since friction model selection is an essential factor in the model-based friction compensation, it is important to find an appropriate friction model that will effectively alleviate the frictional effects in motion control applications. This has been the basis for the continuous search for more efficient and simple model for friction identification and compensation in motion control system.

The recent development in Artificial Intelligent (AI) makes it adaptable for system modeling base on the data training and expert knowledge. It has been shown that the major AI paradigms (Neural Network, Fuzzy Logic, Support vector machine etc.) have the capability of approximating any nonlinear functions to a reasonable degree of accuracy; and hence, have been identified and proposed as appropriate alternatives for friction model and compensation in motion control systems, (Bi et.al., 2004; Kemal and Masayoshi, 2007; Wahyudi and Ismaila, 2008). In addition, the use of artificial intelligence based friction model may also reduce both the complexity and time consumed in the friction modeling and identification.

This chapter first presents an overview of model-based friction techniques which have been used in friction modeling and compensation in motion control systems. Then the application of artificial intelligent based methods in this area is reviewed. The development, implementation and performance comparison of Adaptive Neuro-Fuzzy inference system (ANFIS) and Support Vector Regression (SVR) for non-linear friction estimation in a motion control system so as to achieve high precision performance are described. These two AI techniques are selected based on their unique characterstics over others as discussed latter in this paper. A comparative study on the performance of these two AI techniques in terms of modeling accuracy, compensation efficiency, and computational time is examined. The chapter is concluded with highligths of summary of the results of the study and future directions of research in this area.

2. Review of friction modelling techniques in motion control system

The study of friction is dated back to the work of Leonardo da Vinci (1452-1519) who investigated the nature of friction and proposed the basis for the theory of classical friction. According to da Vinci (1452-1519) theory of friction, and latter work of Amontons (1699), and Charles (1785) friction is proportional to load, opposed motion, and is independent of contact area. With the birth of tribology and its recent advancement, details about the topography of contact between bodies especially at atomic level have been more detailed and investigated by Armstrong (1991) and recently revisited by Farid (2008).

Two main regimes have been identified for friction, namely: pre-sliding and sliding. Pre-sliding regime defines friction at very low velocity prior to sliding motion and is a function of displacement, while sliding regime covers the period when the body is sliding/in motion and during this period friction is a function of velocity of motion. Some of the challenges in friction model includes the merging of friction model in both regimes in order to offer a smooth transition from pre-sliding to sliding regime whichtakes into consideration frictional effects such as: Stribeck, stick-slip, hysteresis, break-away force, nonlocal memory, and friction lag. For motion control applications, friction study is been carried out to compesate its negative effects on control performances.

Several methods have been adopted for friction compensation in the research domain and industry. Detailed review was given by Armstrong (1994). Non-model-based compensation includes the use of stiff proportional-derivative (PD) control, integral control with deadband, dither, impulsive control and joint torque control and nonlinear controllers. Stiff PD approach involves the use of either high derivative (velocity) feedback or high proportional position feedback. This has been shown to be effective for stable tracking and for system designed for high rigidity.

The use of integral control to eliminate the steady state error due to friction is confronted with the problem of limit cycles. This necessitates the introduction of deadband at the input of the integrator control block, thereby limiting the attainable steady state accuracy (Shen and Wang, 1964).

Dither isa high frequency signal added to the control signal to eliminate the effects of the nonlinearities which include friction in the system. The application of dither in aerospace control was reported by Oppelt (1976). The challenges in application of dither lies in its mode of generation and application.

Others form of non-model based techniques include impulsive control, joint torque (Armstrong, 1992; Hashimoto et.al., 1992).

The use of nonlinear controllers has also been reported by many researchers. PD controller plus a discontinuous nonlinear proportional feedback (DNPF) was proposed by Southward et.al.,(1991), while PD plus smooth robust nonlinear feedback (SRNF) was investigated by Cai and Song (1993). A compensation scheme using nominal characteristic trajectory following (NCTF) was presented by Wahyudi et al., (2005) and this has been reported to outperform both the DNPF and SRNF techniques.

The concept of model-based friction compensation is depicted in Figure 1, where the friction signal $\hat{u}_f$ is approximately equal to the actual plant friction u_f, that is $\hat{u}_f = u_f$; u_c is control signal generated by the linear controller G_c; u_{in} is actual input control signal into the plant; θ_r is reference position signal; θ_{out} is output position response of the system; $\dot{\theta}$ is velocity signal; G_c is a linear controller designed with nominal plant model; G_1 is sub-system model 1 and G_2 is sub-system model 2.

Though very simple, the effectiveness of the technique is anchored on the precision of the friction model and the velocity estimation. It is implemented as either feedforward model-based when the desired reference velocity is taken as the input to the model, or feedback model-based when the input velocity is estimated from the sensed output. Both methods of implementation have been adopted by different authors as reported by Armstrong (1994).

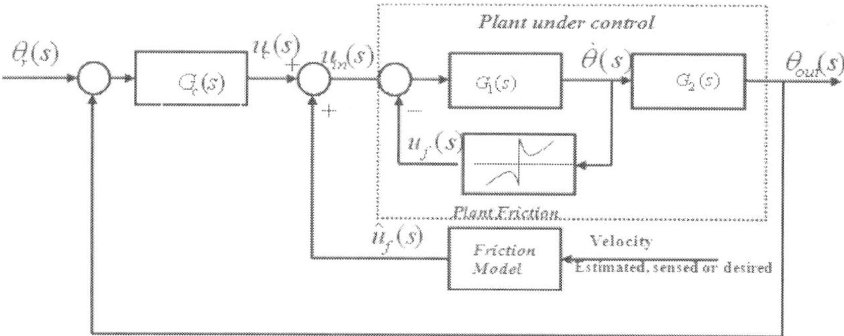

Fig. 1. Block diagram of basic model-based friction compensation**

2.1 Parametric based friction models

Coulomb friction is the earliest physical model of friction based on the work of Da Vinci (1519), Amontons (1699) and Coulomb (1785). It is described as a constant opposing force independent of velocity of motion and is mathematically given by

$$F_f = F_c \, \mathrm{sgn}(\dot{\theta}) \tag{1}$$

and illustrated by Figure by Figure 2a

The viscous friction was developed by Reynold (1866) followed the birth of the theory of hydrodynamics. Viscous friction is proportional to velocity, and it is zero when velocity goes to zero

$$F_f = F_\theta \dot{\theta} \tag{2}$$

This led to the well known combine Coulomb plus viscous static model shown in Figure 2 (b), and represented by

$$F_f = F_c \, \mathrm{sgn}(\dot{\theta}) + F_\theta \dot{\theta} \tag{3}$$

This model has been widely applied in control system due to its simplicity. It has been experimentally proven to be efficient for application above certain minimum velocity (Armstrong, 1991). Canudas et al. (1986) employed Coulomb and viscous model in an adaptive model-based friction compensation and has reported an improved performance in terms of positioning accuracy. Based on its historical place in friction modeling, it is often used for benchmarking the performance of other more complex models (Tjahjowidodo, 2004; Wahyudi and Tijani, 2008). The major problems with this model have been the failure to account for friction at zero velocity and other several friction behaviors especially at low velocity.

Morin (1833) introduced the idea of friction at rest known as stiction or static friction. Stiction friction is defined as the force (torque) requires to initiate motion from rest, and is generally greater than the Coulomb (Kinetic) friction. Friction was then seen to depend not only on velocity but magnitude and rate of the external force. This resulted in a complete model of static friction as shown in Figure 2(c). However, Stribeck (1902) observed a decreasing friction with increasing velocity at low velocity during the transition from stiction to kinetic friction and he proposed the concept of Stribeck friction shown in Figure 2(d). In order to overcome the jump discontinuity of the model at zero velocity, a modification was introduced (Karnopp, 1985) by replacing the jump with a line of finite slope as shown in Figure 2(e). A combination of stiction, Stribeck, Coulomb and viscous friction model is been referred to as Stribeck friction (Armstrong, 1991) or General Kinetic Friction (GKF), (Evangelos et.al, 2002), and is described by

$$F_f = \begin{cases} F_f(\dot{\theta}) & \dot{\theta}(t) \neq 0, \ddot{\theta} = 0 \\ F_e & \dot{\theta}(t) = 0, \ddot{\theta} = 0, |F_e| < F_s \\ F_s \operatorname{sgn}(F_e) & \dot{\theta}(t) = 0, \ddot{\theta} \neq 0, |F_e| > F_s \end{cases} \tag{4}$$

Several variant of Stribeck friction has been reported and evaluated by Armstrong (1991). A general exponential form is given by

$$F_f(\dot{\theta}) = \left[F_c + (F_s - F_c)\exp(-\left|\dot{\theta}/\dot{\theta}_s\right|^\delta) \right]\operatorname{sgn}(\dot{\theta}) + F_{\dot{\theta}}\dot{\theta} \tag{5}$$

where F_f, F_s, F_c, and $F_{\dot{\theta}}$ are the friction force, stiction, kinetic and viscous frictions respectively, $\dot{\theta}$ is the velocity of motion, $\dot{\theta}_s$ is the Stribeck velocity, constant δ is an empirical parameter that determines the shape of the model, in which $\operatorname{sgn}(\dot{\theta})$ is defined as

$$\operatorname{sgn}(\dot{\theta}) = \begin{cases} +1 & \dot{\theta}(t) > 0 \\ 0 & \dot{\theta}(t) = 0 \\ -1 & \dot{\theta}(t) < 0 \end{cases} \tag{6}$$

where values of $\delta = 1$ and $\delta = 2$ indicate the Tustin /exponential model (1947) and Gaussian model respectively.
Hess and Soom (1990) proposed another model of the form

$$F_f(\dot{\theta}) = \left[F_c + \frac{F_s - F_c}{1 + (\dot{\theta}/\dot{\theta}_s)^2} \right]\operatorname{sgn}(\dot{\theta}) + F_{\dot{\theta}}\dot{\theta} \tag{7}$$

which is known as Lorentzian friction model.
Tustin (1947) was the first to make use of a negative viscous friction (stribeck) in the analysis of feedback control. Armstrong (1991) employed exponential, gaussian, Lorentzian together with a polynomial model given by

$$F_f(\dot{\theta}) = F_c + F_2\dot{\theta}^2 + F_3\dot{\theta}^3 + F_4\dot{\theta}^4 + F_5\dot{\theta}^5 + F_6\dot{\theta}^6 + F_7\dot{\theta}^7 + F_{\dot{\theta}}\dot{\theta}^8 \tag{8}$$

for friction identification in a robot arm system. The Lorentzian model gave best performance fit and was later adopted for the friction compensation.
Several other researchers have employed the complete stribeck model both for fixed and adaptive model-based friction compensation (Envangelos, et.al., 2002; and Lorinc and Bela,

2007). Improved performance with respect to tracking and steady state accuracy have been reported by them. A continuous, differentiable friction model with six parameters was recently proposed by Makkar et al., (2005). The performance of the model was evaluated with numbers of simulations and found to account for major friction effects such as Coulomb, viscous, and stribeck. Its experimental implementation for friction compensation has not yet been reported.

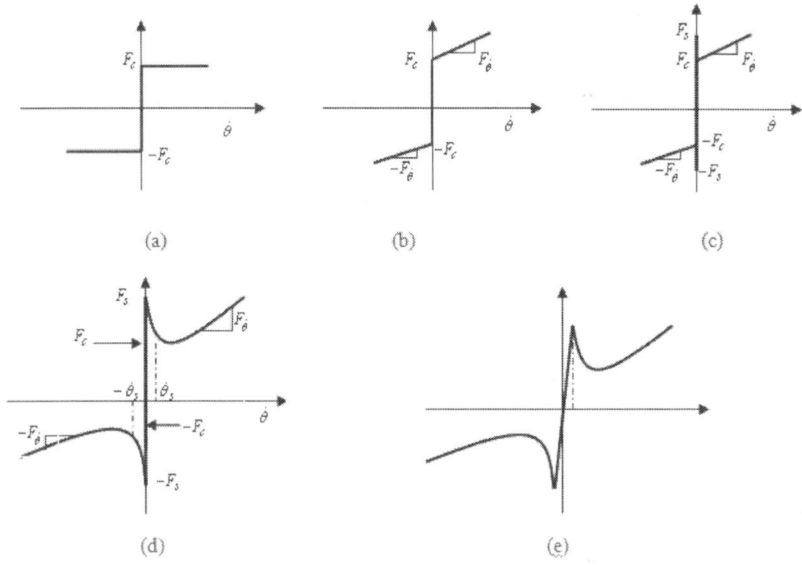

Fig. 2. Static friction models (a) Coulomb friction,(b) Coulomb + Viscous friction (c) Stiction + Coulomb + Viscous friction (d) Stiction + Stribeck + Coulomb + Viscous and (e) Modified Stribeck friction (Karnopp Model)

Though the General Kinetic Friction (GKF) fails to account for pre-sliding friction behaviors and other dynamics characteristics such as friction lag and local memory hysteresis, experimental works have proven that a good static friction model can approximate the real friction force with a degree of confidentiality of 90% (Armstrong, 1991; Lorinc and Bela, 2007). Also, Canudas de Wit et al., (1995) demonstrated that the simulated static friction model and dynamic friction model predicts almost the same limit cycles generated by friction in controlled positioning system. Hence, static friction model-based compensation and identification techniques still have great significant practical applications.

Dynamic friction models have been proposed to account for various pre-sliding friction behaviors and these are becoming essentials for higher precision performance at micro- and nano- scale velocity and positioning control (Yi et. al., 2008). Some of the common dynamic models which have been considered in control applications are Dahl, Lugre, Leuven, and Generalized Maxwell-Slip (GMS). Dahl model (1968) was the first simple dynamic model proposed for simulations of control system with friction. This was used for adaptive friction compensation by Ehrich (1991) and is expressed as

$$\frac{dF}{dx} = \sigma\left(1 - \frac{F}{F_c}\text{sgn}(\dot{\theta})\right)^{\alpha} \tag{9}$$

where F is the friction as a function of displacement x, F_c is the Coulomb friction, $\dot{\theta}$ is the motion velocity and α is empirical parameter which determines the shapes of the model.

It is position dependent model which captures the hysteresis behavior of friction but fails to account for stiction and Stribeck.

Another dynamic model was proposed and implemented by Canudas de Wit et al. (1995). In addition, Canudas de wit et al. (1995) modified the Dahl model to incorporate breakaway (stiction) friction and its dynamics together with Stribeck effect using exponential GFK to give what is been referred to as Lugre friction. This model captures most of the experimentally observed friction characteristics, and is the first dynamic model that seeks to effect smooth transition between the two friction regimes without recourse to switching function. It is mathematically given by

$$\frac{dz}{dt} = \dot{\theta} - \sigma_o \frac{|\dot{\theta}|}{g(\dot{\theta})} z, \tag{10}$$

$$F_t = \sigma_o z + \sigma_1(\dot{\theta})\frac{dz}{dt} + F_{\dot{\theta}}(\dot{\theta}) \tag{11}$$

where z is average of bristle deflection, F_t is the tangential friction force, $g(\dot{\theta})$ is stribeck friction for steady-state velocities, $F_{\dot{\theta}}$ is viscous friction coefficient, while σ_o and σ_1 are dynamic parameters, which are respectively the frictional stiffness and frictional damping.

Lugre model has been employed for friction analysis and compensation in various control systems (Wen-Fang, 2007).

However, Lugre model fails to capture the non-local memory effect of hysteresis. Leuven model proposed by Swevers et. al., (2000) is an elaborate model than Lugre as it incorporating hysteresis function with non-local memory behavior in pre-sliding regime. Apart from its complexity that has rendered it less effective in control system application, Lampaert et. al., (2002) pointed out two major problems associated with Leuven model namely: discontinuity and memory stack algorithm.

GMS is a qualitative new formulation by Lampaert et.al. (2003) based on the rate-state approach of the Lugre and the Leuven models. It is noteworthy that despite the unique advantages of dynamic models, one of the major challenges associated with their practical implementation is the dependency of the models on unmmeasurable internal state of the system and/or availability of very high resolution of (order 10^{-6}) sensing devices (Armstrong, 1991). Hence, many of the reported works employing complex dynamic friction model are based on simulation study.

2.2 Non-parametric based techniques

Due to the complexity and difficulty associated with physical models of friction in terms of model selection, parameters estimation, and implementation, non-parametric based approach using Artificial Intelligent (AI) approach is been alternatively employed in control systems for friction identification and compensation. Neural network (NN), fuzzy logic

(FL)/adaptive neuro-fuzzy inference system (ANFIS), support vector machine (SVM), and genetic algorithm are among the common AI methods that have been reportedly used in positioning control system.

The theory of artificial neural network (ANN) is based on simulated nerve cells or neuron which are joined together in a variety of ways to form network. The main feature of the ANN is that it has the ability to learn effectively from the data, and has been identified as a universal function approximator (Haykin, 1999). ANN with back propagation was proposed by (Kemal M. Ciliza and Masaypshi Tomizukab, 2007; Wahyudi and Tijani, 2008) for friction modeling and compensation with varying structures and applications. The performance of classical friction model was compared with Multilayer Feedforward Network (MFN)-based friction model for friction compensation in (Wahyudi and Tijani, 2008), and MFN was reported to outperform the classical friction model. A hybrid ANN was developed by Kemal and Masayoshi (2007) where static and adaptive parametric models are combined with ANN to better capture the discontinuities at the zero velocity. A radial basis function (RBF) approach was proposed in (Du and Nair, 1999; and Haung et al., 2000) where the center points and variances of the Gaussian functions had to be chosen a priori. Gan and Danai (2000) developed model-based neural network (MBNN), and structured according to linearized state space model of the plant and incorporated into Lugre friction model in a Linear Motor stage.

Despite the extensive use of ANN for friction modelling, no ANN structure has been agreed upon for optimal friction modeling for a varieties of motion control systems. There is need to extend the notion of MBNN for other friction models that are suitable for some motion control systems. Some of the challenges associated with the use of ANN in friction modeling include: selection of appropriate structures (layers, neurons, and models) for a particular application, generalization and local minimal problems.

Though ANFIS has been applied in nonlinear system modeling and control (Stefan, 2000), its application in friction modeling and compensation in motion control has not received much attention in the literatures. ANFIS is a Tagaki Sugeno (TSK) based fuzzy inference system implemented in the framework of adaptive networks (Jang 1995). It has the ability to construct an input-output mapping based on both human knowledge (in the form of fuzzy if-then rules) and stipulated input-output data pairs. Existing work related to the use of Neuro-Fuzzy can be found in many areas such as velocity control in (Jun and Pyeong, 2000), (Chorng-Shyan 2003). In the latter case, fuzzy inference system was introduced to compensate for friction parameter variations. Recently Tijani et.al (2011) reported the application of ANFIS in friction modelling and compensation in motion control system. Their results confirmed that this technique produces better performance in friction modelling than paramteric methods.

Application of Support Vector Regression (SVR) in adaptive friction compensation was recently proposed (Wang et al., 2007, Ismaila et.al. 2009(b)). It is noted that SVR has not been extensively explored as compared to ANN for friction modelling. Also, other forms of SVR such as least square support vector regression regression (LS-SVR) has been proposed as alternative to SVR with a more simplified optimization algorithm (Johan, Van Gestel, De Brabanter and Vandewalle, 2002), however it is yet to be employed in friction identification. In addition, GA was employed for the estimation of optimal parameters for Lugre parametric models by De-peng (2005), while hybrid of ANN and Gafor friction modelling has been reported in (Sung-Kwun et al., 2006).

3. System modelling and identification

Development of an appropriate mathematical model is the first step in order to characterize friction associated with motion control system. Figure 2 shows the experimental set-up of a DC motor-driven rotary motion system which consists of servo motor driven by an amplifier and position encoder attached to the shaft as the feedback sensor. The input to the motor is the armature voltage u driven by a voltage source. The measurable variable is the angular position of the shaft, θ in radian, while the angular velocity of the motor shaft ($\dot{\theta}$ in radian/s) is estimated using an appropriate digital filter. The plant was integrated into MATLAB xPC target environment as shown in Figure 3 for real-time experimental implementation.

Basically, in line with model-based friction compensation approach, the system can be decomposed into nominal (linear model) and non-linear sub-systems as shown in Figure4. The nominal/linear sub-system is obtained from the physics of the system based on first principle approach and system identication process for linear parameters estimation (Tijani et.al,2009). The nonlinear sub-system on the other hand, represents the friction present in the system. The friction occurs between various moving moving parts in the system. For instance, it exists between the motor shaft and bearing, encoder shaft, external shaft, load and associated bearing. As stated in section 2.1, the friction can take different form depending on the geometary of the system and operating conditions. In this study, major sliding friction effects dominating the sliding motion regime are considered. This consists of stiction, Stribeck, and coulomb frcition as shown in Figure 2e. Note that the viscous friction is regarded is included in linear sub-system model and its detailed derivation is reported in (Tijani, 2009). The resulting second order mathematical model is given as

$$G(s) = \frac{\theta(s)}{u(s)} = \frac{K}{s(\tau_p s + 1)}$$

$$(12)$$

where $K = 275$ and $\tau_p = 0.1009$

3.1 Friction identification experiments

Generally, in supervised AI-based modelling the availability of representative data is very important. Two major experiments are required to obtain the velocity to friction relationship for both break-away friction force and Stribeck friction. The major hardware, apart from the Host and Target PC, are the National Instrument (NI) Multifunction input-output (I/O) data acquisition (DAQ) PCI6024E, with BNC-2110 adapter for data acquisition to and from the Target Pc. A Scancon incremental shaft encoder with resolution of $2x10^{-4}$ (in quadrature mode) was used for measuring the position in radian. A current sensor with 0-5Amp current rating which is above the maximum current rating of the motor, 2Amp. was used for measuring the armature current. A simple experiment based on Ohm's Law was carried out to test and model the V-I relationship of the sensor prior to the performance of the experiment. This is required to transform the output voltage of the current sensor to coresponding current.

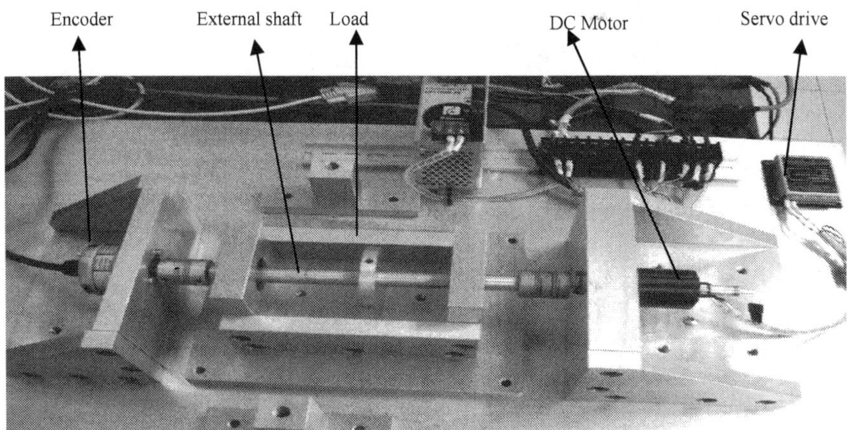

Fig. 2. DC-Motor driven rotary motion systems.

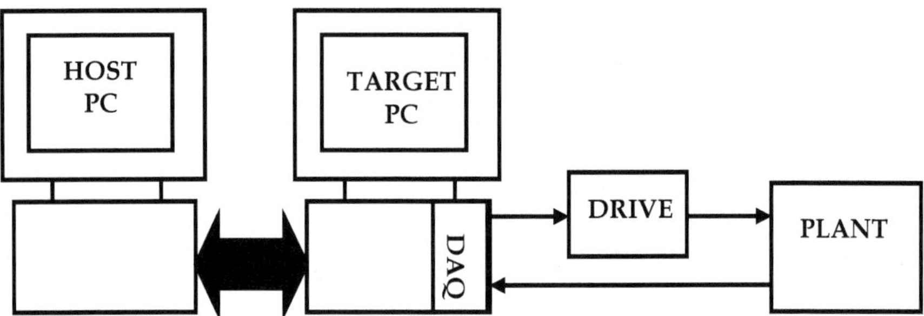

Fig. 3. MALAB xPC target set-up.

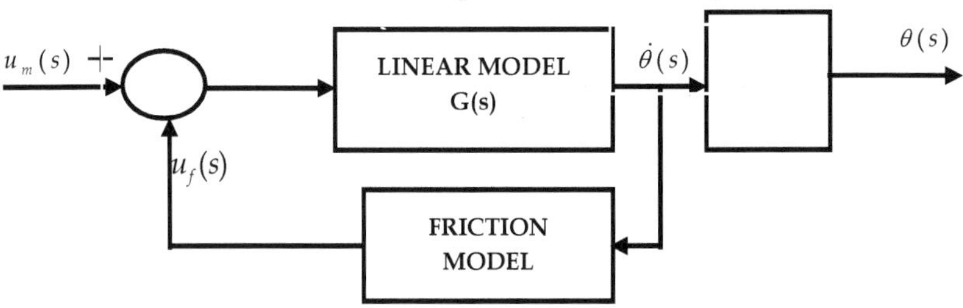

Fig. 4. Complete system model.

The resulting voltage-to-current relationship is given by

$$I_s = 7.8555V_s - 19.6544 \tag{13}$$

where V_s is sensor output in volts and I_s is equivalent current sensor output in amperes.

The first experiment tagged break-away experiment is to yield the break-away friction force (τ_f) in an open-loop mode. The break-away force is the force requires to initiate motion, in other word it represents the stiction friction at zero velocity, i.e. the $\tau_f(\dot{\theta})\big|_{\dot{\theta}=0}$. The systematic steps followed according to (Armstrong, 1991) are:

- "Warming-Up" of the Plant at beginning of each run
- Gradual Increase of the motor Current at steps of 0.001volts command signal in pen loop mode until the shaft moves (or breaks-away), this was taken to be at least 2 encoder counts.
- Repetition of steps 1-2 for several times and Averaging of results in order to guarantee repeatability

The procedures were repeated for both positive and negative directions of motion with 10 time runs for different days with a ramp input. The mean of the resulting values measured by the current sensor in volts is then computed to give the average stiction friction force 2.531volt and 2.475volt for poistive and negative direction of motion respectively. The difference between the friction force values in the poistive and negative directions of motion justifies the asymetric nature of friction.

The second experiment involves identification of steady-state velocity-friction relationship. The direct relationship between the friction torque, τ_f and motor torque τ_m at steady state (i.e when $\dot{\theta} \approx 0$) is explored in this experiment. At steady state, $\tau_f = \tau_m$, and since τ_m is proportional to the armature current i_a, it follows that τ_f is propotional to i_a. The experiement is conducted for a closed-loop system with an appropriate velocity controller. Though any linear controller can be employed, a stiff velocity control scheme such as the pseudo-derivative feedback with feedforward (PDFF) (Ohm,1990) has been shown to give better performance especially at low-velocity control regime (Tijani, 2009). A suitable velocity region is selected for both directions of motion to cover the low and high speed above the region of Stribeck effect. For each constant velocity within this region, the average of armature current and steady state velocity are then computed after the transient period of 0.2 second. Five different runs were carried out for each velocity input, and the overall mean is computed. A total of 108 data sets were obtained for each direction of motion. Figure 5 and Figure 6 show samples of the steady state responses of the plant for positive and negative directions respectively. Finally, the friction data aqcuired in voltage form based on the output of the current sensor is transformed into actual armature current using the V-I relationship in (13). The complete experimental data set for both directions are shown in Figure 7.

4. Artificial intelligent based friction modelling and compensation

The development of Artificial Intelligent (AI) based friction modelling and application of such model in friction compensation in motion control is described in this section. The objective is to demonstrate the suitability of AI techniques in friction compensation in motion control system. Though there exists several AI methods that can be applied based on their approximating capability, the focus in this section is on the ANFIS and SVR based on their unique characteristics over other AI methods.

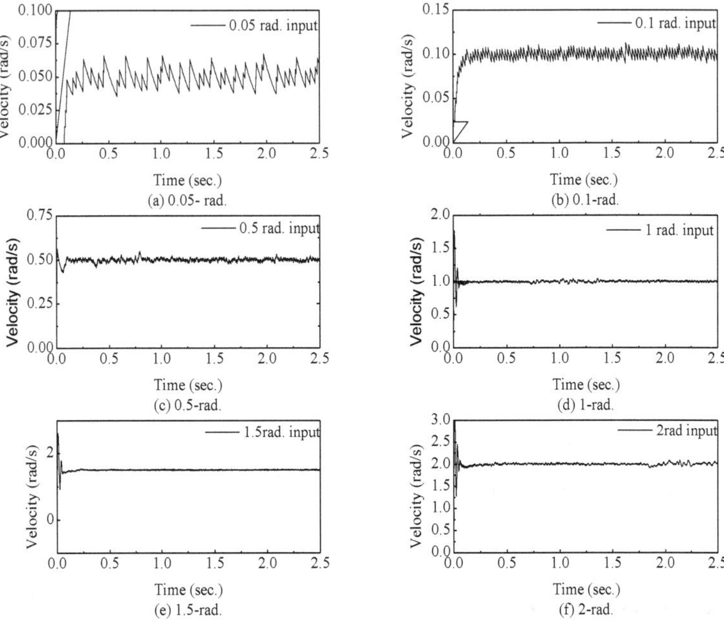

Fig. 5. Samples of positive steady-state velocity responses.

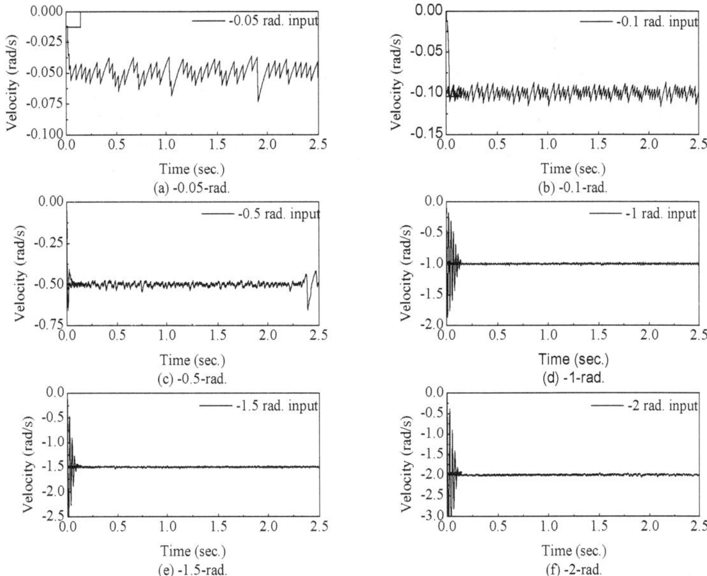

Fig. 6. Samples of negative steady-state velocity responses.

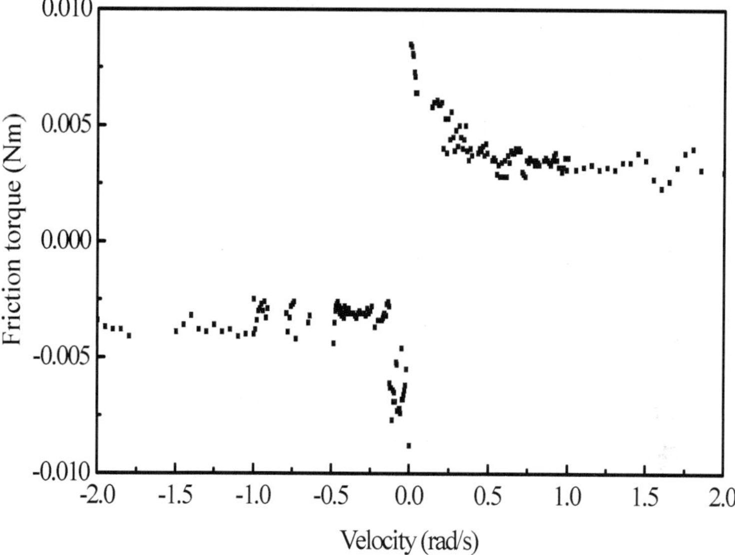

Fig. 7. Complete experimental friction-velocity data set for both positive and negative direction.

4.1 ANFIS and SVR as modeling tools

Both ANFIS and SVR are characterized with unique qualities that make them effective for nonlinear system identification and modeling. ANFIS is an hybrid AI-paradigm, integrating the best features of Fuzzy System (based on expert knowledge) and Neural Networks (based on data mining) in solving the problems of transforming the expert knowledge into fuzzy rules and tuning of membership functions associated with ordinary fuzzy inference system. On the other hand, SVR is an extension of the well developed theories of Support vector machine (SVM) to regression problems with introduction of ε -insensitivity loss function by Vapnik (1995). Unlike traditional learning algorithm for function estimation such as Neural network that minimizes the error on the training data based on the principle of Empirical risk minimization, SVR embodies the principle of structure risk minimization which minimizes an upper bound on the expected risk. Hence, it is characterized by better ability to generalize, and at the same time it is less prone to the problems of overfitting and local minimal. Though initially developed for linear function estimation, the principle of linear SVR was extended to non-linear case by the application of the kernel trick. Due to these unique advantages, SVR has been recently employed for non-linear function approximation and system modeling (Bi etal 2004, Ahmed etal 2008). A brief theoretical overview of the two paradigms are given here while full detail can be obtained in the literatures (Jang, 1993, Tijani et.al., 2011). It should be noted that there are two techniques of SVR namely $\varepsilon-SVR$ and $v-SVR$. The first is based on original concept of ε -insensitivity Vapnik (1995), and it involves the selection of appropriate ε -parameter for the modelling process. The challenges associated with the selection of ε is overcome by the use of $v-SVR$ in which a parameter v is introduced to facilitate the optimal computation of ε-sensitivity function. Tijani (2009) reported a comparison of these two techniques. $v-SVR$ was reporter with both better modelling and compensation accuracy of friction in motion control system. Hence, only the $v-SVR$ is reported in this chapter while the reader is referred to the literature for detailed review of the other two approaches

4.1.1 ANFIS overview

Basically, ANFIS implements Takagi Sugeno Fuzzy Inference System, and consists of five layers minus the input layer O as shown in Figure 8. Besides the input layer O, each other layer performs a specific function based on the associated node function as follows:

Layer 1 is responsible for the fuzzification of the input signal X_1 and X_2 with appropriate membership function. It consists of adaptive nodes in which the parameters of membership function are adjusted during learning process.

Layer 2 compute the firing strength ω_i of each rule using a T-norm (min, product, etc) of the incoming signals.

Layer 3 estimate the normalized firing strength, ϖ_i of each fuzzy rule

Layer 4 also consists of adaptive nodes for computing the consequence parameters Q_i.

Layer 5 compute the overall output, O using a linear combination of all the incoming signals from layer 4 :

Parametrically, ANFIS is represented by two parameter sets: the input/premise parameters and the output/ consequence parameters.

4.1.2 SVR overview

Given a set of N input/output data $\{x_i, y_i\}_{i=1}^{N}$ such that $x_i \in \Re^n$ and $y_i \in \Re$, the goal of $v-SVR$ learning theory is to find a function f which minimizes the regularized risk function(structural risk function) of the form (Sch"olkopf and Smola, 2002):

$$R^v_{reg}[f] := R_{emp}[f] + \frac{1}{2}\|w^2\| + v\varepsilon \tag{14}$$

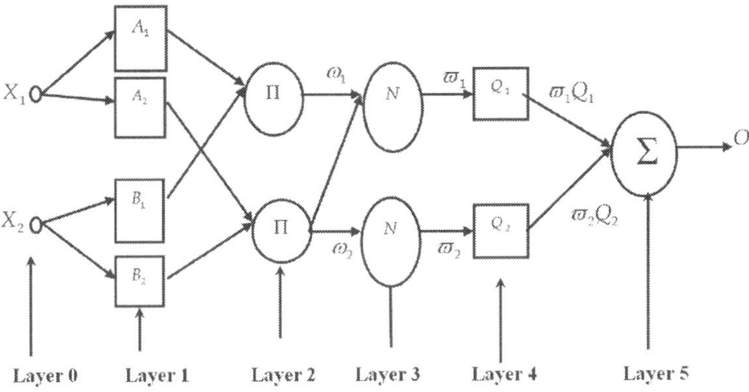

Fig. 8. Two inputs, one output typical ANFIS structure.

where $\dfrac{1}{2}\|w^2\|$ is the regularization term(or complexity penalizer) used to find the flattest function with sufficient approximation qualities, $R_{emp}[f]$ is an empiric risk defined as:

$$R_{emp}[f] := \frac{1}{N}\sum_{i=1}^{N} L(y_i, f(x_i)) \tag{15}$$

and parameter v is for automatic selection of optimal ε and control of number of SVs. For Vapnik's ε-insensitivity, the loss function is defined as :

$$L_\varepsilon(y) = |y - f(x)|_\varepsilon = \begin{cases} 0 & if\ |y - f(x)| \le \varepsilon \\ |y - f(x)| - \varepsilon & otherwise \end{cases} \tag{16}$$

Methodologically, $v-SVR$ processes are similar to that of $\varepsilon-SVR$. It involves formulation of the problem in the primal weight space as a constrained optimization problem by formulating the Lagrangian, then take the conditions for optimality, and finally solve the problem in the dual space of Lagrange multipliers called support values. Though, initially developed for linear function estimation, the principle of linear SVR was extended to non-linear case by the application of the kernel trick. For non-linear regression in the primal weight space the model is of the form

$$f(x) = \omega^T \varphi(x) + b \tag{17}$$

where for the given training set $\{x_i, y_i\}_{i=1}^{N}$, $\varphi(\cdot): \Re^n \to \Re^{n_h}$ is a mapping to a high dimensional feature space by the application of the kernel trick which is defined as

$$K(x_i, x_j) = \varphi(x_i)^T \varphi(x_j) \tag{18}$$

The constraint optimization problem in the primal weight space is

$$\min_{\omega,b,\xi,\xi^*} J_P(\omega,\xi,\xi^*,\varepsilon) = \frac{1}{2}\omega^T\omega + C.\left(v\varepsilon + \sum_{i=1}^{N}(\xi_i + \xi_i^*)\right)$$

Subject to:

$$y_i - \omega^T\varphi(x) - b \le \varepsilon + \xi_i \qquad\qquad i = 1,2...,N$$

$$\omega^T\varphi(x) + b - y_i \le \varepsilon + \xi_i^* \quad i = 1,2...,N \ \ and \ \ \xi,\xi^* \ge 0, \varepsilon \ge 0 \tag{19}$$

where ξ_i, ξ_i^* are the slack variables for soft margin

By defining the Lagrangian and applying the conditions for optimality solution, one obtains the following v-SVR dual optimization problem:

$$\max_{\alpha,\alpha^*} J_D(\alpha,\alpha^*) = -\frac{1}{2}\sum_{i,j=1}^{N}(\alpha_i - \alpha_i^*)(\alpha_j - \alpha_j^*)K(x_i,x_j) + \sum_{i=1}^{N} y_i(\alpha_i - \alpha_i^*)$$

Subject to: $\sum_{i=1}^{N}(\alpha_i - \alpha_i^*) = 0$,

$$0 \le \alpha_i, \alpha_i^* \le \frac{C}{N} \quad \forall \ i = 1,2,...N \quad \text{and} \quad \sum_{i}^{N}(\alpha_i^* + \alpha_i) \le C.v \tag{20}$$

Thus, the regression estimate is given by

$$f(x) = \sum_{i=1}^{N}(\alpha_i + \alpha_i^*)K(x_i, x_j) + b \tag{21}$$

where α_i, α_i^* are the Lagrange multipliers which are the solution to the Quadratic optimization problem, and b follows from the complementary Karush-Kuhn-Tucker(KKT) conditions (Scholkolpf and Smola,2002).

From the foregoing review, it is clear that the choice of Kernel function and the optimization parameters to be selected aprior play important roles in overall performance of the regression process. As previously reported in (Sch¨olkopf and Smola, 2002), the range $0 \le v \le 1$ has been identified as effective range of parameter v for control of errors, thereby simplifying the selection range of parameters combination as compared to ε -SVR.

4.2 Development of ANFIS friction model

The ANFIS-GP model was developed using MATLAB Fuzzy logic toolbox. First the data was partitioned into training (60) and validation (40) data sets, and based on prior information about the friction characteristics, two membership functions were assigned to the input while the value of the premise parameters were initially set to satisfy ε -completeness (Lee,1990) with $\varepsilon = 0.5$. The training was carried out using Hybrid training with 0.0001 error target and 100 epochs. Figure 9 shows the resulting model with Gaussian membership function.

4.3 Development of $v-SVR$ friction model

The SVR-model was developed with reference to the original Matlab toolbox codes by Canu et al (2005). The overall procedures are as follows:

- Partitioning of data into training and validation sets.
- Selection of Kernel function: e.g. Gaussian kernel
- Selection and tuning of the regression parameters: σ -Kernel parameter ($0 \le v \le 1$), and C–Capacity control for optimum performance. Various combinations of these parameters were employed and cross-validated with testing data for both directions of motion.
- Computation of the difference of the Lagrange multipliers $(\alpha_i - \alpha_i^*)$, support vectors (nsv), bias term, b and epsilon, ε .
- Computation of the SVR/decision functions.

The resulting SVR models with training data and associated support vectors (circled 'star data points') are shown in Figure10 (a) and (b) for positive and negative directions respectively.

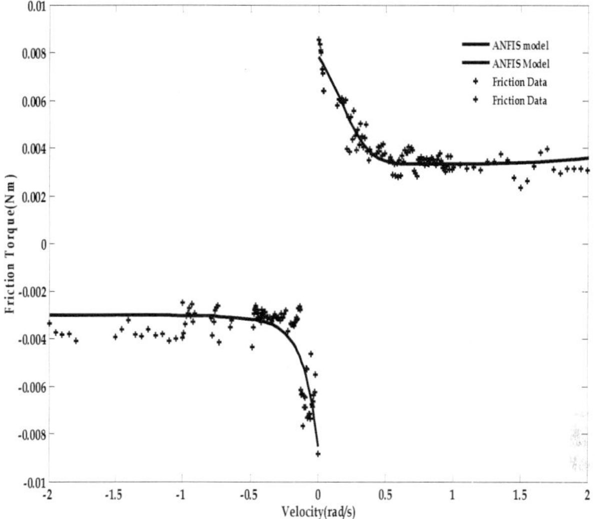

Fig. 9. ANFIS friction model with Gaussian membership function.

4.4 Friction compensation

The developed AI-based friction models are used in model-based friction compensation as shown in Figure 11. The linear PD controller using root-locus technique with nominal plant plant model given in equation (12). The use of PD controller is to enable proper evaluation of the friction model performance since the controller does not have an integral action that has the effect of suppressing the friction effect. The real-time scheme is implemented with the MATLAB xPc target. ANFIS is implemented with the inbuilt MATLAB Fuzzy-Simulink block while the resulting model parameters (difference of Lagrange multipliers and bias) of the v-SVR are integrated to an embedded Matlab function for online real-time friction compensation. Referring to Figure 11, the control law with friction compensation is given as:

$$u_c = u_{in} + \hat{u}_f \tag{22}$$

Hence it can be seen that if $\hat{u}_f(\dot{\theta}) \approx u_f(\dot{\theta})$ and the modeling error is approximately equal

zero, the effect of friction force is effectively compensated and the position accuracy improved.

Figure 12 (a) and (b) show the the comparison of the response of the plant with and without both ANFIS and v-SVR friction compensators for 0.1 and 1 degree step inputs . The tracking errors for 0.1 and 1 degree for 1Hz sine wave input are shown in Figure 13 (a) and (b). These were repeated for 0.5 and 10 degrees step (both directions) and sine wave reference input, and the overall results are reported in Table 2 (a), (b) and Table 3 for point-to-point and tracking control respectively in terms of response time, steady state accuracy and root mean square error(RMSE).

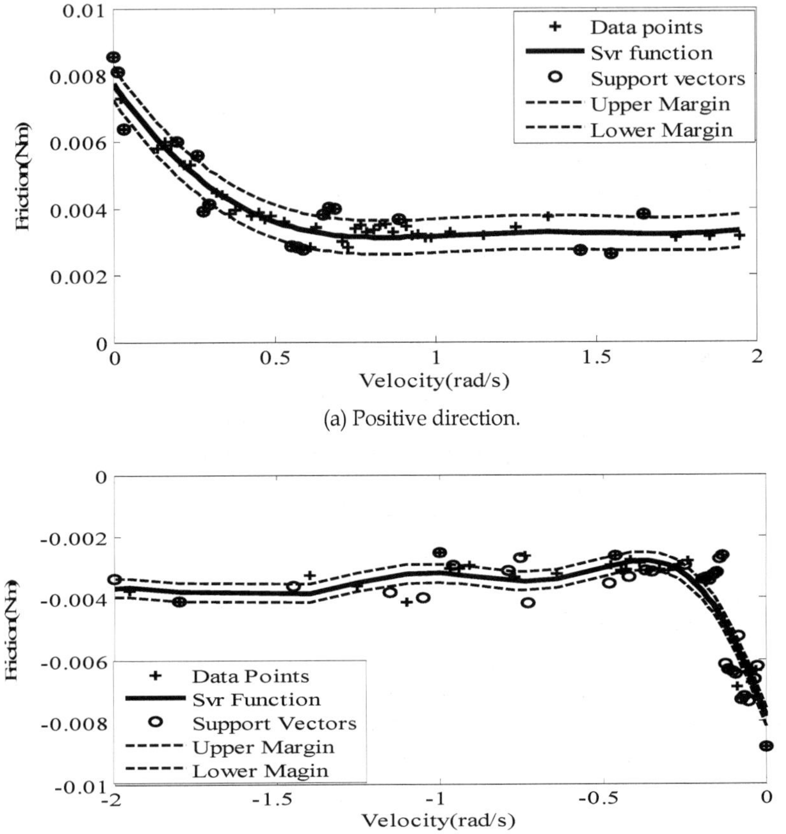

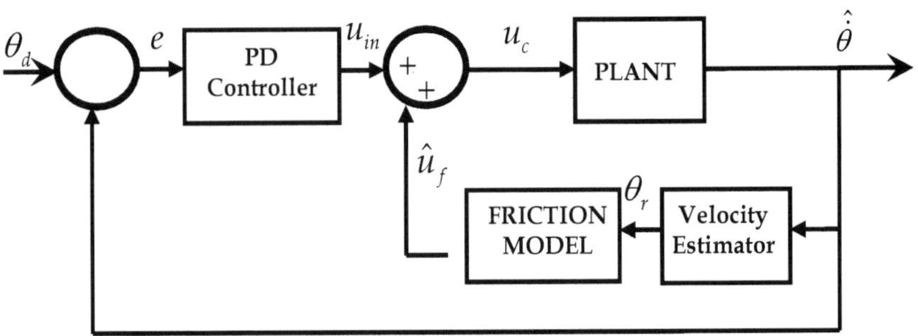

(a) Positive direction.

(b) Negative direction.

Fig. 10. $v-SVR$ friction mod

Fig. 11. Control scheme for the model-based friction compensation.

5. Performance comparison of the proposed AI-models

The performance comparison of the two proposed AI-based friction models is carried out in terms of modeling accuracy, compensation efficiency, and computational time/complexity. The modeling accuracy refers to the performance of the model on training and validation is data. Table 4 gives the comparison of the two models RMSE for both directions of motion. The percentage reduction in both steady state and tracking error for each ANFIS-based and $v-SVR$ compensators was computed so as to compare their compensation efficiency as shown respectively in Figure 14(a) and (b) and Figure 15. Also, the computational time for training and prediction based on the MATLAB resources was computed to examine the complexity of each model as reported in Table 5 .

6. Discussions

The performance improvements recorded with each of the friction compensators over only linear PD controller indicate the importance and requirements of friction compensation for precision positioning control especially at low reference input where the effect of negative friction is highly deteriorating. Comparatively, a better modeling accuracy and compensation efficiency were generally obtained with $v-SVR$ as reported in Table 4, and shown in Figure 14 (a) and (b) and Figure 15. Significant reduction in positioning error over the use of only linear controller was observed in particular up to 90% reduction in steady state error and 60% reduction in root mean square error for PTP and tracking respectively with the v-SVR based friction compensators as against 90% and 50% reduction respectively with ANFIS model. On the other hand, with the MATLAB resources employed, ANFIS is less computational intensive with average computational time of 110ms per training while $v-SVR$ takes 220ms per each iteration in modeling of friction as indicated in Table 5. It should be noted that, many iterative steps are required in SVR development as compared to ANFIS. However, ANFIS is noted to have lesser prediction response with slower time response of 1.6ms as compared to $v-SVR$ with approximately 0.5ms. This implies a tradeoff between desired performance accuracy in favor of SVR and less computational efforts for model development in favor of ANFIS.

The general performance of SVR over ANFIS can be attributed to the fact that SVR algorithm minimizes an upper bound on the expected risk, that is, SVR not only minimizes the error on the training data as in ANFIS modeling but it also minimizes model complexity. So it was able to generalize better than ANFIS on the noisy real-time velocity data during the compensation especially for tracking control.

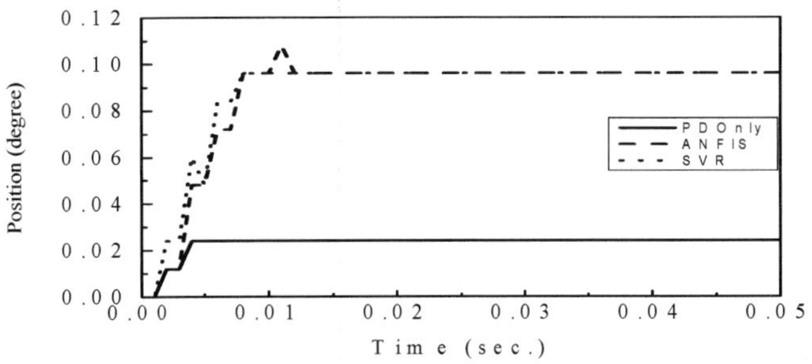

Fig. 12(a). 0.1 deg.

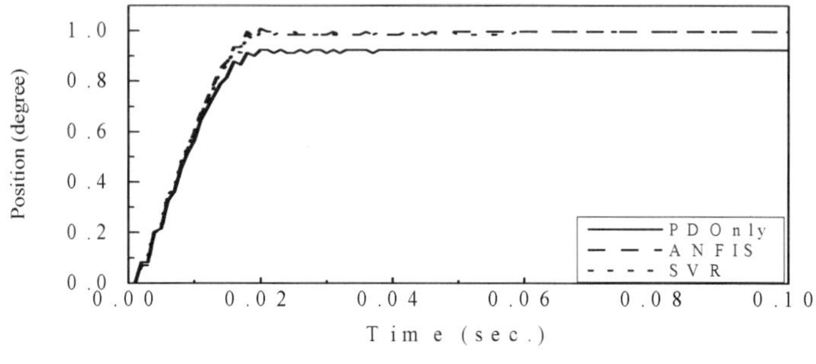

Fig. 12(b). 1.0 deg.

Fig. 12. Step input responses with and without the Friction compensator.

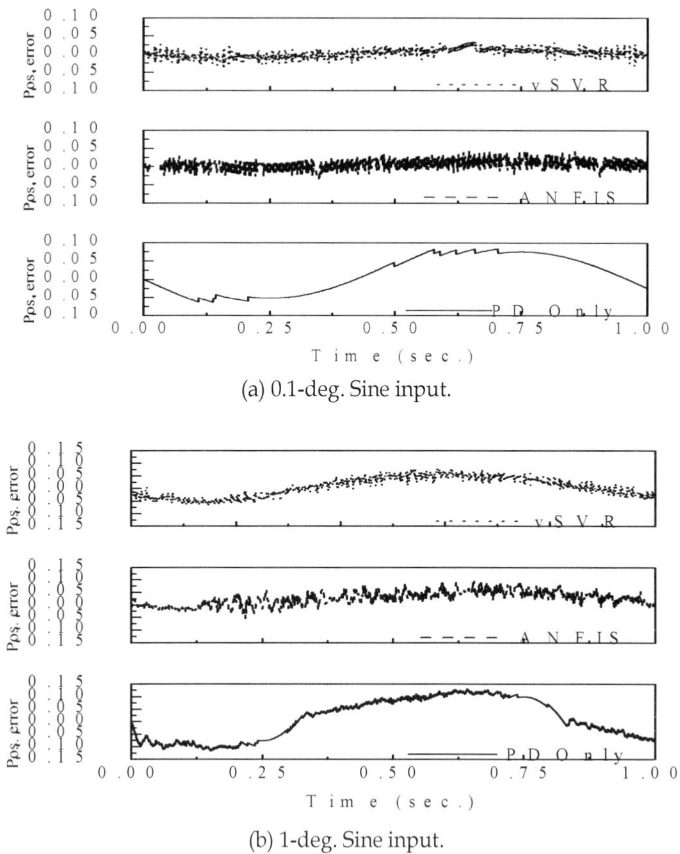

(a) 0.1-deg. Sine input.

(b) 1-deg. Sine input.

Fig. 13(a) and (b). Position tracking error for sinusoidal reference signal.

	POSITIVE STEP INPUTS							
	0.1-deg.		0.5-deg.		1-deg.		10-deg.	
Friction Compensators	ess(%)	Tr(sec.)	ess(%)	Tr(sec.)	ess(%)	Tr(sec.)	ess(%)	Tr(sec.)
No Compensator	75	N/A	37.6	N/A	7.6	0.017	1.8	0.015
ANFIS	4	0.0084	0.8	0.009	0.4	0.015	0.3	0.014
v-SVR	4	0.008	0.8	0.01	0.4	0.015	0.1	0.014

Table 2(a). Performance comparison results for positive PTP positioning control.

	NEGATIVE STEP INPUTS							
	-0.1-deg.		-0.5-deg.		-1-deg.		-10-deg.	
Friction Compensators	ess(%)	Tr(sec.)	ess(%)	Tr(sec.)	ess(%)	Tr(sec.)	ess(%)	Tr(sec.)
No Compensator	76	N/A	44.26	N/A	21	0.017	1.24	0.015
ANFIS	4	0.009	0.8	0.008	0.4	0.012	0.1	0.014
v-SVR	4	0.008	0.8	0.013	0.4	0.013	0.04	0.014

Table 2(b). Performance comparison results for negative PTP positioning control.

Friction Compensators	Root Mean Square Errors (RMSE)			
	0.1-deg.	0.5-deg	1-deg.	10-deg.
No Compensator	0.0355	0.0656	0.0874	0.0959
ANFIS	0.0165	0.0277	0.0380	0.0587
v-SVR	0.0132	0.0255	0.0390	0.0608

Table 3. Performance comparison results for tracking positioning control.

		Training RMSE	Prediction RMSE
ANFIS	Positive Direction	0.000458	0.000443
	Negative Direction	0.000725	0.000744
v-SVR	Positive Direction	0.000408	0.000430
	Negative Direction	0.000690	0.000727

Table 4. Performance comparison in terms of the modelling accuracy.

		Training Computational time(ms)	Prediction Computational time(ms)
ANFIS	Positive Direction	108.581	1.605
	Negative Direction	110.080	1.605
v-SVR	Positive Direction	209.692	0.493
	Negative Direction	224.828	0.493

Table 5. Performance comparison in terms of computational time.

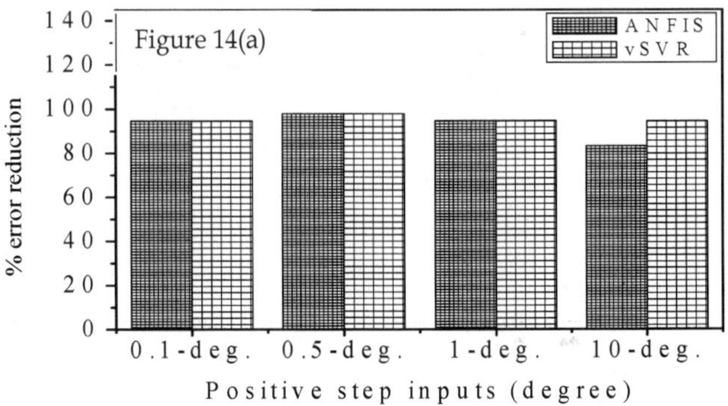

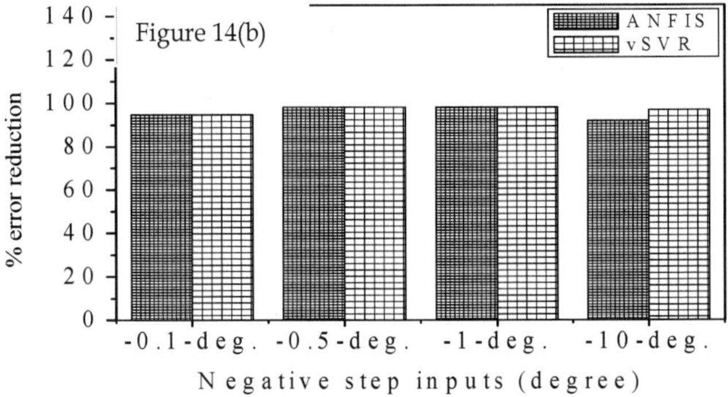

Fig. 14(a) and (b). Comparison of the ANFIS and $v-SVR$ models in terms of %reduction in steady state error over only PD controller for step inputs

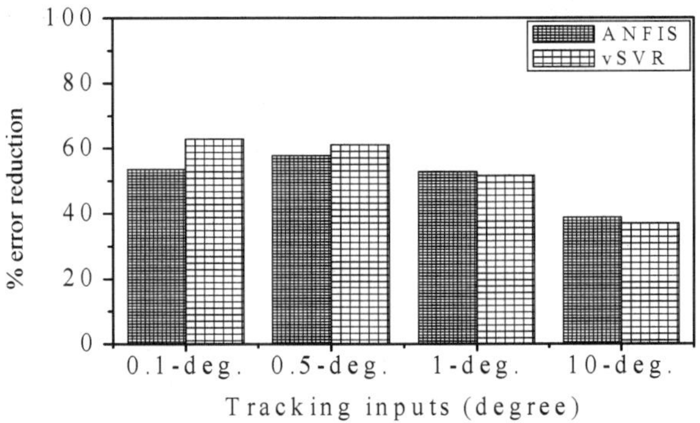

Fig. 15. Comparison of the ANFIS and $v-SVR$ Models in terms %reduction in   tracking error over Only PD controller for tracking control.

7. Conclusion

The application of artificial intelligent based techniques in friction modeling and compensation in motion control system has been presented in this chapter. The chapter focuses on comparative study of the two developed AI-friction models which have been carried out in terms of modeling accuracy, compensation efficiency, and computational time. In comparison, $v-SVR$ outperformes ANFIS both in representing and compensating the frictional effects especially for tracking control at low velocity regime. The results show v-SVR to be better in representing friction than ANFIS with smaller RMSE for both training and prediction of friction. Though, both perform equally in PTP control, v-SVR outperformed ANFIS in tracking control with 60% to 50% reduction in tracking error. Computationally, ANFIS is better with smaller computational processes and time for modeling than SVR, but appears to be poor in prediction than SVR.

It is noted from this study that the performance of the friction model is greatly affected by the precision of the sensor employed. This has limited the minimum velocity that can be controlled to 0.1 degree. Apart from sensor effect, extension of these techniques to micro/nano scale positioning control will required the incorporation of dynamic friction model in the AI-friction model development.

Also, the velocity estimation from the position sensor used introduced noise in the feedback signal. This is responsible for non-smoothness in the tracking responses. This can be avoided either with the use of better position sensor together with more sophisticated velocity filter or by using separate sensor to measure the velocity directly.

8. References

Ahmed G. Abo-Khalil and Dong-Choon Lee, (2008). *MPPT Control of Wind Generation Systems Based on Estimated Wind Speed Using SVR*,IEEE Transactions on Industrial Elecetronics,Vol.55,no. 3,March 2008.

Amontons, G., (1699). *On the resistance originating in machines*, Proc. of the French Royal Academy of Sciences. pp. 206-22.

Armstrong-Helouvry B., Dupont P. and De Wit C., (1994). *A survey of models, analysis tools and compensation method for the control of machines with friction*, Automatica, Vol. 30, No. 7 (1994) pp. 1083-1138.

Armstrong-Helouvry B.,(1992) *Control of Machines with Friction*, Boston, MA, Kluwer, 1991

Bi.D, Y.F.Li, T.S.Tso, and G.L.Wang, (2004), *Friction modeling and Compensation for Haptic display based on Support Vector Machine*, IEEE Trans. Ind. Electron.,vol. 51, no. 2, pp. 491-500,Apr.2004

Cai L. and G. Song, (1993). *A smooth robust nonlinear controller for robot manipulators with joint stick-slip friction*, in Proceedings of IEEE Int. Conference on Robotics and Automation,Atlanta,1993, pp.449-454.

Canudas de Wit, K.J. Astrom and K. Braun,(1986). *Adaptive friction compensation in DC motor drives*, Proceeding of IEEE International Conference on Robotics and Automation, San Francisco, Vol. 3, pp. 1556-1561,1986.

Canudas de Wit, H. Olsson, K. J. Astrom, and P. Lischinsky,(1995). *A New Model for Control of Systems with Friction, IEEE* Transactions on Automatic Control, vol. 40, no. 3, pp. 419-425, 1995

Canu S., Y.Grangyalet , V. Guigue,and A. Rakotomamonjy, (2005). *SVM and Kernel Methods Matlab Toolbox*. PerceptionSystemes et Information, INSA de Rouen, Rouen, France, 2005.

Charles M. Close, Dean H. Frederick, and Jonathan C. Newell, (2002). *Modeling and analysis of dynamic systems*, John Wiley and Sons, Inc.

Chorng-Shyan Lin, (2003). *Recurrent neuro-fuzzy modeling and fuzzy MD control for flexible servomechanisms*, Journal of Intelligent and Robotic Systems 38: 213–235, © 2003 Kluwer Academic Publishers, printed in the Netherlands.

Coulomb, C. A. (1785). *Theorie des machines simples, enayant egard au frottemen de leurs parties, et a la roideur dews cordages"*. Mem. Math Phys., x, 161-342.

Da Vinci, L. (1519). *The Notebooks*, Dover, NY.

Dahl P., (1968). *A solid friction model*, Technical Report TOR-0158H3107–18I-1, the Aerospace Corporation, El Segundo, CA.

De-Peng Liu, (2005). *Research on Parameter Identification of Friction Model forServo Systems Based on Genetic Algorithms*, Proceedings of the 2005 Fourth International Conference on Machine Learning and Cybernetics, Guangzhou.

Dupont P. and B. Armstrong-Helouvry, (1993). *Compensation techniques for servo with friction*, Proceeding of American Control Conference, San Francisco, Vol. 2, pp. 1915-1919, 1993.

Ehrich, N. E., (1991). *An investigation of control strategies for friction compensation*. M.S. Thesis, Dept. of Electrical Engineering, University of Maryland, MA.

Envangelos G., Papadopoulos and Georgios C. Chasparis,(2002). *Analysis and model based control of servomechanisms with friction*, Proc. of the International Conference on Intelligent Robots and Systems (IROS 2002),EPFL Lausanne, Switzerland.

Farid Al-Bender and Jan Swevers, (2008). *Characterization of friction force dynamics*, IEEE Control Systems Magazine, vol. 28, no.6, pp. 64-81.

Hashimoto, M., K. Koreyeda, T. Shimono, H. Tanaka, Y. Kiyosawa and H.Hirabayashi, (1992). *Experimental study on torque control using harmonic drive built-in torque sensors*, Proc. Inter. Conf on Robotics and Automation, IEEE,Nice, pp. 2026-2031.

Haykin S. (1999), *Neural networks: a comprehensive foundation*, 2nd Ed., New Jersey, Prentice Hall.

Jang J.S.R. (1993). *ANFIS: Adaptive-Network-Based Fuzzy Inference System"*, IEEE Trans. Systems, Man, Cybernetics,23 (5/6):665-685, 1993.

Jang J.S.R and N. Gulley, Natick, MA ,(1995). *"The Fuzzy Logic Toolbox for use with MATLAB"*: The MathWorks Inc., 1995

Johan A.K. Suykens, Van Gestel T., De Brabanter J., De Moor B., Vandewalle J., (2002*). Least squares support vector machines*, River Edge, NJ: World Scientific.

Jun Oh Jang and Pyeong Gi Le,(2000). *Neuro-fuzzy control for dc motor friction compensation*, Proceedings of the 39Ih IEEE Conference on Decision and   Control   Sydney, Australia.

Ismaila B. Tijani, Wahyudi M. and Talib H.H., (2011). *Adaptive Neuro-Fuzzy Inference System (ANFIS) for Friction Modeling and Compensation in Motion Control Syste*, International Journal of Modeling and Simulation, Volume 31, No. 1,2011, ACTA PRESS.

Ismaila B. Tijani, Wahyudi M. and Talib H.H., (2009). *A Non-Parametric Friction Model for Accurate Positioning Control using v-Support Vector Regression (v-SVR)*, Proc. of the 2009 IEEE/ASME International Conference on Advanced Intelligent Mechatronics will be held on July 14-17, 2009 in Suntec International Convention and Exhibition Center, Singapore.

Karnopp D, (1985). *Computer simulation of slip-stick friction in mechanical dynamic systems*, Journal of Dynamic Systems, Measurement, and Control, 107H1I:100–103.

Kemal M. Cılıza, Masayoshi Tomizukab (2007), (2007). *Friction modeling and compensation for motion control using hybrid neural network models*, ScienceDirect, Engineering Applications of Artificial Intelligence 20 (2007) 898–911.

Lampaert V., Swevers J., and Al-Bender F.,(2002). *Modification of the Leuven integrated friction model structure*, IEEE Transactions on Automatic Control, 47, 4, pp. 683-687.

Lampaert, V., F. Al-Bender and J. Swevers, (2003). *A Generalized Maxwell-Slip friction model appropriate for control purposes*, Proc. of the 2003   Int. Conference on Physics and Control, Saint-Petersbourg, Russia, 1170-1178.

Ljung, L. (1987). *System Identification: Theory for the User*, Prentice-Hall, Englewood Cliffs, NJ.

Lörinc Márton and Béla Lantos, (2007). *Modeling, identification, and compensation of stick-slip friction*, IEEE Trans. on Industrial Electronics, vol. 54, no. 1.

Makkar C.,W.E.Dixon,W.G.Sawyer,and G.Hu, (2005). *A New Continously Differentiable Friction Model for Control Ssystems Design*, Proceedings of the 2005 IEEE/ASME International Conference on Advaanced Intelligent Mechatronics Monterey,California,USA,24-26 July,2005.

Morin A.J. (1833). *New friction experiments carried out at Metz in 1831–1833*, In Proc. of the French Royal Academy of Sciences, volume 4, pages 1–128.

Ohm D.Y., (1990). *A PDFF Controller for Tracking and Regulation in Motion Control*, Proceedings of 18th Conference, Intelligent Motion,Philadelphia,1990

Oppelt W., (1976). *A historical review of autopilot development, research, and theory in Germany*, Journal of Dynamic Systems, Measurements, and Control,   pages 215–23.

Reynolds, O., (1886). *On the theory of lubrication and its application to Mr. Beauchamp Tower's experiments, including an experimental determination of the viscosity of olive oil*, Phil. Trans. Royal Soc., 177, 157-234.

Rong-Hwang Horng, Li-Ren Lin and An-Chen Lee, (2006). *LuGre model-based neural network friction compensator in a linear motor stage*, International   Journal of Precision Engineering and Manufacturing vol. 7, no.2.

Sch¨olkopf, B. and Smola A. (2002). *Learning with Kernels*,MIT Press,Cambridge MA, 2002

Shen, C. N. and H. Wang, (1964). *Nonlinear compensation of a second- and third-order system with dry friction.* IEEE Trans. on Applications and Industry,*83(71)*, 128-136.

Smola A.J. and B.Scholkopf,(2001). *A tutorial on Support vector Regression*, NeuroCOLT Technical Report NC-TR-98-030, 2001.

Southward S.S., C.J. Radcliffe, C.R.Mac-Cluer, (1991). *Robust nonlinear stick-slip friction compensation*, J.Dyn. syst.,Meas. Control 113,pp. 639-644.

Stefan Brock, (2000). *Application of ANFIS controller for two - mass – system*, ESIT 2000, 14-15, Aachen, Germany.

Sung-Kwun Oh, Witold Pedrycz, Wan-Su Kim, and Hyun-Ki Kim1, (2006). *GA-based polynomial neural networks architecture and its application to multi-variable software process*, Q. Yang and G. Webb (Eds.): PRICAI 2006, LNAI 4099, pp. 834 – 838, 2006.© Springer-Verlag Berlin Heidelberg

Swevers J., Al-Bender F., Ganseman C., and Prajogo T., (2000). *An integrated friction model structure with improved presliding behavior for accurate friction compensation*, IEEE Trans. on Automatic Control, 45, pp. 675-686.

Tjahjowidodo, T., Al-Bender, F. and Brussel, HV, (2004). *Friction Identification and Compensation in a DC Motor*, International Journal of JSPE Vol. 32, No. 3, pp. 200-206.

Tijani Ismaila B., (2009). *Friction Identification and Compensation in Motion Control System using ANFIS and SVM*, Master Thesis, Mechatronics Engineering Department, International Islamic University Malaysia, 2009.

Tustin A. (1947), *The effects of Backlash and of Speed-Dependent Friction on the Stability of Closed-Cycle Control Systems*, IEEE Journal,vol. 94,part 2A,p.143-51.

Wahyudi, Sato K. and Shimokohbe A. (2005). *Robustness evaluation of three friction compensation methods for point-to-point (ptp) positioning systems*, Robotics and autonomous system, Elsevier, Vol. 52, Issues 2-3, pp. 247-256

Wahyudi, (2003). *Friction Identification and Compensation for High Precision Motion Control System - Part 1: Friction Identification"*, Proc. of Industrial Electronics Seminar (IES), 2003.

Wahyudi and Ismaila B. Tijani, (2008). *Friction Compensation for Motion Control System using Multilayer Feedforward Network*, Proceeding of the 5th International Symposium on Mechatronics and its Applications (ISMA08), Amman, Jordan, May 27-29, 2008

Wen-Fang Xie, (2007). *Sliding-Mode-observer-based adaptive control for servo actuator with friction*, IEEE Trans. On Industrial Electronics, vol. 54, no.

Vapnik V., (1995). *The Nature of Statistical Learning Theory*, Springer,New York,1995.

Yi Guo, Zhihua Qu, Yehuda Braiman, Zhenyu Zhang and Jacob Barhen, (2008). *Nanontribology and nanoscale friction*, IEEE Control Systems Magazine,vol. 28,no.6, pp. 92-100.

Zhixiang Hou, Quntai Shen and Heqing Li, (2003). *Nonlinear System Identification Based on ANFIS*, IEEE Int. Conf. Neural Networks & Signal Processing, Nanjing, China, December 14-17, 2003.

4

Mechatronic Systems for Kinetic Energy Recovery at the Braking of Motor Vehicles

Corneliu Cristescu[1], Petrin Drumea[1], Dragos Ion Guta[1],
Catalin Dumitrescu[1] and Constantin Chirita[2]
[1]Hydraulics and Pneumatics Research Institute, INOE 2000-IHP, Bucharest
[2]"Gheorghe Asachi" Technical University, Iasi
Romania

1. Introduction

Vehicle manufacturers are continually concerned with reducing fuel consumption and lowering polluting emissions. (Gauchia & Sanz, 2010). Besides the vehicles which use liquefied gas, methanol, electricity or fuel cells, also, there have been designed and manufactured diferent *hybrid propulsion motor vehicles.* (Toyota, 2008; Permo Drive, 2009; Eaton, 2011).

It is known that during a work cycle of a motor vehicle, which consists of a period of acceleration, another one of running at constant speed and a period of deceleration, the power required during acceleration is much greater than that required while running at constant speed and, in principle, it is this power what determines the size of engine installed on the motor vehicle. Upon vehicle braking, kinetic energy acquired by acceleration of the motor vehicle is converted into heat energy, which is located in the braking system and gets lost, irreversibly, into space, with negative effects on *global warming.* So, rightfully, there has been formulated the technical problem that, during the motor vehicle braking stage, the kinetic energy gained by it to be recovered and stored in power batteries and then used during start-up and acceleration stages. Therefore, vehicle manufacturers consider that one of radical solutions in order to achieve the above mentioned goals is a deep change of motor vehicle propulsion method, promoting *hybrid propulsion systems,* which are considered to be solutions for the near future, for a substantial decrease of fuel consumption and polluting emissions. Propulsion systems that are composed of, besides a conventional propulsion system with an internal combustion engine, at least another one based on another type of energy, capable of providing torque/traction moment at the motor vehicle wheels, form a hybrid propulsion system. If they, along with propulsion, can recover, during braking stage, part of the kinetic energy accumulated in the acceleration stages, and then they are called hybrid regenerative systems. A feature of *regenerative hybrid vehicles* is that they include components that capture and *store kinetic energy* of the vehicle during braking process, for it to be used later, or when accelerating or at constant speed movement. Systems for capturing and storing kinetic energy perform its converting and storing under different forms of energy, namely: as mechanical/ kinetic energy of a flywheel, as potential energy of a working fluid (liquid or gas), as electrochemical energy (Gauchia & Sanz, 2010)), or as electrostatic energy. To restore the recovered and stored energy, drive/propulsion systems are, also, of several types, namely: hydro-mechanical systems (hydrostatic or hydrodynamic), electromechanical systems (direct current or alternating current) and mechanical systems (mechanical or mechanic-inertial). Worldwide, various solutions have been designed for developing hybrid systems, but most common are hybrid systems with thermo-electric drive and hybrid systems with thermo-hydraulic drive. A special competition is under development between the *thermo-electric hybrid system*, (Toyota, 2011; Eaton, 2011), which, in addition to the heat engine, also has an electric propulsion system, and the *thermo-hydraulic hybrid system*, (Permo-Drive, 2011; Eaton, 2011a; Bosch Rexroth, 2011), which, in addition to the driving heat engine, has a hydraulic propulsion system. Compared with electric vehicles, characterized by a reduced autonomy of movement, *hybrid vehicles have many advantages*, Usually, the kinetic energy of the motor vehicle, accumulated in the accelerating phase, in the braking phase is converted in the thermal energy which is, normally and irremediable, wasted in atmosphere. Therefore, the main objectives of the *hybrid systems* are the *recovering kinetic energy* of the road motor vehicles and reducing the *fuel consumption* and *the environment pollution* (Parker Hannefin, 2010).

From the above presented issues, it is clear that hybrid propulsion systems are very complex systems, multidisciplinary and interdisciplinary. Also, they develop dynamic/transient operation modes, with rapid succession of events over time, difficult to drive and control with conventional means. Therefore, for such complex systems, the only technology able to manage, optimize and control in conditions of total safety, is *mechatronics technology*, for which reason *hybrid propulsion systems* represents *a new field of application of mechatronics* (Ardeleanu & al.; Cristescu et al., 2008b; 2007; Maties, 1998).

2. The mechatronic system for kinetic energy recovery at the braking of motor vehicles

Basic solution, adopted to achieve the kinetic energy recovery system for the braking stage, was that of kinetic energy recovery *by hydraulic means*, based on the use of a *hydraulic machine* which can operate both as a pump, during braking, and as an motor, during acceleration/start-up. In the braking stages, the mechanical/kinetic energy of the motor vehicle is *converted* by the hydraulic machine, which is working as a pump, into hydraulic/hydrostatic energy and *stored* at high pressure, in hydro-pneumatic accumulators. In the acceleration/start-up stages, hydrostatic energy, *stored* in hydro-pneumatic accumulators, is *converted back* into mechanical energy by the hydraulic machine, which is working now as a motor and generating acceleration of the motor vehicle, (Cristescu, 2008a).

The *aim* of the designed hydraulic system is the recovery of kinetic energy, in the braking stage of a motor vehicle.

The *technical problem*, which is solved by the energy recovery hydraulic system, is the capturing and storing of the lost energy in the braking stages at medium and heavy motor vehicles.

The *method* consists in using one mechanic and hydraulic module, which is able to capture and convert the kinetic energy into hydrostatic energy and, also, storage and reuse it for acceleration and start-up of the road motor vehicles.

The implementation of a hydraulic system for recovery of kinetic energy, on a motor vehicle, transforms it into a *hybrid motor vehicle* and leads to decreasing of the fuel consumption and, also, to reducing of the environmental pollution.

The *main objectives* of the *hybrid propulsion systems* are *the recovery of kinetic energy* of the road motor vehicles, in order to reduce the fuel consumption and to increase the energy efficiency of the propulsion systems of the motor vehicles.

2.1 Conceptual model and mechatronic configuration of the kinetic energy recovery system

2.1.1 Constructive configuration and implementation of the energy recovery system on motor vehicles

Constructive and functional concept of developing and implementing a system for braking energy recovery is shown, in schematically, in Figure 1, which presents a conceptual model of construction and installation/implementation of the kinetic energy recovery system on a motor vehicle. The energy recovery system consists, in essence, of a hydro-mechanical module which includes a variable displacement hydraulic machine, that can operate both in pump mode, during braking, and in motor regime, during start-up/acceleration of the motor vehicle. The hydraulic machine is driven by a mechanical transmission and is controlled by an electric and electronic control subsystem, which performs, also, the interfacing with the braking and acceleration systems of the basic motor vehicle, operation being controlled through a processor, which provides the information support, specific to mechatronic systems.

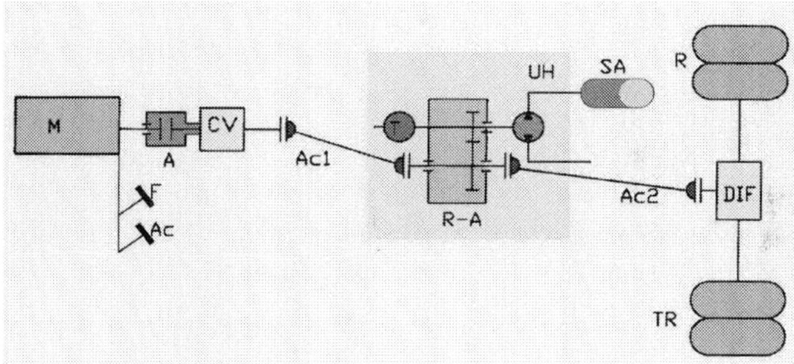

Fig. 1. A conceptual model of construction and installation/implementation of the recovery system on motor vehicles.

Implementation/installation of the energy recovery system can be done on motor vehicles that have a long cardan axle between the gearbox CV and the differential mechanism DIF, by replacing it with two shorter axles. Mechanical connection between the cardan axles Ac1 and Ac2 and the recovery system R-A is permanent and is achieved through a mechanical transmission, which adapts the rotational speed of the cardan axle to the operating rotational speed of the hydraulic machine/unit UH in the system. Depending on the specific conditions provided by the motor vehicle on which the recovery system is installed, the coupling outlet and mechanical transmission can be placed at the end of the cardan axle Ac1 close to the gearbox, at the end of the cardan axle close to the rear drivetrain TR, or between the gearbox CV and the drivetrain TR, by splitting the cardan axle.

Hydraulic unit is a hydraulic machine with variable displacement/geometric volume, which can vary between 0 and a maximum value (V_g=max). Axial piston hydraulic unit can be removed from the zero displacement position, only when the vehicle goes forward. When it goes into reverse, the displacement of the unit remains zero ($Vg = 0$).

Basic schematic diagram of the automatic adjustment system of the motor vehicle hybrid propulsion system, that includes an energy recovery system, is shown in Figure 2. The adjustment system achieves proportionality between the the stroke of the brake pedal, respectively, the stroke of the acceleration pedal, on slowing down, respectively, on starting-up the motor vehicle.

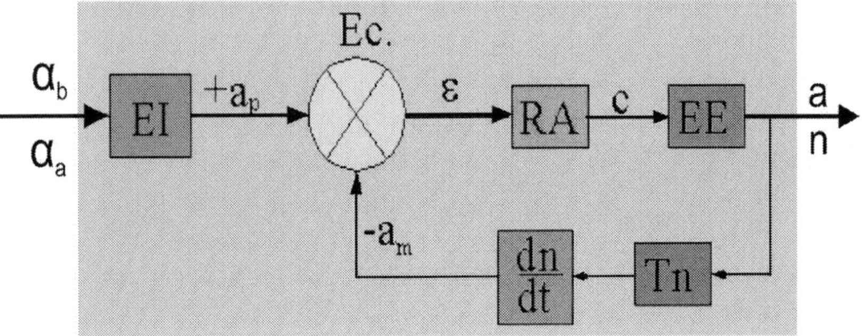

Fig. 2. Automatic adjustment schematic diagram of the hybrid propulsion system of motor vehicles.

According to the adjustment schematic diagram in Figure 2, component elements of the system are the next ones:

EI - the input element, which converts the input parameter of the system, that is the angular stroke of brake pedal α_f, respectively angular stroke of acceleration pedal α_a, into the preset parameter a_p, that is the deceleration, respectively, acceleration, according to the operation stage, braking or acceleration;

EC - the comparison element, which compares the preset parameter a_p with the measured acceleration a_m and transmits to the automatic regulator RA the discrepancy ϵ between the two parameters, in order to operate correction;

RA - the automatic regulator, which determines, depending on the error ϵ, the value of the drive parameter $c_,$, that will work to equalize the preset acceleration a_p with the actual acceleration value a;

EE - execution element, represented by the axial piston hydraulic unit, which determines the value of vehicle acceleration proportional to the received command; this item plays a double part: information and power circulation.

Recovery system also comprises the hydraulic devices to achieve hydraulic circuits, as well as the transducers required for monitoring and automatization of braking and start-up/acceleration processes.

According to the theory of automatic systems, the global systemic model is shown in Figure 3.

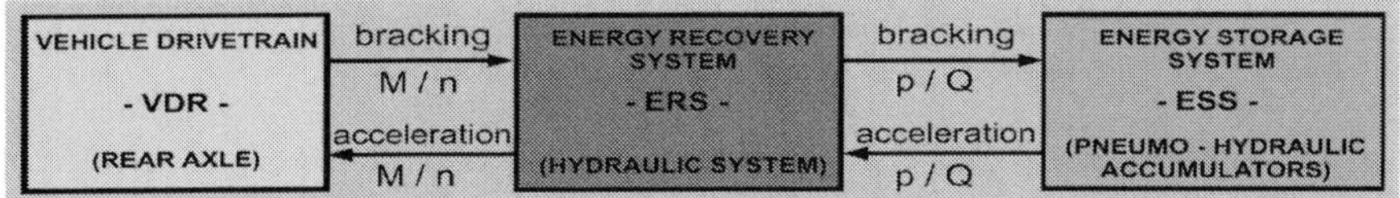

Fig. 3. Global systemic model of a motor vehicle equipped with a kinetic energy recovery system.

During the braking stage, the recovery system ERS captures, from the drivetrain VDR, the vehicle's kinetic energy (with mechanical parameters: torque/moment M and rotational speed n), converts it into hydrostatic energy (with hydraulic parameters: pressure p and flow Q) and stores it inside the storage subsystem ESS. During the start-up stage, the hydrostatic energy (with hydraulic parameters: pressure p and flow Q) is transmitted to the recovery system ERS which converts it into mechanical energy (with mechanical parameters: torque M and rotational speed n), and uses it to add torque/moment to the propulsion and drivetrain of the vehicle, for acceleration or start-up, as appropriate. The general systemic model of interfacing and interconditioning of the energy recovery system with the systems, that command and control motor vehicle movement (braking and acceleration systems), is shown roughly in Figure 4.

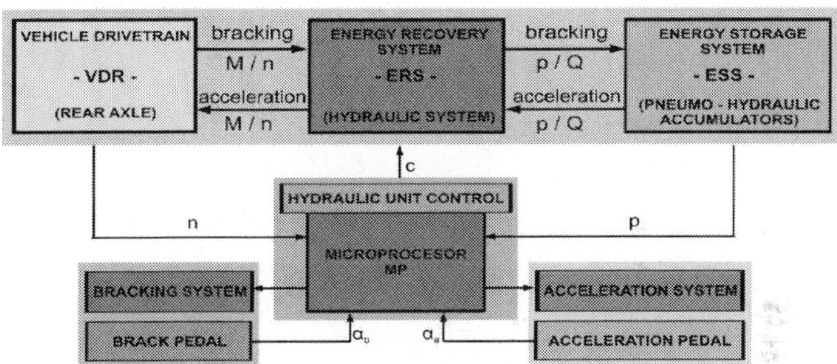

Fig. 4. General systemic model of the command and control system.

As it is shown in Figure 4, the microprocessor MP manages all data of the whole hybrid vehicle, making its operation optimal during the two stages, braking and acceleration. The microprocessor receives information on the braking or acceleration command, rotational speed of drivetrain, pressure inside the storage system, and manages the entire process through commands sent to the energy recovery system and to the conventional braking or acceleration systems.

2.1.2 Mechatronics structure of the kinetic energy recovery system

As one can see in Figure 5, mechatronic model of kinetic energy recovery system in motor vehicle braking has a typical mechatronics structure, see (Maties, 1998; Cristescu et al., 2008b), consisting of the next four main subsystems:

- *mechanical-hydraulic subsystem*, which consists of hydro-mechanical module, hydraulic station, battery of hydro pneumatic accumulators and hydraulic commands pump, installed on a special transmission of the heat engine;

- *electronic drive and control subsystem*, which consists of all electric, electronic and automation elements and components which ensure system operation, including the drive and control panel;
- *subsystem of sensors-transducers*, which consists of all necessary sensors and transducers that provide capturing of evolution over time, of process parameters and conversion into electric parameters, easily processable by the system;
- *computer subsystem* for process control, consisting of user licensed purchased software or software specifically designed and dedicated to the proper functioning and performance of the system, and also the related processor or computer.

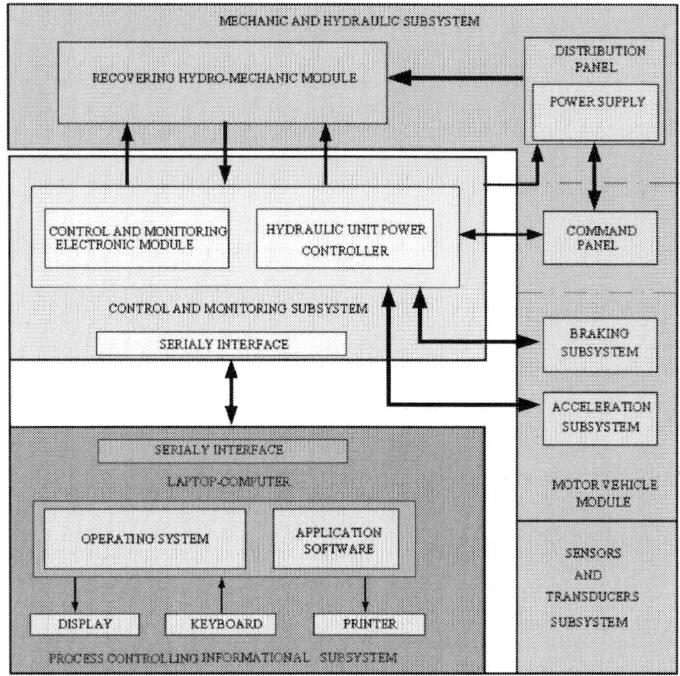

Fig. 5. Mechatronics model of energy recovery system at the braking of motor vehicles.

This structure defines and substantiates the *mechatronic conception* of developing the recovery system. Mechatronic system for recovery of braking energy at motor vehicles operates based on dedicated software, which monitors the system and enables registration of the output parameters and control of the main parameters of the system.

In addition to the specific subsystems of a energy recovery system, mentioned above, mechatronic system monitors and controls, through special interface components, some other subsystems of the basic motor vehicle, on which implementation has been performed, namely: subsystem for interfacing with the classic acceleration subsystem of the motor vehicle and subsystem for interfacing with the classic braking subsystem of the motor vehicle. The energy recovery system is conducted by a computer with specialized software.

2.2 Presentation of the thermo-hydraulic propulsion system

Further on, there is presented a Romanian technical solution for a hybrid propulsion system that has been obtained by implementation of an energy recovery hydraulic system on a medium motor vehicle, which has, already, an existing thermo-mechanical propulsion system. In this maner, the mounting of the hydraulic recovery system, on the motor vehicle with thermo-mechanical propulsion system, leads to transformation of the vehicle into a thermo-hydraulic hybrid vehicle. Entire hybrid propulsion system has been conceived as a mechatronic system, see (Cristescu, 2008a).

2.2.1 The conceptual model of the thermo-hydraulic hybrid vehicle

In Figure 6 is presented the conceptual model of the Romanian technical solution for a hybrid propulsion vehicle, which consists in a energy recovery hydraulic system that has been implemented on a medium motor vehicle.

The *conceptual model* illustrates a thermo-hydraulic parallel hybrid motor vehicle, as the energy recovery hydraulic system implemented does not interrupt the thermo-mechanical direct driveline to the motor vehicle wheels.

This hybrid vehicle has resulted after the implementation of kinetic energy recovery system with hydraulic drive on the vehicle type *ARO-243*, with thermo-mechanical propulsion. Basic motor vehicle allows discontinuity of the thermo-mechanical driveline of the rear bridge, by removing the appropriate cardan axle, thermo-mechanical drive remaining only on the fore bridge, which is exactly the *thermo-mechanical* propulsion subsystem of the vehicle. By mounting *the energy recovery hydraulic system* on the rear bridge of the vehicle, there is created a second drive subsystem namely the *mechanical-hydraulic subsystem* that drives the rear bridge; thus there is made a *parallel hybrid thermo-hydraulic propulsion system* of the motor vehicle, these subsystems being able to propel either separately or together, (Cristescu, 2008a).

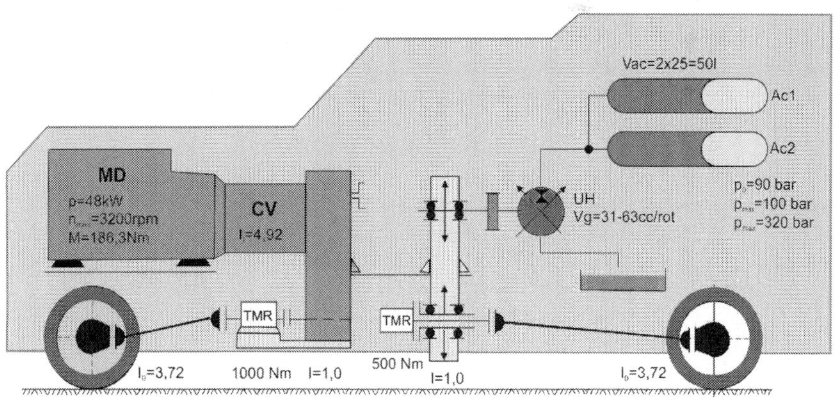

Fig. 6. The conceptual model of the thermo-hydraulic hybrid vehicle with energy recovery hydraulic system.

The recovery hydraulic system of kinetic energy has been designed to be implemented on a Romanian automotive, well-known as *ARO 243* type, which has a *4x4 driving* system. In the conceptual model of the hybrid propulsion vehicle, presented in Figure 6, can be distinguished the Diesel engine MD, the gearbox CV and the gear transmission to the front wheels, through one torque transducer (TMR) and one cardan axle. There can be seen the mechanical transmission to the hydraulic machine/unit UH, the tank for low pressure LT and the storing system for height pressure, which consists of the two hydraulic and pneumatic accumulators AC1 and AC2. The hydraulic power is transmitted, to the breech wheels, through the torque and rotation transducer (TMR) and a cardan axle. The hydraulic machine can be connected, in parallel, anywhere in the driveline, but, generally, it is mounted between the gearbox and differential mechanism. The main part of the recovery system is the hydraulic machine with variable geometrical volume, that can work both as a pump, in the braking process, and, also, as a hydraulic motor, in the start-up process of the motor vehicle.

The hydraulic machine is driven through a gearbox transmission, being assisted by an electro-hydraulic system, which is interfaced with the subsystems for braking and acceleration of the vehicle, all controlled by a processor. Operation of the recovery system has a lot of sensors and transducers, for monitoring and controlling the evolution of parameters.

The hybrid propulsion system, which contains the energy recovery hydraulic system, has been developed in a *mechatronic conception* (Maties, 1998). The system contains: mechanical and hydraulic subsystem, drive and control electronic subsystem and the data management informatic subsystem. The interface of the first two subsystems is the subsystem of sensors and transducers, which provides information on the evolution of the main parameters of the kinetic energy recovery mechatronic system. The sensors and transducers subsystem allows data acquisition from the torque, temperature, flow and pressure transducers (Calinoiu, 2009). The mechatronic system is working on basis of dedicated software, which allows monitoring and recording the evolution of output and control parameters of the system. This component defines the mechatronics basis for the system design and development.

2.2.2 The main physical modules of the energy recovery hydraulic system

In essence, by mounting of the kinetic energy recovery system, Figure 7, on the motor vehicle *ARO-243*, presented in Figure 7(a) and Figure 7(b), transforms it in a *hybrid motor vehicle*, which have now, besides of the existing thermo-mechanic propulsion subsystem, an supplementary propulsion system, named hydro-mechanic propulsion subsystem.

The main parts/subassemblies of the kinetic energy recovery mechatronic system are:

- *hydro-mechanical module*, Figure 7(c) , is composed of a chain transmission, equipped with a torque and rotation transducer TMR, and a hydraulic unit/machine UH, *serving* as a pump, during braking, and as an motor, during start-up. The hydraulic machine is a variable-displacement one, manufactured by the company *Bosch Rexroth Group* (Bosch Rexroth Group, 2010), where flow control is performed electronically, through an automatic control closed loop;
- *hydraulic station* SH itself, Figure 7(d), represents the subassembly connecting the hydro-mechanical transmission and the hydro pneumatic accumulators battery, where hydrostatic energy is stored. Hydraulic station consists of oil tank with its specific elements, and of hydraulic blocks with equipment necessary to perform the functions;

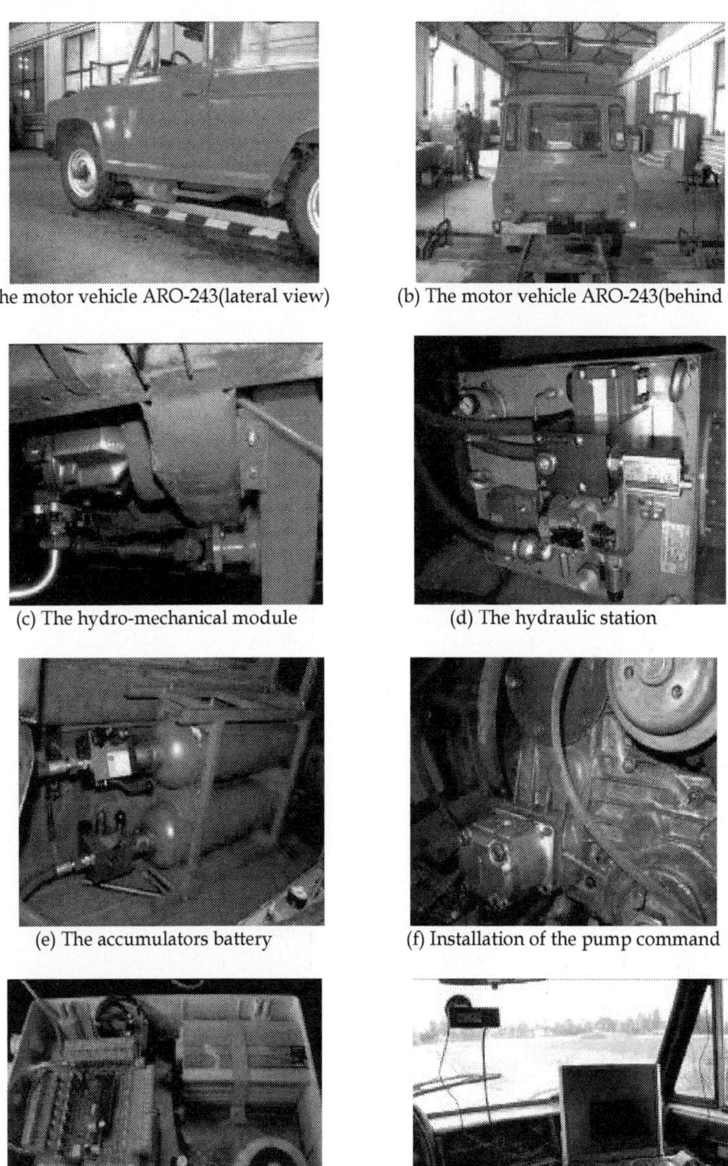

(a) The motor vehicle ARO-243(lateral view)　　(b) The motor vehicle ARO-243(behind view)

(c) The hydro-mechanical module　　(d) The hydraulic station

(e) The accumulators battery　　(f) Installation of the pump command

(g) Electronic drive and control subsystem　　(h) Informatics subsystem

Fig. 7. The main parts/subassemblies of the kinetic energy recovery mechatronic system.

- *hydro pneumatic accumulators battery, Figure 7(e), is a unit consisting of two hydro* pneumatic accumulators, enabling hydrostatic energy storage, during braking stage, and supply of hydraulic motor with potential hydrostatic energy, during start-up or acceleration of the motor vehicle;
- *pump command*, Figure 7(f), is mounted to the power outlet of the heat engine and serves to hydraulically drive the hydraulic machine and unlockable valves for hydrostatic power supply of hydraulic machine.

In addition to the presented subsystems, the system has, also, an electronic drive and control subsystem, Figure 7 (e), and an *informatics management subsystem*, Figure 7 f), all designed and developed in a unitary mechatronic conception.

2.3 Some theoretical results obtained by mathematical modeling and numerical simulation

Motor vehicle dynamic behavior is determined by the size, direction and way of forces acting on it. They are classified into two broad categories: active forces or *traction forces*, which cause motor vehicle movement, and *resistance forces*, which oppose its movement. *Resistant forces* are given by the resistance to running on the road, the resistance of air to movement, additional resistance opposed to running on a ramp, as well as inertial forces that appear on accelerating or stoping a motor vehicle. To overcome these resistance forces, energy consumed to propel the motor vehicle fall into:

- *irreversible consumed energy*, for overcoming all resistance to forward (rolling, aerodynamics, losses in transmission) and which are due, first, to internal and external friction of the motor vehicle;
- *recoverable energy*, used for accelerating or climbing a ramp, in this case the kinetic energy and potential energy, which can be recovered. This recoverable energy can be partially accumulated, instead of being dissipated through braking system, if the motor vehicle is equipped with energy recovery, storage and reuse system.

Therefore, as a first step, preliminary theoretical research has been conducted, based on mathematical modeling and numerical simulation, in order to know the dynamic behavior of motor vehicle *ARO 243*, intended to be equipped with a hydraulic system for kinetic energy recovery at braking. For mathematical modeling and computer simulation of dynamic behavior of experimental motor vehicle there have been used mathematical relations in the specialized literature and *MATLAB with Simulink* software package, (The Math Works Inc., 2007), which is dedicated to numerical calculation and graphics in science and engineering. Some theoretical results obtained are presented below.

2.3.1 Dynamic behavior of the motor vehicle with thermo-mechanic propulsion system

To model the start-up of the motor vehicle *ARO 243* with thermo-mechanical propulsion system, when propulsion is provided exclusively by a 48 kW Diesel heat engine, there has been conducted, first, mathematical modeling and developed a sub-software for simulation of the external feature of heat engine, i.e. of variation diagram of moment/torque Me and engine power Pe, depending on engine rotational speed n_{mot}. This simulation sub-software will be included, as a subroutine, in the general software for simulation of starting the heat propulsion motor vehicle. After numerical simulation, using the data about the engine, we obtained the diagram in Figure 8.

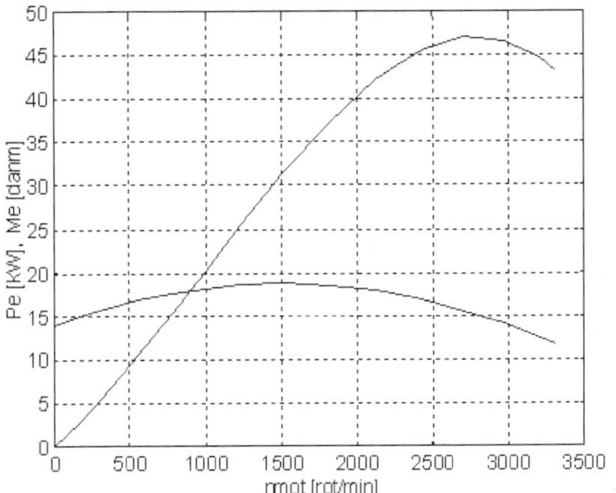

Fig. 8. External feature of a 48 kW Diesel engine.

Mathematical modeling of motor vehicle start-up stage is performed on the basis of relations known in specialized literature, which are based on the principle of *D'Alembert*, according to which equation of movement is written as:

$$\delta \frac{G_a}{g} \frac{dv_a}{dt} = \left(F_R - \sum F \right),$$ (1)

where: v_a is the motor vehicle velocity; G_a is the motor vehicle total weight; g is gravitational acceleration; F_R is the traction force at drive wheels, and $\sum F$ is the sum of resistances to advance that do not depend on acceleration. Coefficient δ is the inertial coefficient of rotating masses, which takes into consideration their influence on motor vehicle movement, with values in the range 1.2 ÷ 1.4, for speed step I, see (Untaru et al., 1974).
It can be written that the sum of resistance forces is given by the next relation:

$$\sum F = G_a \left(f \cdot \cos\alpha + \sin\alpha \right) + K \cdot S \cdot v_a^2$$ (2)

where: f is the coefficient of resistence to running; a is the ramp angle; K is the aerodynamics resistance coefficient; and S is the frontal surface of motor vehicle. With this notations, the equation becomes:

$$\frac{dv_a}{dt} = \frac{g}{\delta \cdot G_a} \cdot \left(F_R - G_a\, f \cdot \cos\alpha - G_a \cdot \sin\alpha - K \cdot S \cdot v_a^2 \right)$$ (3)

If it is considered that the movement is done in a horizontal plane, and starting of the motor vehicle is done at low velocity, equation can be simplified more. Based on the relationship known in classical literature, there has been developed a complete mathematical model for the starting-up stage and, based on this one, there has been developed a numerical simulation software, which allowed, based on structural and functional features of the vehicle, to obtain some theoretical results of interest in the dynamic evolution of the motor vehicle. Some of these preliminary theoretical results are shown in the figures below. Thus,

Figure 9 shows the variation of kinematics parameters and traction force at the wheels of the investigated motor vehicle. The variation of stroke on start-up is shown in Figure 9(a) and the variation of velocity on start-up in Figure 9(b). The variation of acceleration on start-up it can see in Figure 9(c) and the variation of traction force at the wheel is done in Figure 9(d).

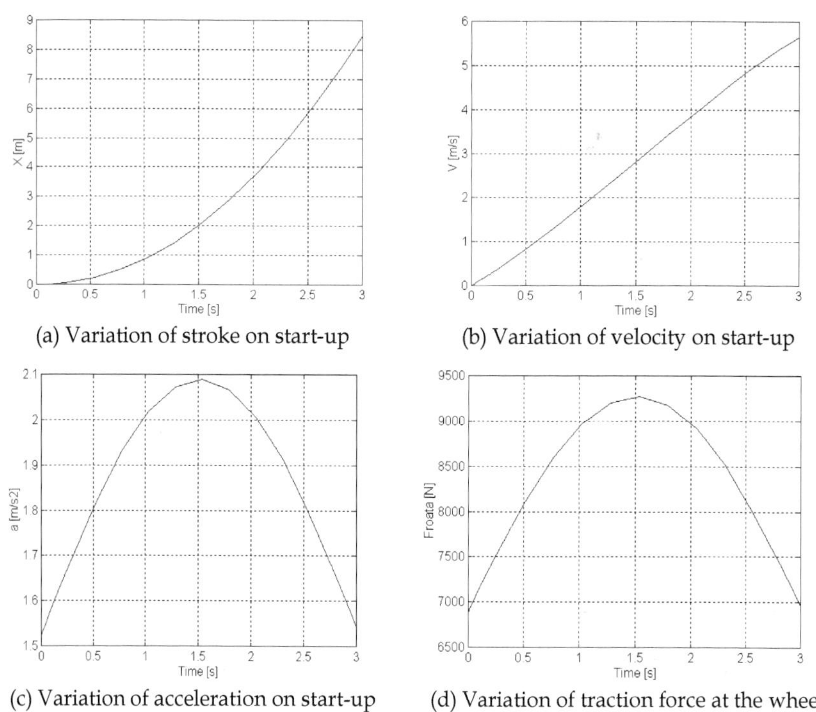

(a) Variation of stroke on start-up   (b) Variation of velocity on start-up

(c) Variation of acceleration on start-up   (d) Variation of traction force at the wheel

Fig. 9. Variation of kinematics parameters and traction force at wheels on thermal starting of the motor vehicle.

2.3.2 Dynamic behavior of the motor vehicle with hydro-mechanic propulsion system

As mentioned above, through implementation, on the motor vehicle ARO 243, of the hydraulic system for the recovery of kinetic energy during braking, it became a *parallel thermal-hydraulic hybrid motor vehicle,* which can be powered exclusively by the heat engine, analyzed in section 2.3.1, exclusively by hydraulic means, which will be studied in this subchapter, or combined, using both sources of power, being a *hybrid propulsion system.* To concretize the way of transmission of energy flow and to highlight the main subsystems participating in the starting process, there has been developed a conceptual model of the hydro-mechanic system, shown in Figure 10. At this stage, it was envisaged that the flow of hydrostatic energy comes from the hydro pneumatic accumulators, where it is stored for reuse, through the hydraulic station of the system, reaching the hydraulic machine which, operating as a hydro motor, converts the hydrostatic energy into mechanical energy and directs it, by means of the cardan axle and differential mechanism, towards the rear axle to drive wheels of the motor vehicle.

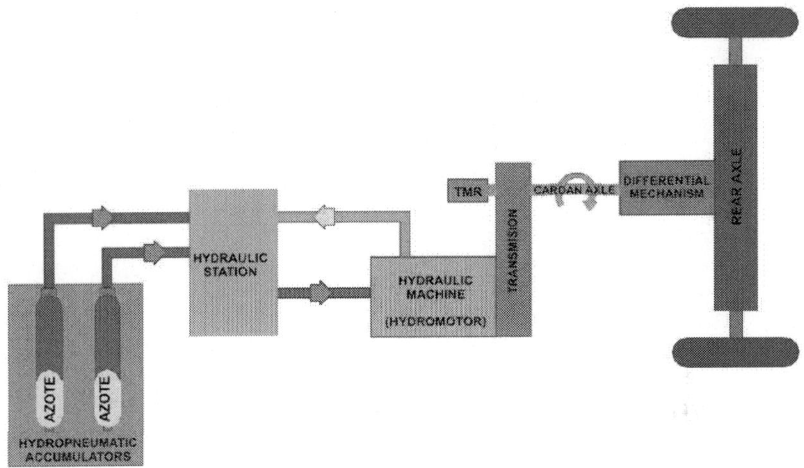

Fig. 10. A conceptual model of the hydro-mechanic starting system of the motor vehicle.

The sstudy upon the dynamic behavior of the motor vehicle equipped with hydraulic system for energy recovery, during starting, propelled, exclusively, by a hydraulic system, also, has been achieved through mathematical modeling and numerical simulation, and it has enabled knowledge of evolution of the main kinematic and dynamic parameters of the motor vehicle. Mathematical modeling of the motor vehicle powered exclusively by hydrostatic energy, supplied by hydro pneumatic accumulators, started from the known equation of motion of the motor vehicle, but, first, there was necessary mathematical modeling of the *polytrope decompression* process of azote inside the accumulators, Figure 11, to assess/evaluate the variation of pressure of the oil that actuates the hydraulic motor, see (Cristescu, 2008).

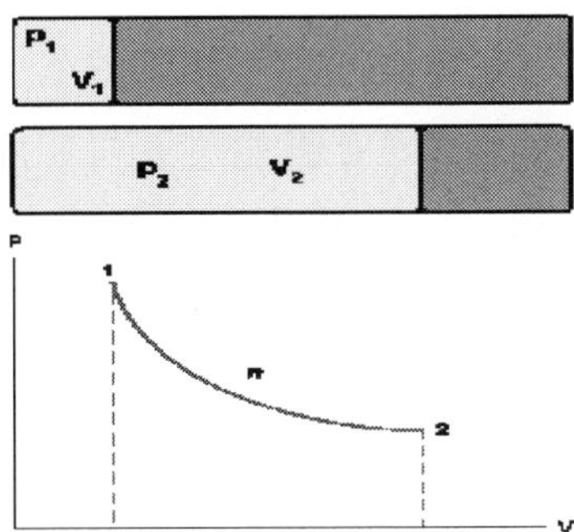

Fig. 11. Polytrope transformation of azote between the initial state 1 and final state 2 during the starting process.

In the assumption that there is no loss of fluid along the hydraulic circuits and the liquid is incompressible, and if it is marked with θ_{MH} rotation angle of the hydraulic motor shaft and with ω_{MH} its angular velocity of rotation, then it is obtained the pressure variation law for the oil inside the accumulators, according to the relation (4). With the relations known in classical literature (Untaru et al., 1974), there has been developed a mathematical model for the hydraulic starting-up stage and, after mathematical modeling and numerical simulation, have been obtained the variations of main parameters of dynamic behavior of the motor vehicle with hydraulic propulsion, presented in Figure 12.

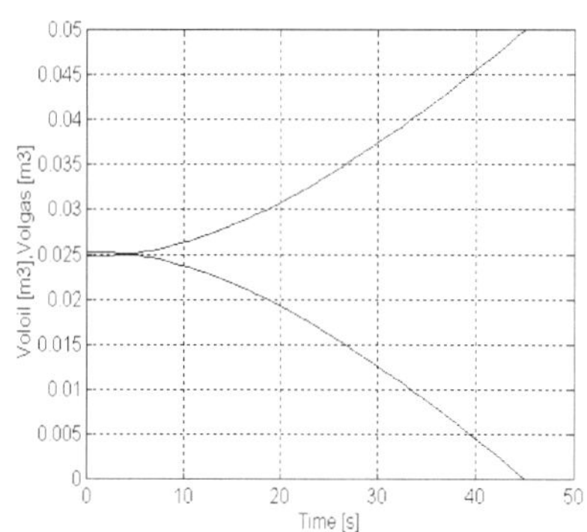

(a) Variation of oil and gas volumes

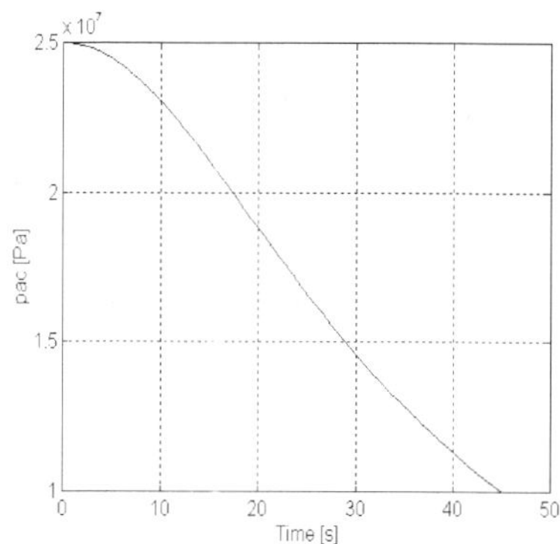

(b) Variation of pressure inside the accumulators

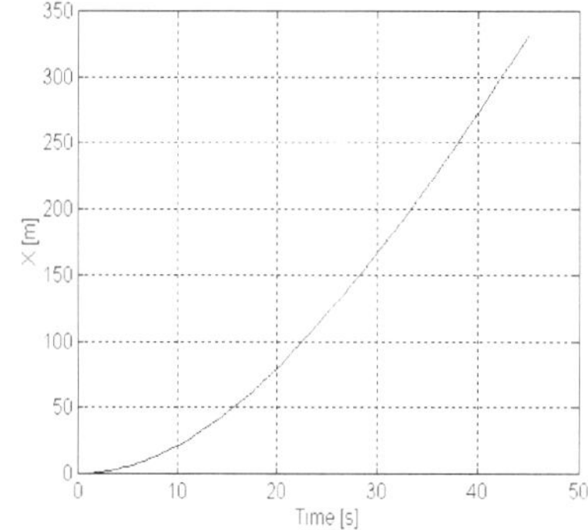

(c) Variation of start-up stroke

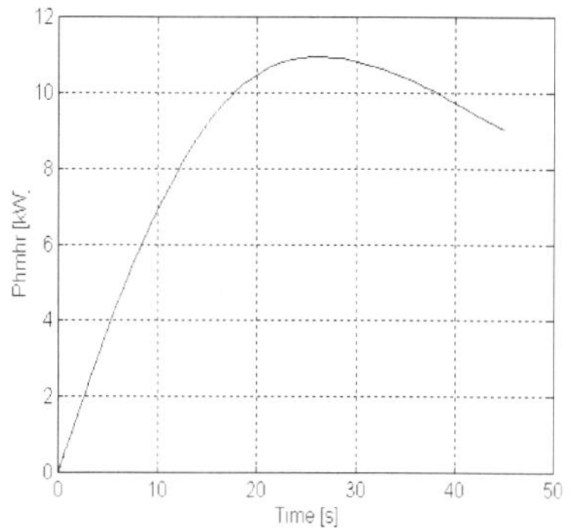

(d) Variation of power of HM

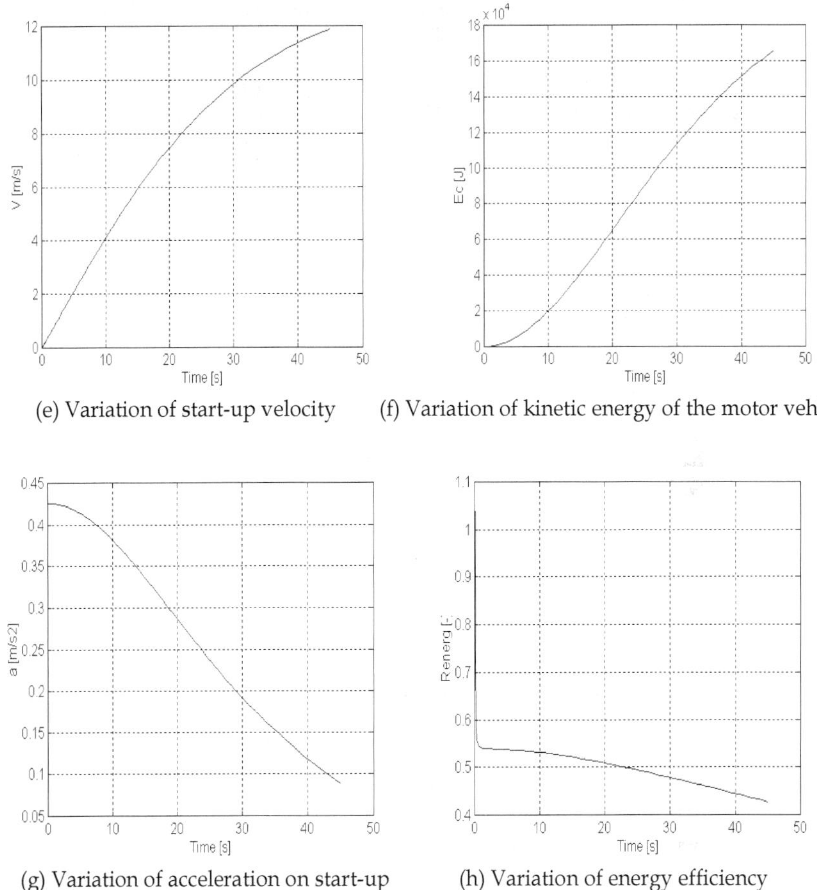

(e) Variation of start-up velocity   (f) Variation of kinetic energy of the motor vehicle

(g) Variation of acceleration on start-up   (h) Variation of energy efficiency

Fig. 12. Variation of the main parameters of hydraulic starting process of a motor vehicle.

In the Figures 12, there are presented some theoretical results of interest regarding the dynamic behavior of the motor vehicle at its hydraulic start-up. Thus, the variation of oil and gas volumes are shown in Figure 12(a), where it can see that the oil volume is in decreasing and the gas volume is in continuous increasing. The pressure in the accumulators is in continuous decreasing, as see in Figure 12(b). The variation of start-up stroke is shown in Figure 12(c). The Figure 12(d) highlights the existing of a maximum value of the power at e hydraulic motor (HM). The variation of start-up velocity is done in Figure 12(e) and this corresponds with the variation of kinetic energy of the motor vehicle, which is shown in Figure 12(f). The variation of acceleration on start-up of the vehicle is presented in Figure 12(g). The variation of energy efficiency of hydraulic propulsion is shown in Figure 12(h) and is around of 60%. The pressure variation in accumulators is done by the next relation (4):

$$p_{ac} = \frac{p_0}{\left(\left(\dfrac{p_0}{p_1}\right)^{\frac{1}{n}} + \dfrac{V_g}{V_0}\cdot\dfrac{\theta_{MH}}{2\pi}\right)^{x}} = \frac{p_0}{\left(\left(\dfrac{p_0}{p_1}\right)^{\frac{1}{n}} + \dfrac{V_g}{V_0}\cdot\dfrac{\omega_{MH}}{2\pi}\cdot t\right)} = p \qquad (4)$$

In the above relation, *state 0* is the state of preloading the battery with azotes, characterized by azotes loading pressure p_0 and their maximum volume V_0. State 1 is the *initial state* of the decompression process, characterized by the maximum pressure p_1 and minimum volume of gas $V_{1,}$, and *state 2* is the *final state* of the start-up process, when the minimum allowable pressure p_2 is reached and, also, the minimum volume of gas V_2. Based on this relation and on those known from the technical literature (Calinoiu et al., 1998)) there has been developed a mathematical model and a numerical simulation software in *MATLAB with Simulink* graphical environment, (The Math Works Inc., 2007), which allowed to obtain graphs of variations of the main parameters of interest, describing the dynamic behavior of the motor vehicle propelled exclusively by a hydraulic system.

2.3.3 Dynamic behavior of the motor vehicle at braking with kinetic energy recovery

To know the dynamic behavior of the hybrid motor vehicle, during braking with recovery of the kinetic energy available/accumulated at the beginning/before of the braking, there is made the *assumption* that, in this stage, the heat engine is operating at ralanty rotational speed and is disconnected from the transmission, being precluded the use of engine braking. Assuming the above, all available *kinetic energy* is taken by the running system and sent to the mechanical hydro pneumatic system of energy recovery at a motor vehicle through the rear axle, where it is mechanically coupled, by means of the differential mechanism and cardan axle. The *kinetic energy* taken from the drivetrain is then *converted* by the hydraulic machine, which operates in *pump mode* during this stage, into *hydrostatic energy* that is *stored* in the battery of accumulators. To concretize the way of transmission of energy flow and to highlight the main subsystems participating in the braking process with kinetic energy recovery, there has been developed a conceptual model of hydraulic braking process with kinetic energy recovery, shown in Figure 13.

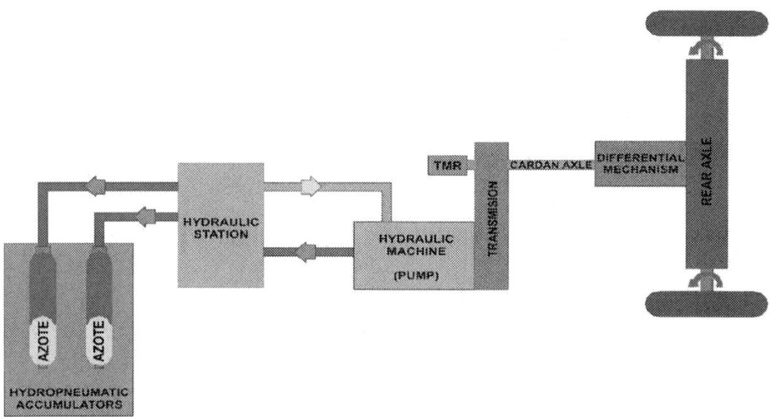

Fig. 13. A conceptual model of the hydraulic braking process with kinetic energy recovery.

At this stage, it was envisaged that the flow of mechanical/kinetic energy comes from the rear axle and drive wheels of the motor vehicle, by means of the differential mechanism and cardan axle, reaching the hydraulic machine which, operating as a pump, converts it into hydrostatic energy and directs it, through the hydraulic station of the system, towards the hydro pneumatic accumulators, where it is stored for reuse. Based on this conceptual model, there has been developed the *physical model* of braking system with energy recovery in, which lies at the basis of mathematical modeling. In Figure 14 is presented the *physical model* of the brake process with recovery of kinetic energy.

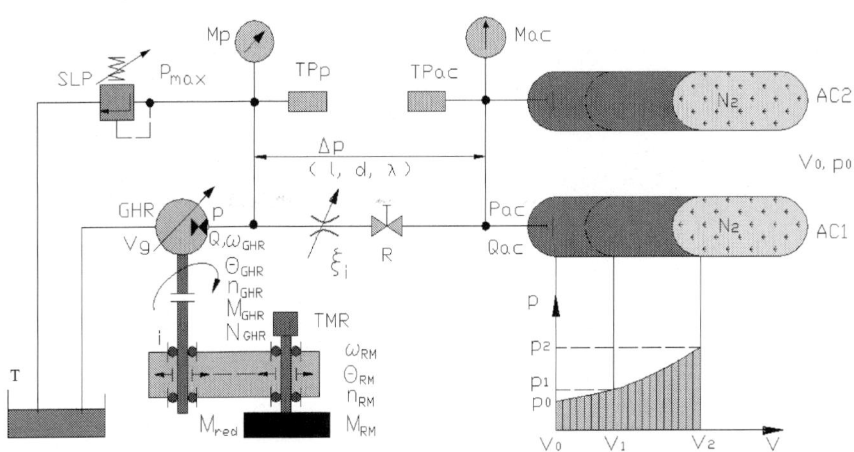

Fig. 14. The physical model of hydraulic braking system with kinetic energy recovery.

The kinetic energy *Ec*, accumulated by the motor vehicle before beginning of braking, impresses on the reduced masses M_{red} a translational motion, respectively a rotational motion, with reduced kinematic parameters at the axis of drive wheels, as indicated in the figure 7: θ_{RM} ω_{RM} , n_{RM} , respectively angular stroke, angular velocity and rotational speed at drive wheels, and, also, θ_{GHR} , n_{GHR} şi ω_{GHR}, representing angular stroke, rotational speed and angular velocity at the axis of hydraulic rotary generator (pump) GHR with displacement V_g and flow Q. Reduced torque at drive wheels M_{RM} , actuates the hydraulic machine (pump) GHR with torque M_{GHR}. The pump discharges the fluid flow Q, through a pipe with diameter *d*, length *l*, with local ζ and linear λ resistance, producing, on the route, a pressure drop Δp, before it can be compressed from a pressure p_1 or p_0, , to the pressure p_2, inside the accumulators AC1 and AC2. In the meantime, oil volume increases from V_0 or V_1 to V_2. The pump limit discharge pressure is read from a manometer M_P, controlled by the pressure limiting valve SLP and taken over electronically from the pressure transducer TP_p. The pressure inside the hydropneumatic accumulators p_{ac}, is read from the gauge M_{ac} and taken over electronically from the pressure transducer. Given the length of the braking process, which is a few tens of seconds, it is considered that the *compression process of azote* inside the accumulators is *polytrope*, with heat exchange with the environment, and must be properly modeled mathematically. Mathematical model of the hybrid motor vehicles, during braking with recovery of the kinetic energy, can, also, be obtained based on the principle of $d'Alembert$, with the equation of motion,\ of the following form:

$$M_{red}\frac{dv}{dt} = F_{act} - F_{rul} - F_{rezh} - F_{reza}$$

(5)

In the above equation, we have made the following notations:

-   F_{act} is the sum of active forces that generate or sustain motion,
-   F_{rul} is the *resistance force at running on a ramp of angle* α ,
-   F_{raer} is the *aerodynamics resistance force,*
-   F_{rezh} is braking hydraulic resistance force, reduced at drive wheels

Resistant hydraulic brake torque, produced by the hydraulic generator of displacement V_g, reduced at driving wheels, M_{RM}, generates a *resistance hydraulic brake force* at driving wheels, which is determined by the relation:

$$F_{rezh} = F_{roata} = \frac{1,59Vg \cdot (p_{ac} + \Delta p) \cdot i_o \cdot i_t}{\eta_{mh} \cdot R} ,$$

(6)

where: $p = p_{ac} + \Delta p$ is pressure of the fluid discharged by pump, and Δp pressure drop along the hydraulic circuit; i_0 is the transmission ratio of the differential mechanism; i_t – the transmission ratio of the mechanical transmission from hydraulic generator to cardan axle; η_{mh} - mechano-hydraulic efficiency, and R is the running radius of drive wheels.

Given the above, as well as other parameters known from the previous section, the equation of motion of the hybrid motor vehicle, during the braking stage with recovery of the accumulated *kinetic energy,* becomes like in (7). In Figure 15 is shown variation of the main parameters of *dynamic behavior* of the motor vehicle with energy recovery system in the braking process with kinetic energy recovery, obtained after mathematical modeling and numerical simulation.

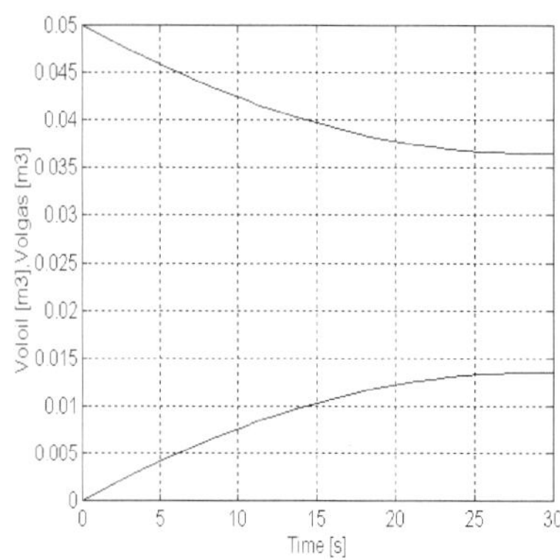

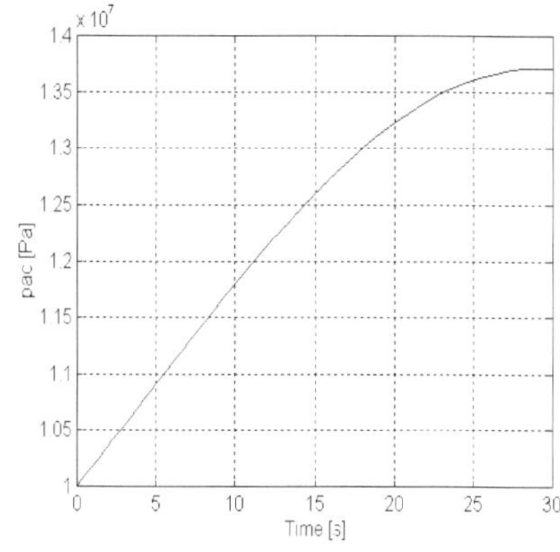

(a) Variation of volumes of oil and azotes inside the accumulators

(b) Variation of pressure inside the accumulators

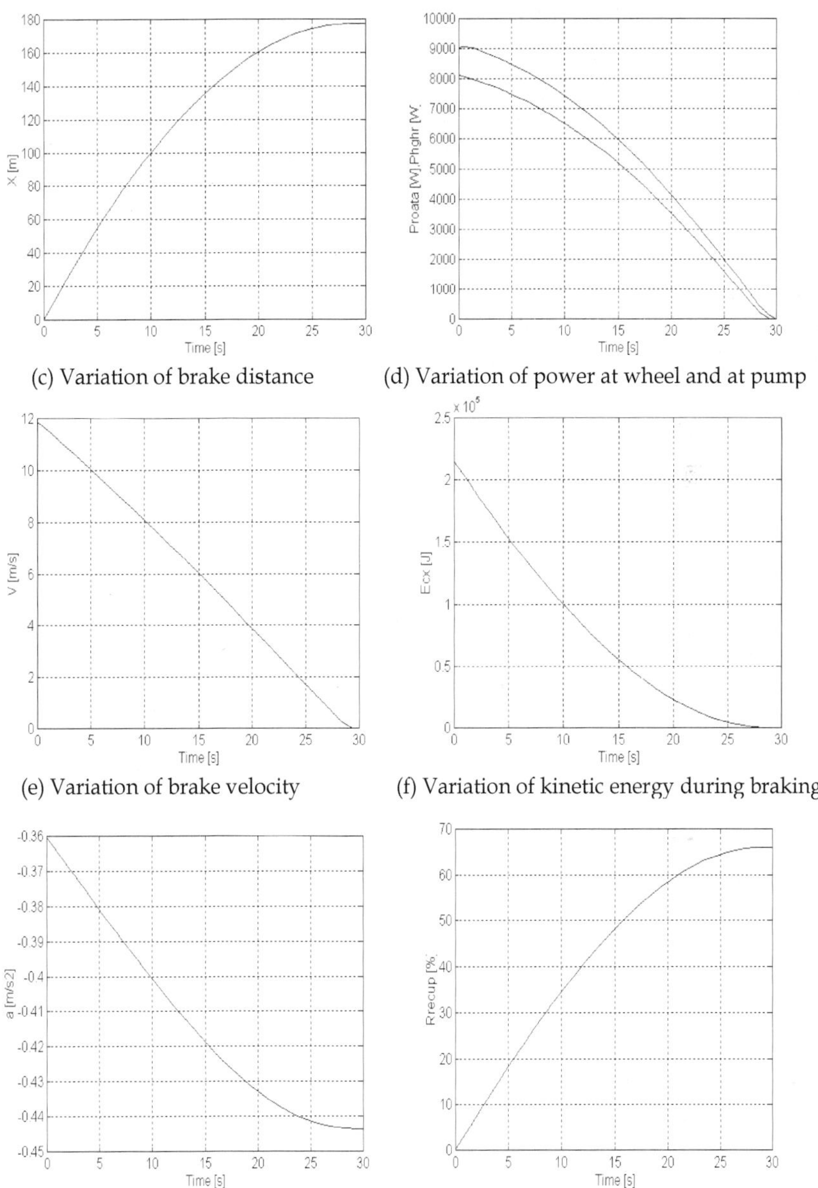

(c) Variation of brake distance   (d) Variation of power at wheel and at pump

(e) Variation of brake velocity   (f) Variation of kinetic energy during braking

(g) Variation of brake acceleration   (h) Variation of braking energy recovery coefficient

Fig. 15. The variation of the main dynamic parameters of the braking process with energy recovery.

$$M_{red} \cdot \frac{dv}{dt} = F_{act} - G_a \left(f \cdot \cos\alpha + \sin\alpha \right) - \frac{1,59 V_g \cdot \left(p_{ac} + \Delta p \right) \cdot i_o \cdot i_t}{\eta_{mh} \cdot R} - K \cdot S \cdot v^2 \qquad (7)$$

Since research on braking with kinetic energy recovery is conducted on a horizontal track, at speeds below 40km/h, there can be neglected the parameters corresponding to the ramp and air and there can be obtained the simplified form of the equation of motion of a motor vehicle. Reduced mass M_{red} is considering the cumulative effect of the actual mass of the motor vehicle (G_a/g), which is in translation motion and that of the masses in rotating motion. A special problem is modeling the compression of azote inside the accumulators, but based on specific assumptions, (Cristescu, 2008), one gets, in the end, to an expression similar to that in the start-up stage (relation 4). Using the above equation of motion, and the other relations known from literature, there is obtained a complete mathematical model, which, by numerical simulation, allowed obtaining variations of dynamic parameters specific to the braking process with energy recovery. The above figure presents the main parameters of *dynamic behavior* of the motor vehicle with energy recovery system. in the braking process with kinetic energy recovery. Thus, the variation of oil and gas volumes are shown in Figure 15(a), where it can see that the oil volume is in increasing and the gas volume is in continuous decreasing. The pressure in the accumulators is in continuous increasing, as see in Figure 15(b). The variation of braking stroke is shown in Figure 15(c). The Figure 15(d) shows the variation of power at wheel and at pump, in the braking phase. The variation of braking velocity is done in Figure 15(e) and this corresponds with the variation of kinetic energy of the motor vehicle during the braking, which is shown in Figure 15(f). The variation of acceleration on braking of the vehicle is presented in Figure 15(g). The variation of kinetic energy recovered at braking of vehicle and the evolution of coefficient of braking energy recovery is shown in Figure 15(h). His maximum is around of 65%.

2.4 Dynamic behavior of the motor vehicle with hybrid propulsion system

The hybrid system, studied in this section, is a mechatronic system, with the next specifical components: mechanical subsystem, drive and adjustment electrohydraulic subsystem, electronic interfacing component and computer component for "governance" of the process. Mechatronics is an interdisciplinary field of science and technology generally dealing with problems in mechanics, electronics and informatics. However, several areas are included in it, which form the basis of mechatronics, and cover many known disciplines, such as: electro technique, energetic, encryption technology, information micro processing technology, adjustment technique, and others. Among these, a special place is held by the electro hydraulic adjustment systems, which are very complex systems, within them interfering phenomena specific to fluid flow in the field of hydraulic volumetric transmissions and to automatic adjustment processes, (Drumea & al., 2010; Popescu & al.., 2011). Due to the complexity of these phenomena, determining the optimal solutions for their design and implementation is performed iteratively. Meeting the required performances involves the use of mathematical modeling and numerical simulation processes of these systems, together with validation of the achieved results by experimental means. For the system analyzed, was followed the next working procedure: mathematical modeling and numerical simulation of mechatronic system (first, for thermo-mechanic system and then for thermo-hydraulic hibrid system) and, finally, testing the energy recovery system in laboratory conditions.

2.4.1 Simulation of dynamic behavior of the motor vehicle with thermo-mechanic propulsion system

Simulation networks presented in this section have been developed and analyzed by modules using *AMESim* numerical simulation software, (LMS IMAGINE SA 2009). The final model used for simulation of HIL in the stage of tests was made using the models developed during the unfolding of research activity upon the system. The first model developed was the model of the motor vehicle with thermo-mechanic propulsion system, figure 16. Input data into the model are: aerodynamics coefficient of the vehicle and torque at the drive wheels, and output data – rotational speed at its wheels.

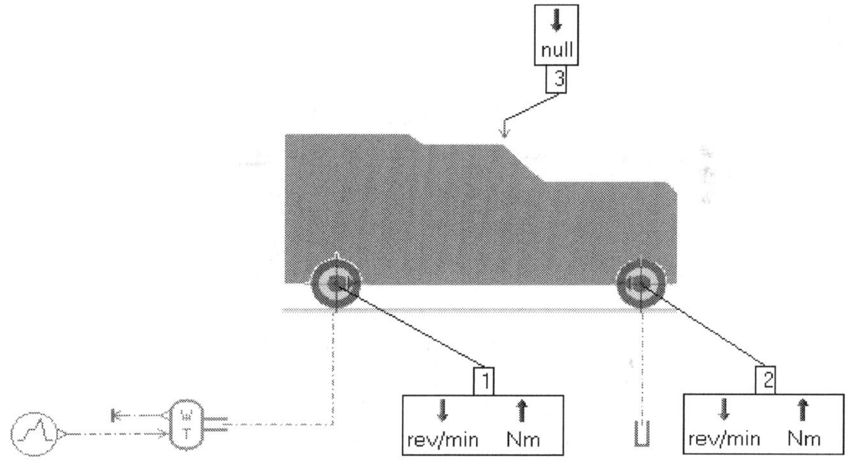

Fig. 16. The model of the motor vehicle with thermo-mechanic propulsion system.

To achieve the simulation network of the motor vehicle with thermo-mechanic propulsion system, the next models have been used: the model of the heat motor vehicle, the models of the elements that convey energy from the vehicle to the ground (drive wheels and free wheels), the model of the differential mechanism, the model of the gearbox, the model of the clutch and the model of heat engine. For the modeling of heat engine, there has been used a simulation network of the external feature of heat engine, using technical data from the table 1. The diagram of relationship between rotational speed and drive torque is presented in Figure 17. This technical feature, from table 1, corresponds to an *Andoria 4CT90 TD* engine, which was part of motor vehicle endowment in some ARO models.

The simulation network of the motor vehicle with thermo-mechanic propulsion system is presented in Figure 18

Rotation al speed	[rpm]	1000	1500	2000	2500	3000	3500	4000
Torque	[Nm]	170	183	186	183	178	168	158

Table 1. Table with technical data of heat engine.

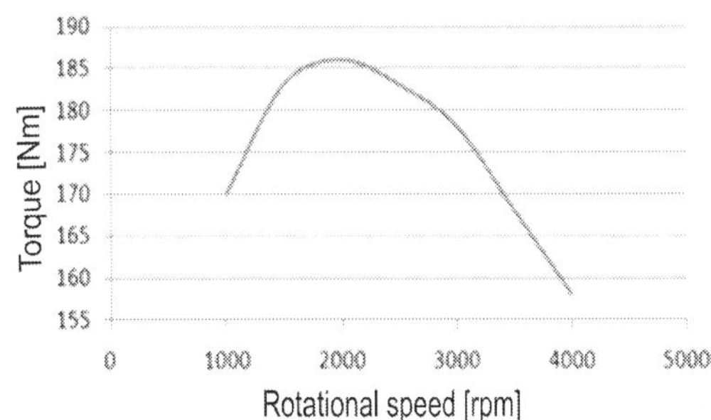

Fig. 17. External feature of the heat engine.

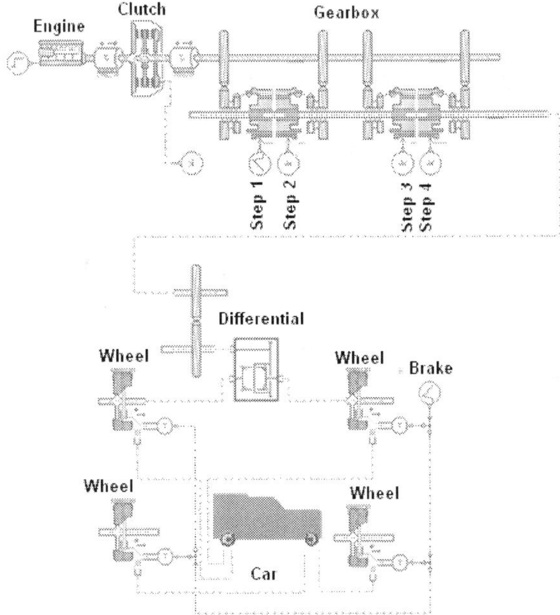

Fig. 18. The simulation network of the motor vehicle with thermo-mechanic propulsion system.

Data about the drive module used to define the models of the simulation network: transmission with 4 speeds (with the next transmission ratios: step I 4.92; step II 2.682; step III 1.654; step IV 1); mechanical switch box with 2 steps; differentials on the front and back bridges, with transmission ratios of 3.72:1; diameter of the wheel D = 736 mm; rolling radius R = 350 mm; cross surface St = 3.57 m²; motor vehicle weight: own weight 1680 daN; total weight 2500 daN; rolling resistance coefficient f = 0.02; ramp angle α = 0º; gravitational acceleration g = 9.81 m/s²; aerodynamics resistance coefficient K = 0.0375 daN/m²; efficiency of the transmission η = 0.9.

Simulation network was run under the next conditions: at the input of the heat engine has been forced a control signal (acceleration pedal), corresponding to the torque/rotational speed dependence curve in Figure 17. The grafical results ar presented in Figure 19. It was maintained constant (100%) for a period of 40 seconds, as is shown in Figure 19(a). At the moment t = 40 s, full closure was ordered to supply no longer the heat engine. The aim of this simulation was to register the evolution of dynamic parameters of the motor vehicle, in the stage of running on energy received from the heat engine and during movement due to inertia of the system, sees Figure 19, namely: the *variation over time of control signal of heat engine* (0..1 corresponds to 0..100%), see Figure 19(a), the *eevolution over time of displacement of motor vehicle*, see Figure 19(b). Evolution of running velocity of motor vehicle, see Figure 19(c), Evolution over time of acceleration of vehicle, see Figure 19(d), variation of torque at the heat engine shaft, see Figure 19(e), Variation of rotational speed at the heat engine shaft, gearbox and differential mechanism, see Figure 19(f).

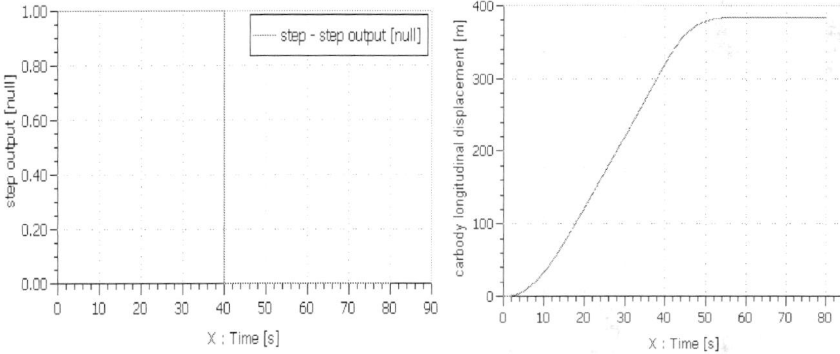

(a) Variation over time of control signal of heat

(b) Evolution over time of displacement of vehicles

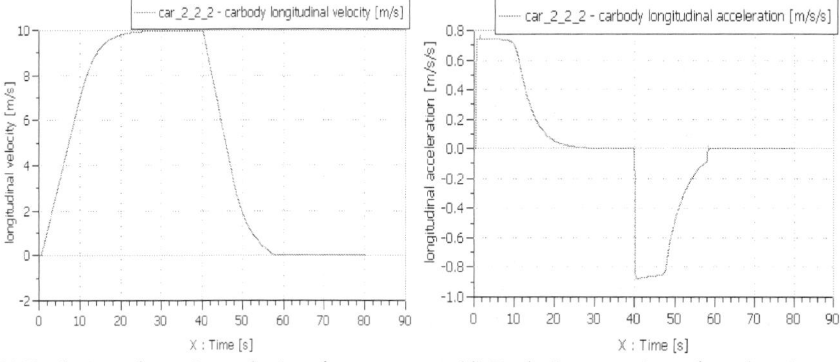

(c) Evolution of running velocity of motor vehicle

(d) Evolution over time of acceleration of vehicles

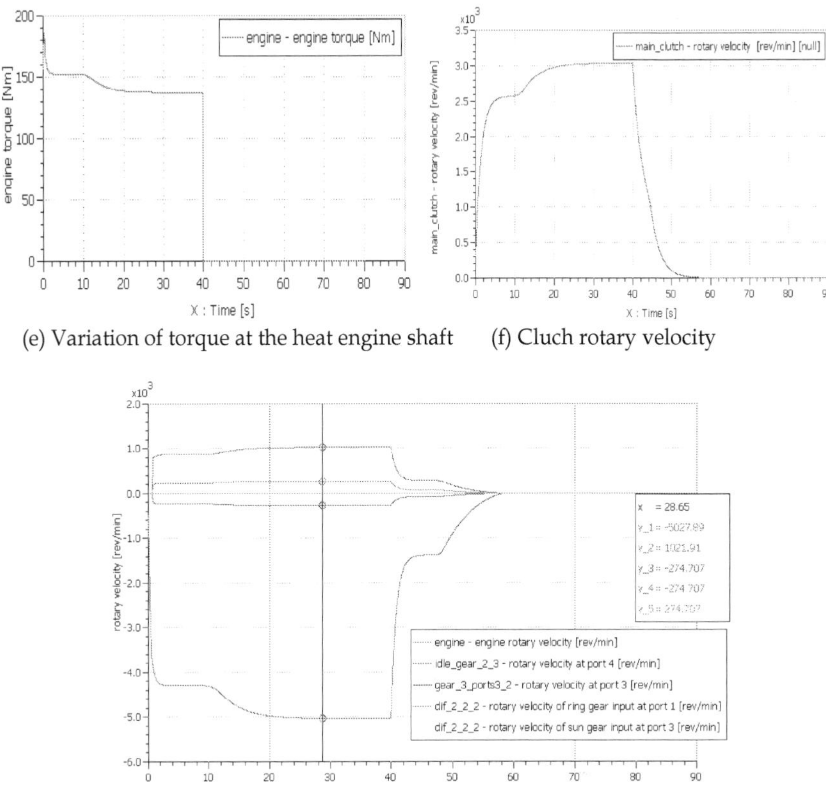

(e) Variation of torque at the heat engine shaft   (f) Cluch rotary velocity

(g) Variation of rotational speed at the heat engine shaft, gearbox and differential

Fig. 19. Variation of the dynamic parameters of the motor vehicle with thermo-mechanic propulsion system.

2.4.2 Simulation of dynamic behavior of the motor vehicle with thermo-hydraulic propulsion hybrid system

The motor vehicle with thermo-mechanic propulsion system has been analyzed with the simulation network shown in Figure 18. The simulation network of dynamic behavior of the motor vehicle with thermo-hydraulic propulsion hybrid system includes the simulation network of thermo-mechanical system, shown in Figure 18, to which was attached the components of energy recovery hydraulic system, to storage and to use of recovery energy achieved at the braking of motor vehicle. Hydrostatic component attached to the thermo-mechanic model is a basic one, greatly simplified for the reason to have an overview of the simulation network. Full schematic diagram includes a series of other elements of hydrostatic instrumentation absolutely necessary for the development of such a system. As it can be seen, in the Figure 20, the most important elements of the hydrostatic system are: bidirectional and reversible hydrostatic unit, battery of oleopneumatic accumulators and mechatronic system for control and adjustment of capacity of the hydrostatic unit.

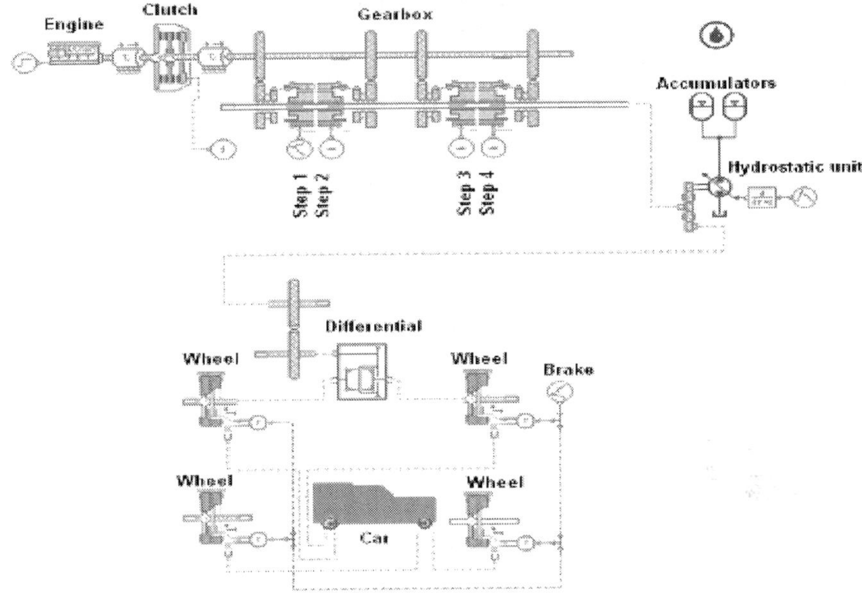

Fig. 20. The simulation network of the dynamic behavior of the motor vehicle with thermo-hydraulic propulsion system.

Data about the hydrostatic drive module used to define the simulation network are the next: capacity of the hydrostatic unit: 45 cm^3; volume of the oleopneumatic accumulators: 25 liters; system which conveys mechanical energy between the hydrostatic unit and gearbox with transmission ratio: 1:1; density of working oil 850 kg/m^3; oil elasticity module: 16000 bar; gas pressure inside accumulators: 100 bar. The ssimulation network of the dynamic behavior of the motor vehicle with thermo-hydraulic propulsion hybrid system has been similarly to the previously presented network, to determine the evolution of dynamic parameters of vehicle. The conditions, under which the model has been run, were the next:

-    at the input of the heat engine has been forced a control signal (acceleration pedal) corresponding to the torque/rotational speed dependence curve in Figure 17. It was maintained constant (100%) for a period of 40 seconds (Fig. 19a). At moment t = 40 s full closure was ordered to supply no longer the heat engine.
-    at moment t = 40 s hydrostatic unit was ordered with a control signal corresponding to its operation in pump mode, with capacity varying after a ramp-step-ramp signal 0 .. 100%, for 10 seconds. During this period the energy recovery function is performed (loading of oleopneumatic accumulators).
-    during time span t1 = 40 seconds t2 = 60 seconds the hydrostatic drive has capacity of 0 cm^3, the energy recovery system is "decoupled" from the mechanical system.
-    at moment t = 60 s hydrostatic unit was ordered with a control signal corresponding to its operation in motor mode, with capacity varying after a ramp-step-ramp signal 0 .. 100%, for 20 seconds. During this period the use of recovered energy function is performed (discharge of oleopneumatic accumulators).

The graphical results, recorded from simulation process, are shown in Figures 21, where it can see: the evolution over time of displacement of motor vehicle, in Figure 21(a), the evolution over time of running velocity of motor vehicle and control signal of hydrostatic unit, in Figure 21(b), the evolution over time of acceleration of vehicle, in Figure 21(c), the variation of torque at the heat engine shaft, in Figure 21(d), the variation of force at the drive wheel, in Figure 21(e), the evolution of pressure inside of accumulators, in Figure 21(f), and, finally, the evolution of the oil flow inside the accumulators depending on control signal of the hydrostatic unit capacity, which can be seen in Figure 21(g),

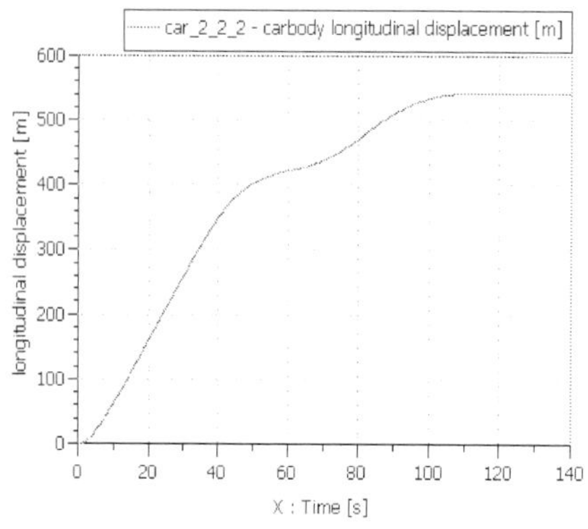

(a) Evolution over time of displacement of motor vehicle

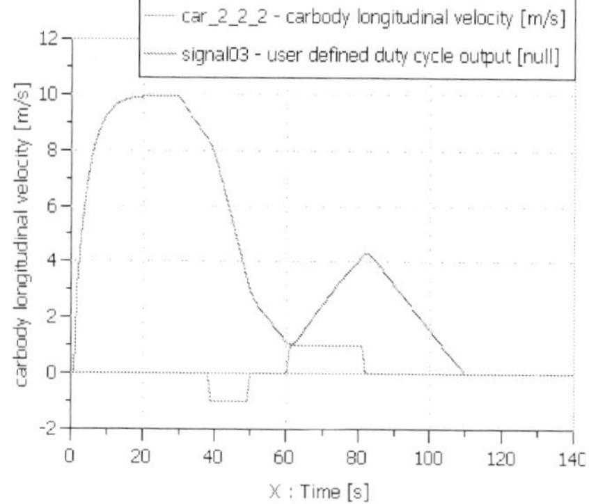

(b) Evolution over time of running velocity of motor vehicle and control signal of hydrostatic unit

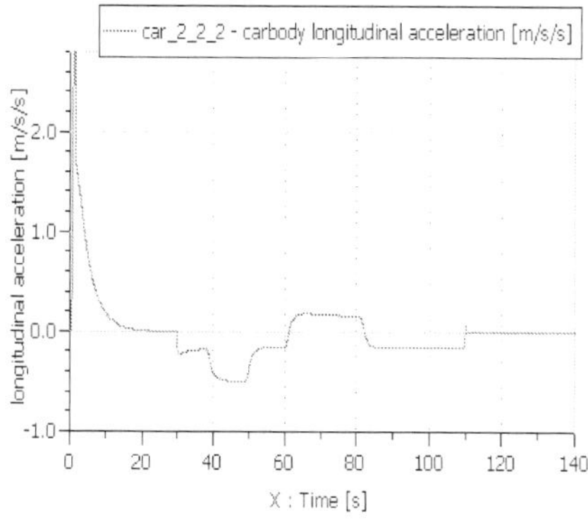

(c) Evolution over time of acceleration of vehicle

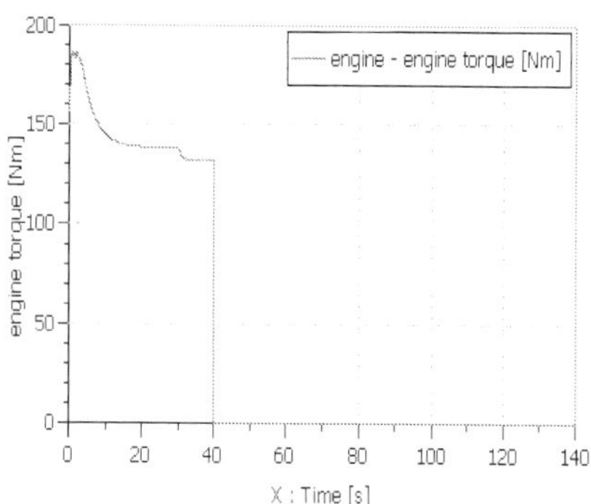

(d) Variation of torque at the heat engine shaft

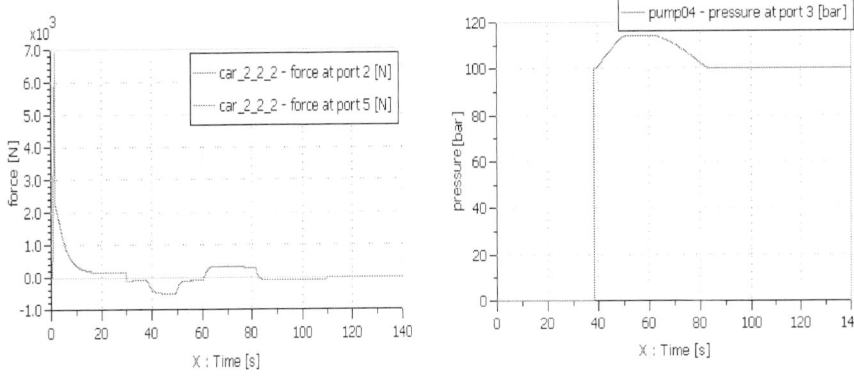

(e) Variation of force at the drive wheel

(f) Evolution of pressure inside of accumulators

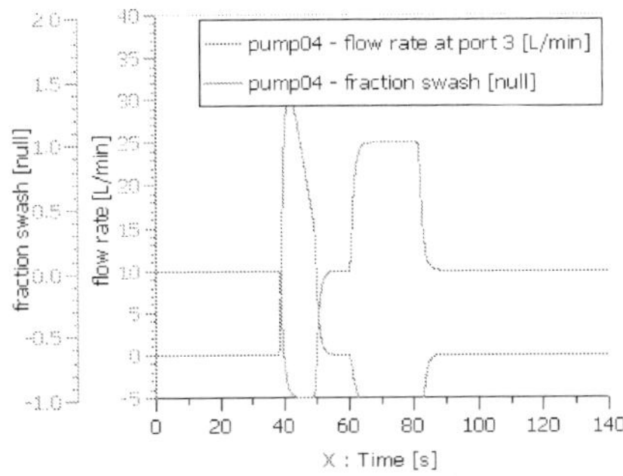

(g) Evolution of oil flow inside the accumulators depending on control signal of the hydrostatic unit capacity

Fig. 21. The variation of the dynamic parameters of the motor vehicle with thermo-hydraulic propulsion system.

3. The mechatronic stand for testing the kinetic energy recovery system

For testing, in laboratory conditions, of the energy recovery mechatronic system, there was necessary to design and physically develop a test stand, able to reproduce the characteristic working modes of a hybrid motor vehicle with the ability to recover kinetic energy during braking. The stand, in itself, is conceived also as one *mechatronic system*.

The goal of stand design and development was to create the possibility of putting the developed mechatronic system for kinetic energy recovery under a series of tests, conducted during all the working modes/stages, before being implemented on a motor vehicle, in order to *understand its dynamic behavior* and the genuine abilities of the system, and, also, to detect early any gaps or shortcomings and new needs, to improve the system on the fly. The stand, also, allows the development of complex experimental research and minimizes the risks borne by a project of this complexity, in case of its direct implementation on the vehicle, without testing in laboratory conditions, (Cristescu, 2008a).

3.1 The technical solution adopted for designing of test stand

The technical solution adopted, in principle, for design and implementation of the test stand of mechatronic system for braking energy recovery, was that of *simulation*, in laboratory conditions, of the *transitional working regimes* for starting and braking the motor vehicles, based on the use of specific equipment only with *electric and hydraulic* drive and control, monitoring the evolution of parameters within the system and managing the processes by computer, using some dedicated software. For *simulating* the operation of the heat engine of the motor vehicle, a combined solution was chosen, based on hydraulic electro-pump, composed of an electric motor and a high pressure hydrostatic pump, which drives a hydraulic motor (or the acceleration module), together simulating the *thermal power*, torque and rotational speed source, parts of the normal equipment of a motor vehicle. The second source of power, *hydraulic power*, characteristic to the energy recovery system, is represented exactly by the *hydro-mechanical module* of the energy recovery system tested on stand, composed of a hydraulic machine and the chain or gear transmission, shown in Figure 25 (a). One load module gathers/integrates, on its input, the two powers, simulating thus the *thermo-hydraulic hybrid propulsion system* of motor vehicles. In this way, *3 propulsion systems* of the motor vehicle can be simulated on stand:

- thermo-mechanical propulsion, based on the heat engine of the motor vehicle;
- mechano-hydraulic propulsion, based on the hydraulic recovery system;
- thermo-hydraulic hybrid propulsion.

Technical solution adopted allows simulation of braking modes with kinetic energy recovery system, namely:

- braking with recovery of kinetic energy impressed by the thermo-mechanical system;
- braking with recovery of kinetic energy impressed by the hydraulic propulsion system

3.2 The general assembly and the structure of the mechatronic test stand

General assembly of mechatronic stand, designed to test the kinetic energy recovery system, is shown in Figure 22, and the physical development of the stand is shown in Figures 23 and Figure 24.

The structure of mechatronic test stand consists of the following modules, which can be seen in Figure 25:

1. *hydro-mechanical module* of the tested mechatronic system for energy recovery, as a source of hydraulic power of the hybrid drive system, consisting of a hydraulic machine and a mechanical chain or gear transmission, fitted with a torque and speed transducer, to monitor the main parameters: torque and speed, shown Figure 25(a);

2. *test module or loading module*, comprising a load device, with a frame containing a torque transducer, having coupled, at its output, a hydraulic unit, and at its input, the hydro-mechanical module of the enrgy recovery system, subjected to testing, shown in Figure 25(b);

3. *module of the electropump*, with variable rotational speed and displacement, which forms together with the acceleration module (hydraulic motor), the subsystem for simulation of the drive engine, shown in Figure 25(c);

4. *acceleration module,* comprising a hydraulic motor, torque and speed transducer, and cardan shaft that connects mechanically the two drive systems simulated, *heat* and *hydraulic,* shown in Figure 25(d);

5. *module for storage of the fluid* under pressure or *battery of accumulators,* comprising a supporting frame on which two hydropneumatic accumulators are mounted, as well as the related security devices, shown in Figure 25(e);

6. *module of the hydraulic station,* with working fluid conditioning subsystem, consisting of an oil tank equipped with temperature control system, drive pump and hydraulic blocks, shown in Figure 25(f);

7. electrical, *electronic and automation subsystem,* with an electrical and electronic subsystem for actuation and control of stand operation and with a *subsystem of sensors and transducers* for monitoring parameters, Figure 25(g);

8. *informatic and control subsystems,* for monitoring and control of stand operation, shown in Figure 25(h);.

The first six modules represent the mechano-hydro-pneumatic subsystem of the test stand, which, toghether with the electronic subsystem and the informatic and control subsystems, create a typical structure of one mechatronic system, (Maties, 1998).

The main modules of the mechatronic test stand were presented in Figure 25.

The stand allows to do testing in the field of hydrostatic transmissions, in order to optimize them functionally and to improve their energy efficiency. The stand is proper for rotary hydrostatic transmissions, with or without energy recovery systems, which are part of fixed (industrial) and mobile (towed vehicles and motor vehicles) equipment, including their subsystems, for functional tests and to establish performance parameters.

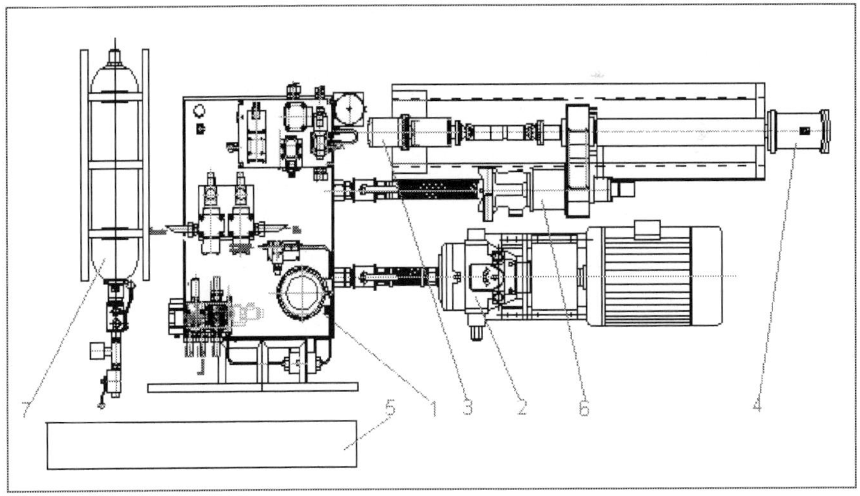

Fig. 22. General assembly of the mechatronic stand for testing of the kinetic energy recovery system.

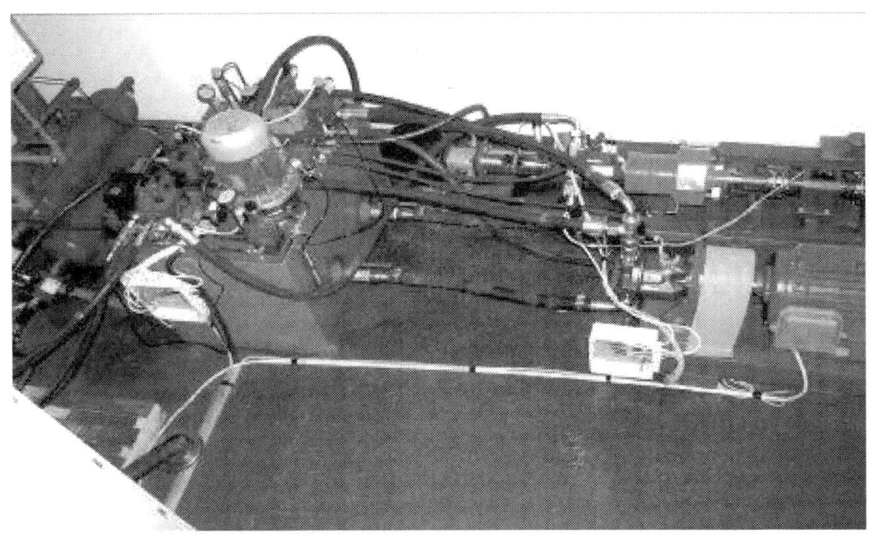

Fig. 23. Mechatronic stand for testing the kinetic energy recovery system – overview.

Fig. 24. Mechatronic stand for testing the kinetic energy recovery system – frontal view.

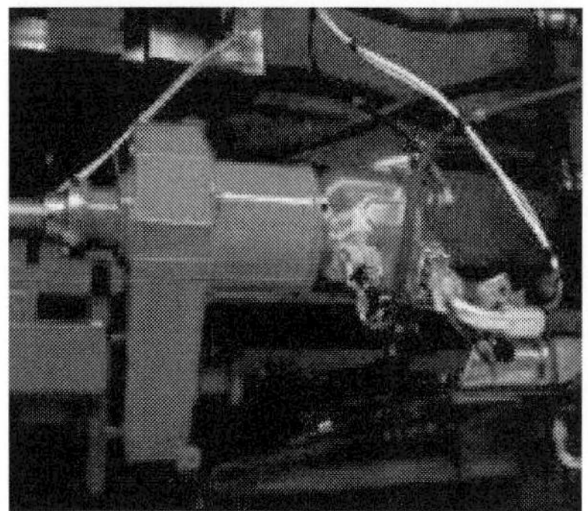

a) The hydro-mechanical module of the recovery system

b) The testing module

c) Module of the electropump

d) The acceleration subsystem

e) The accumulating subsystem

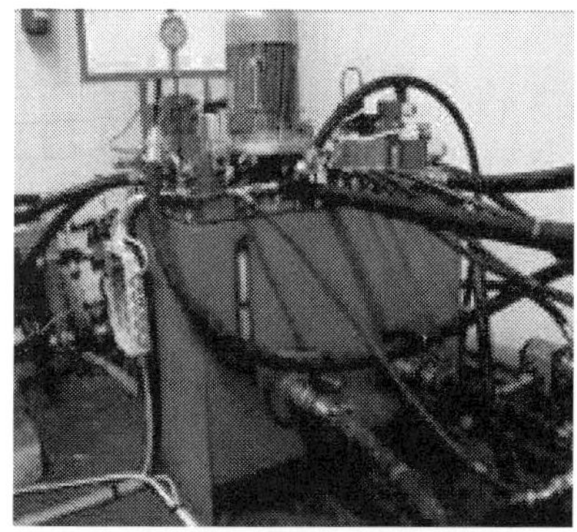

f) The hydraulic station

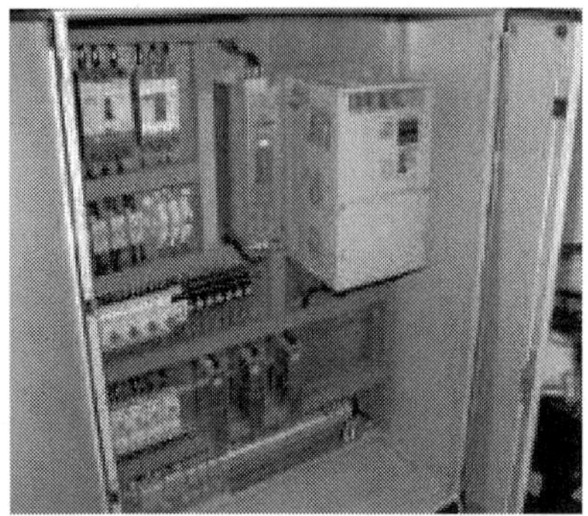

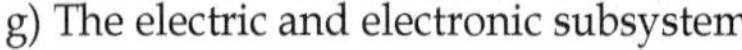

g) The electric and electronic subsystem   h) Informatic and control subsystems

Fig. 25. The main modules of the mechatronic test stand.

3.3 Testing of dynamic behavior of the hybrid motor vehicles by using of the real-time simulation network

The analyzed system has been studied both by means offered by conventional methods of mathematical modeling and numerical simulation and, also, by using the hybrid networks of *real-time co-simulation* and simulation (Ion Guta, 2008).

In order to testing of dynamic behavior of the hybrid motor vehicles by using of the *real-time simulation network*, is necessary to do this in two steps. For developing the real-time simulation the first step is the creating of the co-simulation subsystem, which will be presented in the next subchapter. In the second step, it will be used the hybrid simulators, which connect in terms of information the mathematical models and components of physical systems

3.3.1 The creating of the co-simulation subsystem

For achieving the co-simulation networks, there have been used the above presented models, developed by means of *AMESim* software, (LMS IMAGINE SA, 2009). These were coupled to a simulation supervisory application, developed by the authors, of this work by means of *LabVIEW* programming language, (LabVIEW, 1993). This was a first step for developing the real-time simulation application presented in the experimental section of this work. In Figure 26 can be seen the co-simulation subsystem, the process model being coupled to the application developed in *LabVIEW* and loaded on a NI PXI industrial computer, through the communication process implying sharing of memory (shared memory). For communication between the two systems, there can also be used TCP/IP sockets or TCP/IP protocol.

Application developed using *LabVIEW* language, seen in Figure 27(a), has an operator interface that allows governing of the simulation process and visualization of data obtained during simulation, Figure 27(b). The application contains an automation component which controls the hydrostatic equipment within the simulation network, by adjustment of hydrostatic unit capacity, opening and closing of way directional control valves, comprised in the hydrostatic subsystem.

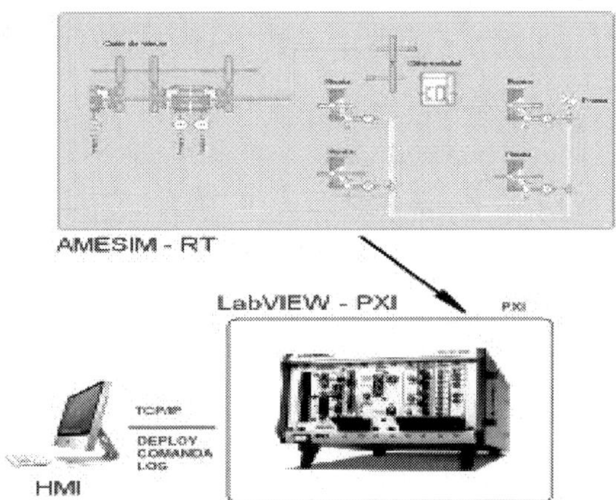

Fig. 26. Co-simulation subsystem.

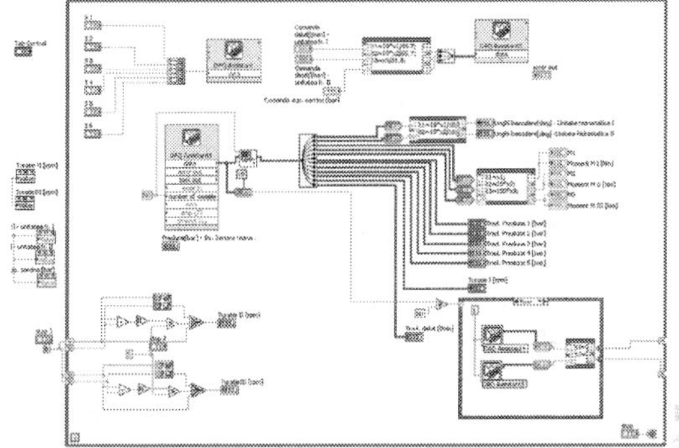

(a) Block diagram of data acquisition module

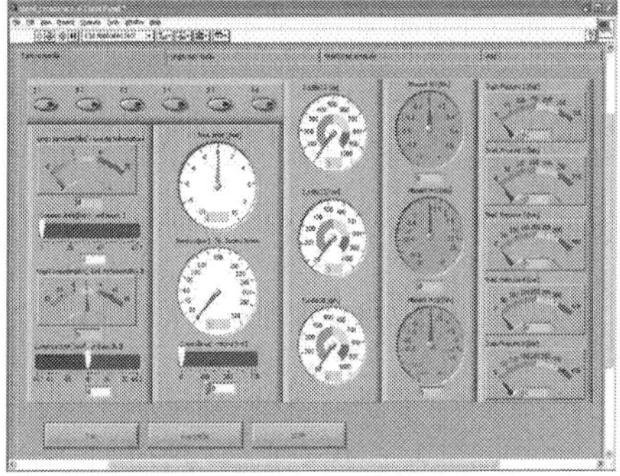

(b) Interface VI of stand functioning

Fig. 27. The application developed in LabVIEW language.

3.3.2 Testing energy recovery system by using the hybrid networks of real-time co-simulation and simulation

The solution adopted to achieve the hydrostatic transmission testing system for the energy recovery systems, was that of simulation of operating, braking and start-up modes of motor vehicles, based on electrohydraulic actuation equipment and systems for simulation and numerical modeling specific to the field of hydrostatic drive. The simulations and the experiments have been achieved in the laboratory of hydrostatic transmissions of the institute INOE 2000 – IHP, where. are conducted experimental research in the field of hydrostatic rotary transmissions, in order to optimize them functionally and improve their energy efficiency. To know the dynamic behavior of the energy recovery system, in laboratory conditions, it was used the concept of *"real-time simulation"* of a system, or *"Harwar-in-the loop"* (HIL), involving the simultaneous use of a mathematical model and a *physical part* of the system, see (Gauchia & Sanz, 2010).

The introduction of computers in monitoring and control of industrial process, led to change of technological systems. Great flexibility offered by these systems allows "software" optimization of complex systems. In this scenario it is rational the *use of hybrid simulators*, which connect in terms of information the mathematical models and components of physical systems. This concept has been established in the specialized literature as *"real-time simulation"* or *"numerical simulation with control loop equipment"*. (Ion Guta, 2008). Modern methods of experimentation, in the field of hydraulic and pneumatic drive systems, imply the existence of at least one numerical calculation equipment. The necessity of using electro-hydraulic converters, for control and adjustment of various physical parameters such as force, displacement, together with the exponential growth of digital electronics, confirms this. Digital equipment can be found in the structure of sensors and transducers, numerical displays, electronic servo-amplifiers (compensators) or process computers. As part of the endowment of any modern laboratory of electro hydraulic drives there are not lacking *sensors and transducers* with electronic communication interface, adjustment systems (proportional electro hydraulic directional control valves, hydraulic or pneumatic servo pumps/ motors etc.) with analog/digital control ports and electronic adjustment blocks. The ability to "load" the numerical calculation systems, with "virtual models" of systems developed using advanced modeling languages, increases even more their flexibility, as it can be seen in Figure 28.

The system includes a numerical model simulating the dynamic behavior of a motor vehicle with thermo-mechanic propulsion, a process computer of PXI (from *National Instruments*) family, an experimental stand and a system for regular acquisition of data in the analyzed process. The purpose of this analysis is to be excited correspondingly, based on specific input data into the mathematical model, the *power components of the experimental stand* by means of the process computer, in order to be quantified the *amount of energy* that it can recover under simulated operation conditions.

To perform experiments in the simulation model (Figure 20) has been removed simulation of the electro hydraulic subsystem. In place of this component, there has been introduced into the model, information gathered from the testing stand, which contains the physical component of the electro hydraulic subsystem. The next technological parameters on the stand have been introduced into the model: rotational speed at the shaft of hydrostatic unit and torque obtained at the shaft of the unit. From the simulation model, a command has been sent to the physical unit on the stand, by means of which has been emulated the heat engine. The command has been sent so, that rotational speed achieved at the shaft of

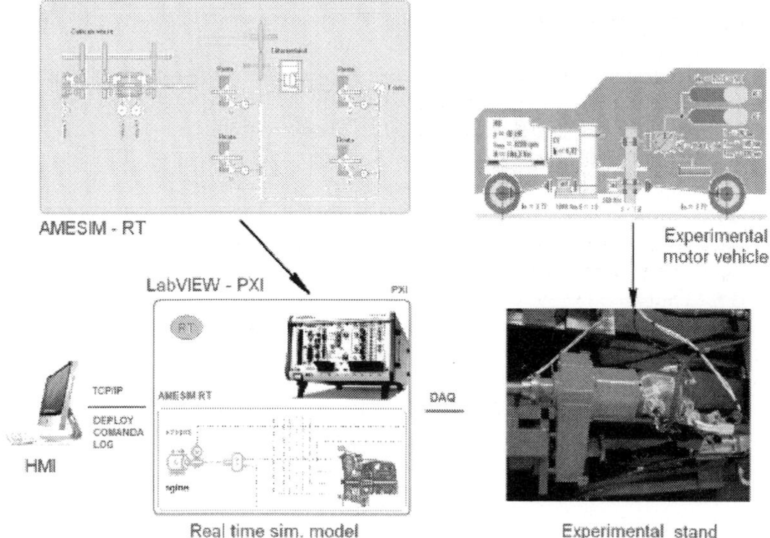

Fig. 28. The hibride network of real-time simulation for testing energy recovery system, in laboratory conditions.

hydraulic motor (which emulates, on the testing stand, the real heat engine) to be dependent on its torque, according to the torque/rotational speed functional curve imposed in the simulation model. Adjustment of rotational speed at the drive shaft has been performed by appropriate variation of the hydrostatic unit capacity. In parallel, computer component of the mechatronic stand, for recovery of braking energy of a motor vehicle, has controlled the devices on the stand so that the simulation model on the PXI industrial computer to record the next cyclogram:

- drive of clutch (coupling of the heat engine to the motor vehicle gearbox) at t = 0 seconds;
- drive of gearbox accordingly to speed step 1 at t = 0 seconds;
- drive of acceleration of the engine till achieving a running velocity of the vehicle of 10 m/s at t = 0 .. 30 seconds;
- drive of clutch (decoupling of the heat engine from the motor vehicle inertial load) at t = 30 .. 70 seconds;
- drive of hydrostatic unit capacity of the energy recovery system, corresponding to its operation in pump mode (working with energy recovery) at t = 32 .. 50 seconds;
- free operation till the motor vehicle stops;
- drive of clutch (coupling of the heat engine to the motor vehicle gearbox) at t = 70 seconds;
- drive of gearbox accordingly to speed step 1 at t = 70 seconds;
- drive of engine acceleration simultaneously with drive of capacity of the system hydrostatic unit corresponding to its operation in motor mode (use of hydrostatic power available in the mechatronic recovery system) till achieving a running velocity of the motor vehicle of 10 m/s at t = 70 seconds;

- drive of clutch (decoupling of the heat engine from the motor vehicle inertial load) at t = 100 seconds;
- drive of hydrostatic unit capacity of the energy recovery system corresponding to its operation in pump mode (working with energy recovery) at t = 105..118 seconds;
- free operation till the motor vehicle stops.

Data obtained from experiments of real-time simulation for testing of energy recovery system are shown in Figures 29, where it can see the evolution over time of displacement of motor vehicle, in Figures 29(a), the evolution over time of running velocity of motor vehicle, in Figures 29(b), the vvariation of torque at the shaft of the system equivalent to a heat engine and at the shaft of the hydrostatic unit, in Figures 29(c), the evolution over time of acceleration of motor vehicle, in Figures 29(d). Finally, the comparative study on the evolution of torque at the drive shaft, with and without contribution of mechatronic system for energy recovery in the braking phase ,is presented in Figures 30.

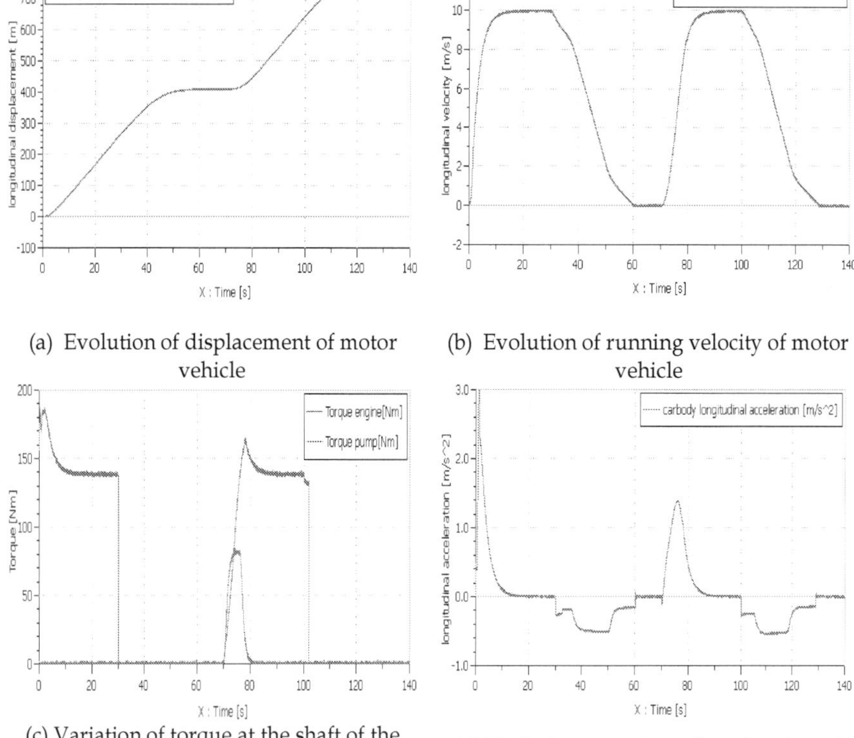

(a) Evolution of displacement of motor vehicle

(b) Evolution of running velocity of motor vehicle

(c) Variation of torque at the shaft of the system equivalent to a heat engine and at the shaft of the hydrostatic unit

(d) Evolution over time of acceleration of motor vehicle

Fig. 29. Data obtained from experiments of real-time simulation for testing of energy recovery system.

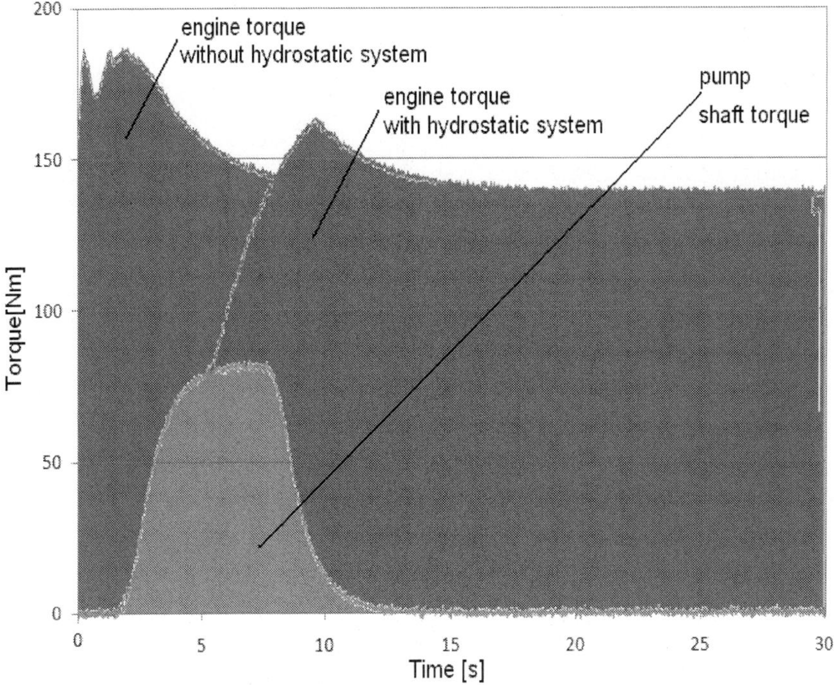

Fig. 30. Comparative study on the evolution of torque at the drive shaft with and without contribution of mechatronic system for recovery of the motor vehicle braking energy.

4. Conclusions

Given the necessity of finding alternative solution to reduce consumption of fossil combustible, being now in exhaustion, and to mitigate the negative impact of emission on the environment, vehicle manufacturers have indicated that an effective solution, could be the development of hybrid propulsion systems, in particular those regenerative propulsion systems, which can recover a portion of the kinetic energy of the vehicle, accumulated before braking.

In this context, the chapter presents some specific problems concerning the complexity of the hybrid propulsion systems of the road vehicles and points out that, indeed, this is a new area suitable for the application of mechatronics, where it is the only technology able to monitor, to manage and to optimize the transient regimes specific for this systems.

By addressing the problem of recovering kinetic energy, when road vehicles are at braking, the authors have reached automatically and at the issue of the hybrid propulsion systems, and they gained o good theoretical and practical experience, which is communicate in this chapter and which can be a point start-up for other researches.

In the first part, the paper presents the general problem of the energy recovery systems and makes a brief presentation for one Romanian mechatronic hydraulic system for energy recovery, which transforms one motor vehicle, where it is implemented, into motor vehicle with hybrid propulsion system, including the main modules of the system.

There are presented some theoretical results obtained by mathematical modeling and numerical simulations, in frame of a preliminary research, which allowed to be chosen some basic components of mechatronic system of energy recovery.

The complexity of issues required by a hybrid propulsion system with energy recovery, have imposed, on the one hand, the choice of mechatronic technology like modality to conceive and to design and, on the other hand, has led to designing and manufacturing of a stand for testing of kinetic energy recovery system, stand which is presented in the second part of the chapter. Also, are presented some graphical results obtained by *real-time simulation*, this new research technology used and by others researchers, which involves the simultaneous use of a mathematical model and a physical part of the studied system. The obtained graphical results confirm, generally, the preliminary theoretical results.

The chapter presents and demonstrates the possibility to design, manufacturing and implementing the energy recovery systems on medium and heavy road motor vehicles, in order to increasing the energy efficiency. The solution allows the extrapolation to different sizes of vehicles and can be mounted on new motor vehicles, as well as on old cars, in the framework of a rehabilitation.The hydraulic and electric necessary components are available on the market

Also, the chapter demonstrates that the only technology which can control and monitories the energy recovery systems, especially the hibrid propulsion systems, is the mechatronics technology.

5. Acknowledgement

This chapter presents some results obtained on a research project conducted under the Romanian *Excellence Research Program* and funded by *National Authority for Scientific Research-ANCS from Romania*.

6. References

Adelt, Ph., Esau, Nat., Hölscher Ch., Kleinjohann B. Kleinjohann L., Krüger M. & Zimmer D. (2011). Hybrid Planning for Self-Optimization in Railbound Mechatronic Systems, In: *Intelligent Mechatronics*, Ganesh Naik (Ed.), pp. 169-194, ISBN: 978-953-307-300-2, InTech, Available from:
http://www.intechopen.com/articles/show/title/hybrid-planning-for-self-optimization-in-railbound-mechatronic-systems

Ardeleanu, M., Gheorghe, Gh. & Matei, Gh. (2007). *Mecatronics. Principles and Applications*, Publishing House AGIR, ISBN: 973-720-142-3, Bucharest, Romania

Bosch Rexroth Group, *Hydraulic Hybrids from Rexroth: Hydrostatic Regenerative Braking System HRB*, Available from:
http://www.boschrexroth.com.cn/country_units/america/united_states/en/Company/Press/trade_show_information/a_downloads/Hydrostatic_Regenerative_Braking_Brochure.pdf and www.boschrexroth.com

Calinoiu, C., Vasiliu, N. & Vasiliu, D. (1998). *Modeling, Simulation and experimental Identification of the Hydraulic Servomechanisms* Technical Publishing House, ISBN 973-31-1315-8, Bucharest, Romania

Cristescu, C. (2008), *The recovery of kinetic energy at braking of motor vehicle*, Publishing House AGIR, ISBN 978-973-720-219-2, Bucharest, Romania

Cristescu, C., Drumea, P., Ilie, I., Blejan, M. & Dutu, I. (2008). Mechatronics system for recovering braking energy conceived for the medium and heavy motor vehicles, *Abstract Proceedings of The 31th International Spring Seminar on Electronics. Technology-Reability and Life Time Prediction*, pp. 192-193, ISBN 978-963-06-4915-8, Budapest, Hungary, 7-11 May, 2008. http://www.isse-eu.net. , www.ett.bme.hu/isse/isse2008/

Drumea, P.; Popescu, T.C.; Blejan, M. & Rotaru, D. (2010). Research Activities Regarding Secondary and Primary Adjustment in Fluid Power Systems, *7th International Fluid Power Conference Aachen Efficiency through Fluid Power, Scientific Poster Session*, Aachen, Germany, 22-24 March 2010,

Eaton (2011). *Electric Hybrid* Available from http://www.eaton.com/Eaton/ProductsServices/SystemsOverview//ElectricHybrid/index.htm

Eaton (2011). *Hydraulic Hybrid*, Available from: http://www.eaton.com/Eaton/ProductsServices//ProductsbyCategory/HybridPower/SystemsOverview/HydraulicHybrid/index.htm and www.eaton.com

Gauchia, L. & Sanz J. (2010). Dynamic Modeling and Simulation of Electrochemical Energy Systems for Electric Vehicles, In: *Urban Transport and Hybrid Vehicles*, Seref Soylu (Ed.), pp. 127-150, ISBN: 978-953-307-100-8, Sciyo, Available from: http://www.intechopen.com/articles/show/title/dynamic-modelling-and-simulation-of- electrochemical-energy-systems-for-electric-vehicles

Ion Guta Dr., Vasiliu, N., Vasiliu, D. , & Calinoiu C. (2008) *Basic concepts of real-time simulation (RTS)*, U.P.B. Scientific Bulletin, series D: Mechanical Engineering, Vol. 70/2008, No.4, pp. 291-301, ISSN 1457-2358

LabVIEW (1993). *Basics Course Manual. National Instruments Corp.*, Austin, SUA

LMS IMAGINE SA (2009). *Advanced Modelling And Simulation Environment, Release* 8.2.b., *User Manual*, Roanne, France

Maties, V. (1998). *Mechatronics.* Publishing House Dacia, Cluj-Napoca, Romania

Math Works Inc. (2007). *Simulink R5*, Natick, MA, U.S.A

Permo-Drive (2009). *What is the Regenerative Drive System (RDS^{TM})?* Available from: www.permo-drive.com and http://www.permodrive.com/tech/index.htm

Parker Hannifin (2010). *Fluid power: Saving Energy with Efficient Fluid Power /Hydraulic Hybrid Drive System*, Available from: www.parker.com and http://www.mobilehydraulictips.com/tag/parker-hannifin/page/2/

Popescu, T. C., Vasiliu, D. &Vasiliu N. (2011). *Numerical Simulation - a Design Tool for Electro Hydraulic Servo Systems, Numerical Simulations - Applications, Examples and Theory*, Lutz Angermann (Ed.), ISBN: 978-953-307-440-5, InTech, Available from: http://www.intechopen.com/articles/show/title/numerical-simulation-a-design-tool-for-electro-hydraulic-servo-systems/

Toyota Prius Hybrid (2008) Available from:

> http://www.autostalk.com/general-motors/toyota-prius-hybrid-443231-30.html

Untaru, M., Cîmpian, V., Hilohi, C. & Vărzescu, V. (1974). *Construction and calculation of automotive (Construcţia şi calculul automobilelor) Technical Publishing House*, Bucharest, Romania

5

Integrated Mechatronic Design for Servo Mechanical Systems

Chin-Yin Chen[1], I-Ming Chen[2] and Chi-Cheng Cheng[3]
*[1]Taiwan Ocean Research Institute, National Applied
Research Laboratories, Kaohsiung, Taiwan R.O.C.
[2]School of Mechanical & Aerospace Engineering, Nanyang
Technological University, Singapore
[3]Department of Mechanical and Electro-Mechanical Engineering,
National Sun Yat-Sen University, Kaohsiung, Taiwan R.O.C.
[1,3]Taiwan
[2]Singapore*

1. Introduction

Mechatronic systems typically exhibited high a degree of complexity due to the strong cross coupling of the involved different engineering disciplines such as mechanical, electronic and computer. This complexity originates from the large number of couplings on various levels of the contributing elements and components, coming from different disciplines. The difficulty for the design engineer in his daily work is that these couplings have to be considered in an early phase of the design process. With shortening product lift cycle, design managers are consistently trying to identify means for producing a better product in a shorter period of time.

Therefore, the realm of Mechatronics is high speed, high precision, high efficiency, highly robust. The difficulty in the Mechatronic approach is that it requires a system perspective: system interactions are important, system modeling is required, and feedback control systems can go unstable. Mechatronic design concepts include direct-drive mechanisms, simple mechanics, system complexity, accuracy and speed from controls, efficiency and reliability from electronics, and functionality from microcomputers. Starting at design and continuing through manufacture, Mechatronic designs optimize the available mix of technologies to produce quality precision products and systems in a timely manner with features the customer wants. The real benefits to industry of a Mechatronic approach to design are shorter development cycles, lower costs, and increased quality, reliability, and performance [25].

Additionally, in order to evaluate concepts generated during the design process, without building and testing each one, the mechatronics engineer must be skilled in the modeling, analysis, and control of dynamic systems and understand the key issues in hardware implementation. Thus, as the Fig. 1 shows, the essential characteristic of a mechatronics engineer and the key to success in mechatronics system is a balance between two sets of skills [22]:

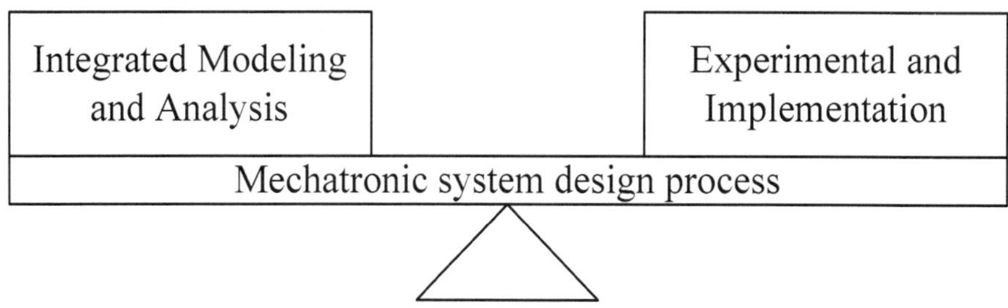

Fig. 1. Balance of mechatronic design process [22].

1.1 Integrated modeling and analysis of dynamic mechatronic systems

During the design of mechatronic systems, it is important that changes in the mechanical structure and the controller be evaluated simultaneously [24]. Although a proper controller enables building a cheaper mechatronic system, a badly designed mechanical system will never be able to give a good performance by adding a sophisticated controller. Therefore, it is important that during an early stage of the design a proper choice can be made with respect to the mechanical properties needed to achieve a good performance of the controlled system. On the other hand, knowledge about the abilities of the controller to compensate for mechanic imperfections may enable that a cheaper mechanical structure be built. This requires that in an early stage of the design a simple integrated model is available, that reveals the performance limiting factors of the mechatronic system.

Consequently, in order to help mechanical structure and controller of mechatronic system modeling simultaneously, the mechatronic system design methods must be integration. Accordingly, some of numerical based integrated design strategies for mechatronic system were proposed to some fields such as: aerospace [1-3], robotics [4-6] and manufacturing systems [7-8] in the early years. However, the dynamic models derived with the above integrated methods typically have a high order. A critical issue in the mechanical structure and control modeling with the integrated design approach is difficulty from each domain.

Therefore, for complex multibody systems of mechatronics, graphical modeling software is helpful to formulate automatically the equations of motion from a high-level description. Among the computer modeling methods, symbolic methods allow to build the equations of motion in symbolic format, whereas numerical methods produce the equations of motion as complex numerical procedures. The symbolic format has the advantages of portability and efficiency, and it provides interesting insights in the analytical structure of the equations. However, numerical methods are able to deal with a more general class of problems, and they are especially suitable to model the dynamics of a flexible mechanism with complex topology in a systematic way. After this clarification, let us further characterize the modeling requirements in the design procedure, which are directly associated with the objectives of this research

1.2 Experimental validation and hardware implementation of designs

In an industrial process, design of controllers involve formulation of reasonably accurate models of the plant to be controlled, designing control laws based on the derived models and simulating the designed control laws using available simulation tools such as MATLAB/Simulink. Whereas implementation is accomplished by converting the designed control laws to the native code of target systems, most commonly embedded microprocessor based architecture or personal computer with analog and digital interfaces. Controllers can be designed in the continuous, discrete or hybrid time domain whereas implementation is accomplished mostly in discrete time domain as most of the present day controllers are being implemented in digital machines. Presence of the vast difference in design and implementation of control applications is inherent due to different concepts in the field of control engineering and computer science. Thus, transformation of controller designs to implementation induces possibilities of errors and unreliable behaviors. In some cases, these errors cannot be identified by rigorous tests of the implementation thus these errors results in failure of the system causing serious and even catastrophic disaster.

Furthermore, the typical controller design task requires selection of controller strategies, structures and parameter values. Before implemented engineers should be tested using actual plant data or in prototype implementation with physically measured inputs and generated outputs. This phase is necessary for experimental validation of model simplifications and other assumptions made when designing the controller. On the other hand, real-time simulation provides the best conditions for performance tuning. However, sometimes the reverse situation occurs when plant model is substituted for the actual plant while the controller might be fully implemented. This approach is called Rapid Controller Prototyping (RCP) simulation [9–11]. For this technique, engineers have actuators, sensors and other physical components interfacing with real-time simulation. Furthermore, RCP techniques allow implementing and validating control strategies during the development process that users can work within the same environment from the requirement analysis to the controller design and implementation phase.

According to those two sets of skills, in the mechatronic design process, it can be broadly categorized into three stages in a computer-supported design environment, namely, the design problem of understanding behavior of mechatronic system through an analysis of need, initial solution generation through conceptual design, and solution refinement and finalization through multi-discipline detailed design. In computer support for engineering design, there is little support for the first two stages in the design process, primarily due to the complexity and diverse needs of these design activities during the three stages. The final stage in the design process is currently the main area that has reasonable computer support, and can be used to assist engineer designers to improve their designs or products. This stage of computer support can be further decomposed into component modeling, component matching and sizing, and behavior simulation and comparison for informative decision-making. This decomposition facilitates further investigation of the constituents of each design support activity.

Notably, one typical problem with many current computer-modeling methods is that they are extremely domain dependent. In the mechatronic system design processes, which include structural design, controller design and implementation in three domains, also consider interactions among multiple domains, such as integrated design, rapid prototyping and animation technology (Fig. 2). Therefore, mechatronic design engineers must to be trained to use in different application domains such that they would be competent in using all these domain-dependent technologies. This task itself is very challenging. Consequently, to solve dependent problems for mechatronic systems, mechatronic engineers always use a dynamic equation that includes all parameters in the structural and control domains. Unfortunately, one of the most significant problems when using equation-based mechatronic modeling is the amount of modeling data that must be analyzed during the process. Because of this enormous amount of data and the numerical algorithm that must also be utilized, this method of modeling and simulation is typically very slow and prone to errors. Additionally, this method requires excellent knowledge of numerical solution methods and programming principles.

Based on those reasons, in order to easy integrated design and simulation of mechatronic system for different concept domains, in this study the graphical environment called Computer Aided Rapid System Integration (CARSI) technology will be developed to achieve the structure design, controller design and implementation in the same design environment.

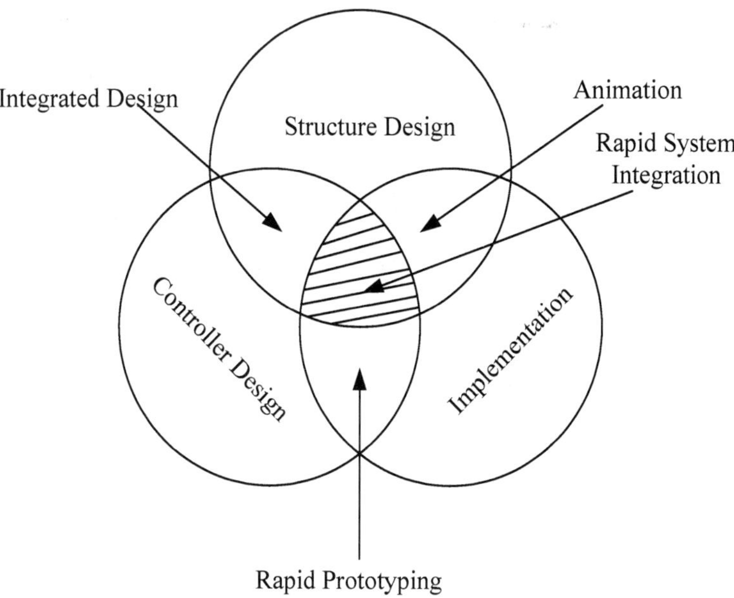

Fig. 2. Skills for mechatronic system design.

In this chapter, next section will describe the integrated design strategy using the sequential, iterative, and simultaneous methods. In Section 3, the integrated design method DFC will employ to develop a legged mechatronic system. Followed by section 4 and section 5 will present CARSI technology to put together the design, simulation and implementation at same environment. The end is concludes the work in this chapter.

2. Integrated design strategy

With a multilevel decomposition approach [12], a large complex optimization problem is broken into a hierarchy of smaller optimization sub problems. This hierarchy can be thought as levels of increasing details. At the upper level, the sub problem is formulated in terms of global quantities, which describe the overall behavior of the entire system. On the lower level, the sub problems are stated in terms of local quantities and local constraints, which have only a small impact on the entire system. Each sub problem uses local design variables to reduce the violation of constraints, which are unique to that sub problem. Each level is a multi-objective optimization problem characterized by a vector of objective functions, constraints and design variables. So considering the structure and control two-level problem for a mechatronic system, the multilevel decomposition procedure can be written as below. At structure level,

$$\text{min.} \quad Y_{Ni}(X_N), i = 1, ..., n_N$$
$$\text{s.t.} \quad g_{Nk}(X_N) \leq 0, k = 1, ..., nc_N$$
$$\sum_{i=1}^{NDV_N} \frac{\partial Y_{Rj}}{\partial X_{Ni}} \Delta X_{Ni} \leq \varepsilon_{2j}, j = 1, ..., n_R \tag{1}$$
$$X_{Ni}^L \leq X_{Ni} \leq X_{Ni}^U, i = 1, ..., NDV_N$$
$$X_{Rj}^L \leq X_{Rj}^* + \sum_{i=1}^{NDV_N} \frac{\partial X_{Rj}^*}{\partial X_{Ni}} \Delta X_{Ni} \leq X_{Rj}^U, j = 1, ..., NDV_R$$

where Y_N and Y_R are the objective function vectors at the structure level and the control level, respectively; g_N and g_R are the corresponding constraint vectors; X_N and X_R are the corresponding design variable vectors, ε_{2j} is a tolerance on the change in the jth objective of control level during optimization at the structure level; L and U are lower and upper bounds of design vectors, $\partial Y_{Rj}^*/\partial X_{Ni}$ and $\partial X_{Rj}^*/\partial X_{Ni}$ represent the optimal sensitivity parameters of the control level objective function and design variable vectors, respectively, with respect to the structure level design variables. n_N and n_R denote numbers of objective functions for each level; nc_N is the number of constraints for structure level; NDV_N and NDV_R are numbers of design variables for the structure and the control levels.

Similarly, the process of control level becomes

$$\text{min} \quad Y_{Rj}(X_N^*, X_R), j = 1, ..., n_R$$
$$\text{s.t.} \quad g_{Rk}(X_N^*, X_R) \leq 0, k = 1, ..., nc_R \tag{2}$$
$$X_{Ri}^L \leq X_{Ri} \leq X_{Ri}^U, i = 1, ..., NDV_R$$

Where X_N^* is the optimum design variable vector from the structure level and must be fixed during optimization at the control level.

Following (1) and (2); the integrated design methodology can be broken into sequential, iterative, and simultaneous three strategies:

In the sequential strategy, the mechanical structure is usually designed first (Eq. 1). It is then fitted with off-the-shelf electric motors and drive electronics. Finally, a controller is designed and tuned for the existing physical system until the goal is archived (Eq. 2); therefore, it is called Design Then Control (DTC) strategy. In this method, the structure is assumed to be fixed and cannot be changed by excluding considerations from a dynamics and control point of view. Consequently, this approach leads to a system with non-optimal dynamic performance.

Based on this reason, in order to improvement system performance, the iterative strategy is discussed. For this method, the structural design is also first performed based on loading considerations (Eq.1). Sizes and masses of mission-related components are estimated and a structure that maintains the desired component relationships during operations is designed. Next, a controller is designed for the fixed structure to obtain the required dynamic performance (Eq.2). The control design must also provide satisfactory closed-loop stability and robustness properties. If the nominal system does not provide an adequate performance, the design process must return to the structural discipline for modification (Eq. 1). After modification, the structure parameters are returned to the control discipline for redesign (Eq.2). This iterative process continues until a satisfactory compromise is found between the mission and control requirements. Now suppose that it is desired to simplify the (1) and (2) formulation as much as possible. One could presumably simplify the problem by assuming that all the objective functions and constraints are convex within both the structure and controller design subspaces. In other words, one could presumably assume that, when the structure design variables X_N are fixed, all the objective functions and constraints in the above problem will be convex and vice versa. However this assumption is not a sufficient guarantee for the system level optimization problem to be convex [7]. Thus, in order to achieve the optimization problem into the system level, the simultaneous design strategy must be considered.

As (1) and (2), given a combined structure and controller optimization problem for mechatronic system, the system level is often nonconvex, even if the individual structure and control optimization sub-problems are convex (individual design problem for (1) and (2)). The main reason is easy involved the static and variation optimization problem during iterative design process. Thus, some of researchers were used closed-loop eigenvalues [2][3], Design For Control (DFC) [5][6][23], and convex integrated design [8] to improve structure and control problem simultaneous.

Therefore, as Fig.3 shows, comparing above three strategies, even system performance will be increased during sequential, iterative, and simultaneous strategies, but

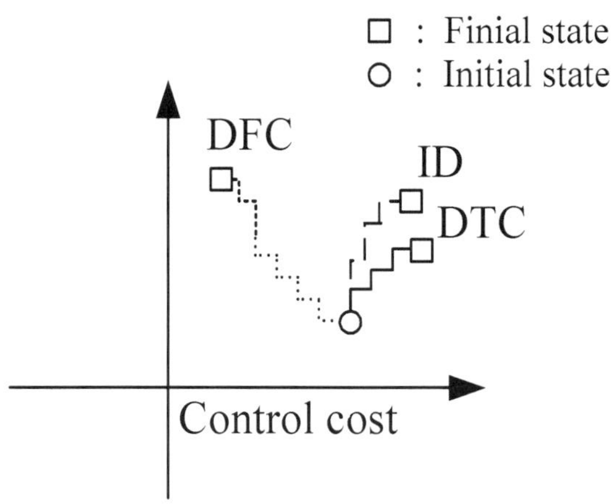

Fig. 3. Control cost in iterative process.

3. Legged mechatronic system design

Most mobile robots are equipped with wheels. A wheel is easy to control and direct, provides a stable base on which a robot can maneuver and is easy to construct. However, one major drawbacks of a wheel is the limitation it imposes on the terrain the can be successfully navigated. Therefore, research into legged locomotion is important as legs can overpass rough terrain. Thus, create a leg mechanism that walks has becomes a central goal in the field of robotics [13-15]. Based on this reason, in this study, the CARSI will be used in rapid legged mechatronic system design process, and the flow chart shows in Fig. 4.

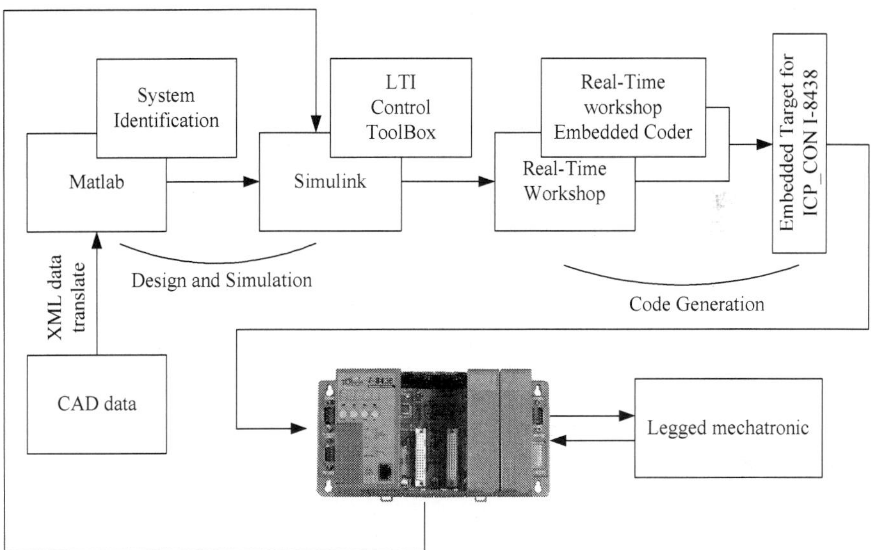

Fig. 4. Legged mechatronic system design flow chart.

3.1 Legged structure

Basic considerations for a leg design for a walking machine are as follows: the leg should generate an approximately straight-line trajectory for the foot with respect to the body; the leg should have a simple mechanical design; and, when specifically required, it should have the minimum number of DOFs to ensure motion capability. Therefore, the basic principle in this study is to create a walking machine via the linkage method with symmetrical coupler curves to combine the functions of a four-bar linkage and a pantograph into one leg structure [16][17].

Based on the embedded-type leg mechanism (Fig. 5), an embedded trajectory P is first designed via a four-bar linkage, and then magnified by a scale ratio n (B0E=nB0D) to obtain the gait profile G. Therefore, according to design specifications (Table 1), the parameters of the embedded four-bar linkage are obtained. Moreover, all design processes are based on the following assumptions:

1. No transmission loss exists between the input and end effect of this mechanism.
2. Ground reaction force on the end effect is constant.

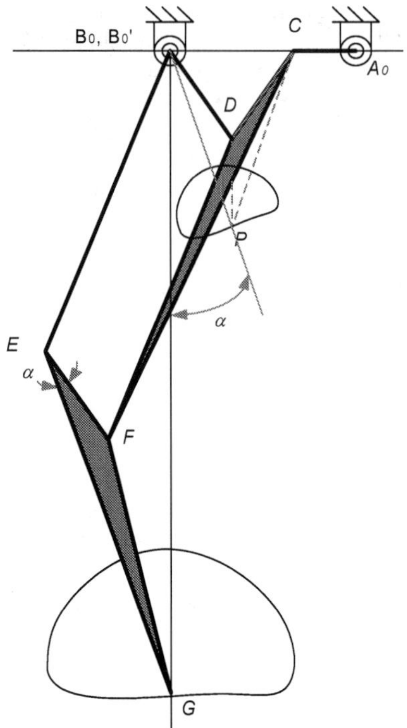

Fig. 5. Legged structure.

3.2 Optimal multivariable design for gait profile

As discussed, gait profile can be designed using an embedded four-bar linkage, and magnified using a pantograph to satisfy the target. Additionally, to decrease leg size (or minimize scale ratio n) and obtain an enhanced footpath height, the design objective function can be formulated as (3).

$$I_1 = \min \quad (\beta(l_s)^{-1} + \gamma(l_h)^{-1}) \tag{3}$$

s.t.

$$\phi \geq 2\cos^{-1}\sqrt{\cos\beta_1 \sin\beta_2}$$

$$12cm \leq \overline{A_0 B_0} \leq 14cm$$
$$2cm \leq \overline{A_0 C} \leq (2\overline{CD} - \overline{A_0 B_0})cm$$
$$\pi < \mu + \phi' < 2\pi$$
$$45° \leq \mu \leq 135°$$

where:

β, γ : weighting factor, β =1, γ =0.2

l_s : stride length of the embedded four-bar linkage

l_h : foot-path height of the embedded four-bar linkage

α : skew angle

$$\angle CDF = \phi + \alpha = \phi'$$

Table 2 lists optimal results based on (3) and those constrains. Additionally, Fig. 6 shows the six-bar walking machine gait profile and the embedded four-bar linkage profile.

Parameters	DTC	DFC	Var. %
Structure Parameters			
$\overline{A_0B_0}$ (cm)	12.8	12.8	-
$\overline{A_0C}$ (cm)	2.6	2.6	-
$\overline{B_0D} = \overline{EF}$ (cm)	8	8	-
$\overline{B_0E} = \overline{DF}$ (cm)	30.8	30.8	-
Mass of $\overline{A_0C}$ (kg)	0.050	0.080	60
Mass of $\overline{B_0D}$ (kg)	0.035	0.07	100
Mass of $\overline{B_0E}$ (kg)	0.134	0.134	-
Mass of ΔCDF (kg)	0.368	0.215	-41.5
Mass of ΔEFG (kg)	0.316	0.177	-44.0
r2(cm) / δ_2 (deg)	1.3 / 0	0 / 0	- / -
r3(cm) / δ_3 (deg)	16.0 / 35.6	14.6 / 38.0	-8.6 / 6.3
r4(cm) / δ_4 (deg)	4 / 0	0 / 0	- / -
r5(cm) / δ_5 (deg)	12.5 / 42.0	11.6 / 36.6	-7.5 / -13
r6(cm) / δ_6 (deg)	15.4 / 0	15.4 /0	- / -
α (deg)	51.2	51.2	-
n	3.8	3.8	-
ϕ' (deg)	128.8	128.8	-
Controller Parameters			
K_p	4.3	3.5	-18.6
K_i	4000	4800	20
K_{pp}	150	180	20
Max $|\tau|$ (N-m) without acc/dec (0.5 step/s)	0.14	0.12	-10
Max $|\tau|$ (N-m) without acc/dec (2 step/s)	0.5	0.2	-60

Table 2. Integrated Design Results.

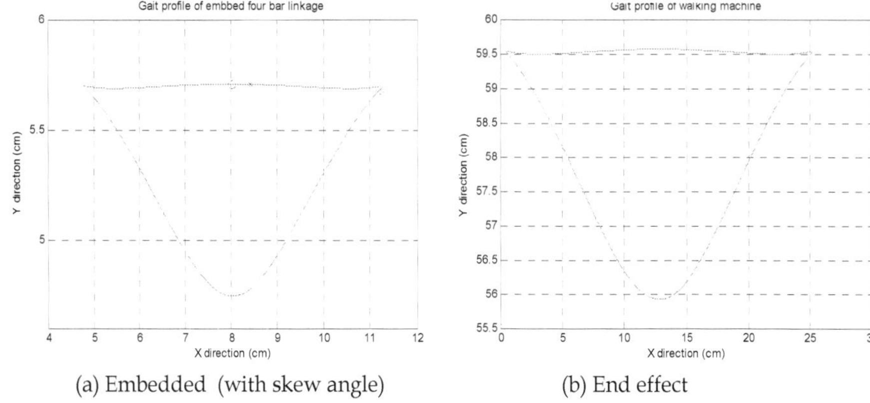

(a) Embedded (with skew angle)   (b) End effect

Fig. 6. Gait profile.

3.3 Controller design

When kinematic design of the walking machine was complete, controller design was considered. Therefore, to integrate and model the mechatronic system of a walking machine in the design process, Lagrange's equation, which formulated as (4), is applied to derive the all parameters in this controller design process.

$$\frac{d}{dt}\frac{\partial K}{\partial \dot{\theta}_2} - \frac{\partial K}{\partial \theta_2} + \frac{\partial P}{\partial \theta_2} = \tau \tag{4}$$

where K is kinetic energy, P is potential energy, τ is control torque, and θ_2 is angle of the input link. Fig. 7 presents the detailed parameters of a system dynamic for a walking machine. Thus, the primary parameters K and P can be expressed by (5) and (6), and control torque τ was re-formulated as (7).

$$K = \sum_{i=2}^{6} \left[\frac{1}{2} m_i (V_{ix}^2 + V_{iy}^2) + \frac{1}{2} J_i \dot{\theta}_i^2 \right] \tag{5}$$

$$P = (m_2 r_2 \sin(\theta_2 + \delta_2) + m_3 (L_2 \sin\theta_2 + r_3 \sin(\theta_3 + \delta_3))$$

$$+ m_4 r_4 \sin(\theta_4 + \delta_4) + m_5 (L_2 \sin\theta_2 + L_3' \sin(\theta_3 + \rho_3) + r_5 \sin(\theta_5 + \delta_5) + m_6 r_6 \sin(\theta_6 + \delta_6))g \tag{6}$$

where m_i is mass of each linkage; V_{ix} and V_{iy} are the velocity in the x and y direction, respectively, for each linkage; r_i is the length from the central mass to a reference point; L_i is the characteristic length for each linkage; δ_i is the angle of central mass for each linkage; g is gravity; $L_3' = \left(L_3^2 + L_6^2 - 2L_3 L_6 \cos\phi' \right)^{0.5}$.

In the other hand, use of simple controllers, such as PD/PID controllers, for industrial manipulators and servo system applications are well known which works on the basis of position loop control. In this work, in order to improve tracking performance for velocity and position simultaneously, the IP controller was employed in the velocity loop, and the P controller was used in the position loop. The equation of control power τ can be formulated in (7).

$$\tau(t) = K_i \int (e_\theta K_{pp} - \omega_m)dt - K_p \omega_m \qquad (7)$$

where e_θ, K_{pp} K_p and K_i are position tracking error, position loop proportion gain, velocity loop proportion gain, and velocity loop integration gain, respectively. Following the Integral Time Absolute Error (ITAE) criteria, the design objective for the controller was written as (8), where η and ς are weight factors, and $\eta =1$, $\varsigma =0.1$, the results are listed in Table 2.

$$I = \min(\eta \int t|e_\theta(t)|dt + \varsigma|\tau(t)|)$$
$$s.t \quad \tau(t) \le 5 \quad Nm \qquad (8)$$
$$e_\theta \ge 0$$

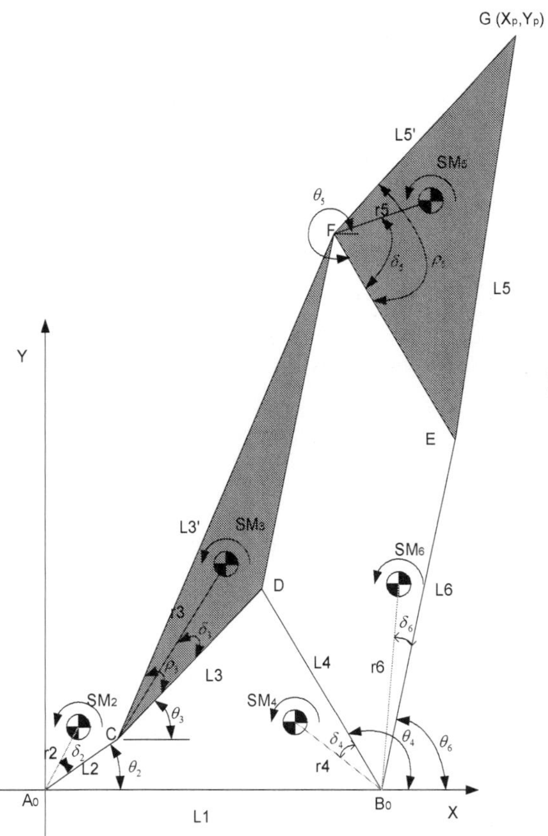

Fig. 7. Dynamic model of linkage for Leg system.

4. System optimization using the Design For Control (DFC) approach

As Fig. 8 shows the DFC iterative process [5][6][23], if system performance is unsatisfactory, design process is returned to structural domain. The structural modification process will go out of used the single domain constrains, and overall system dynamic conditions will be replaced with original conditions, and pass into control domain to acquire a new controller solution. Hence, DFC is not only used a concurrent (parallel) integrated design process to achieve system performance, but also to enhance the control requirements to easy control system in the design approach.

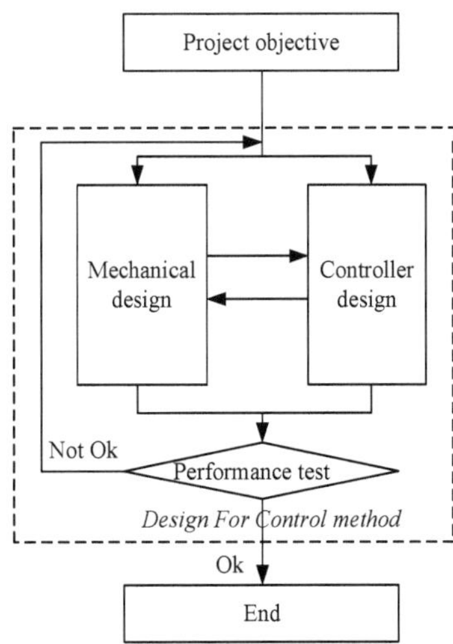

Fig. 8. Integrated design of mechatronic system using DFC.

Following use of the DFC concept and Lagrange's equation, the aim of the design process is to decrease potential and kinetic energy first during these interactive design processes. Thus, modifying system parameters of the leg linkage for the walking machine must be considered, and system performance is based on structural results (5) and (6) in tuning the controller parameters (7) at the same time. Therefore, following (5) and (6), two methods can be utilized to improve system performance, namely, variable input speed [18], which reduces kinetic energy for (5), and mass redistribution [5-6], which decreases both terms (5) and (6) simultaneously. Thus, the "complete force balancing" method based on mass redistribution was applied to enhance system performance. Hence, the primary objective in this interactive process can be formulated as (9).

Based on this objective, when (9) equals zero, the dynamic equation for legged mechatronic can be re-formulated as (10). From this equation, the dynamic behavior for this mechatronic leg can be also reduced to a simple equation, i.e., control power is only considered as kinetic energy and near constant potential energy during this interactive control design process. As mentioned earlier, the basic idea of the DFC approach is to spare control design effort and improve real-time performance by providing a simple dynamic model through judicious mechanical design. Consequently, key parameters of the internal moment, δ_i , r_i , and m_i , were improved (Table 2). Additionally, Table 2 also lists optimal control gains.

$$I = \min \frac{\partial P}{\partial \theta_2} \tag{9}$$

$$\frac{d}{dt}\frac{\partial K}{\partial \dot{\theta}_2} - \frac{\partial K}{\partial \theta_2} = \tau \tag{10}$$

4.1 Multi-domain graphical model integration

With the rapid developments in computer science over the last 20 years, computer-aided engineering software, such as Pro/Engineer, Solidworks, Ansys and Matlab, have been widely utilized in structure and control fields. Therefore, file format standards, such as the Initial Graphics Exchange Specification (IGES), STEP (ISO-10303) and DXF, were developed to address the incompatibility issue of various CAD/CAM systems. This standard allows for efficient and accurate exchange of product definition data across almost all CAD/CAM systems.

As each computer-aided engineering software package using a unique method of describing geometry both mathematically and structurally, some information is always lost when translating data from one system data format to another. Intermediate file formats are also limited in what they can describe, and can be interpreted differently by both the sending and receiving systems. When transferring data between systems, identifying what needs to be translated is important. Additionally, translating intermediate files always focuses on the same engineering domain. Therefore, in the mechatronic system, intermediate file formats or parameters must be considered in detail to be accepted by each domain. That is, modeling of different system domains in the same model is possible when the language used for describing the model is extensible and includes several standard libraries for different domains; this helps users because they can use modeling tools with which they are familiar for different tasks.

XML (eXtensible Markup Language) is a World Wide Web Consortium (W3C) recommended general-purpose markup language that supports a wide variety of applications [19]. The XML language and its 'dialects' can be designed by anyone and can be processed using appropriate software. Notably, XML is also designed to be reasonably human-legible.

According to kinematic synthesis results for the leg linkage (Table 2), the 3D graphical model for the mechatronic leg was first designed using Pro/E (Fig. 9(a)). Moreover, based on the (4), (5), (6), parameters of linkages, such as mass, length, position, center of gravity, unit, volume, and constrain (or joint type), were obtained from this CAD data. Following this step, the graphical model, based on XML syntax, in reference to control requirement parameters was obtained. According to this model and parameters (Table 2), the embedded controller was also created using this graphical model (Fig. 9(b)).Consequently, to simplify modeling and simulation for the mechatronic system in this study, a graphical environment called CARSI technology was employed for structural design, controller design and implementation in the same design environment.

Based on CARSI method, Fig. 10(a) and 10(b) present results obtained by the DTC and DFC methods, respectively. Comparing the simulation results obtained with DTC and DFC, the performance of the mechanism after applying the mass-redistribution scheme was significantly improved; the maximum control torque at low and high speeds was reduced by 10%, from ± 0.14N-m to ± 0.12 N-m and 10%, from ± 0.5N-m to ± 0.2 N-m, respectively.

According to these analytical results, once the mass and mass center are fixed during machine walking, the potential energy term will have almost no influence even when speed is changed. Conversely, based on the DTC result, when walking speed changed, control power increases.

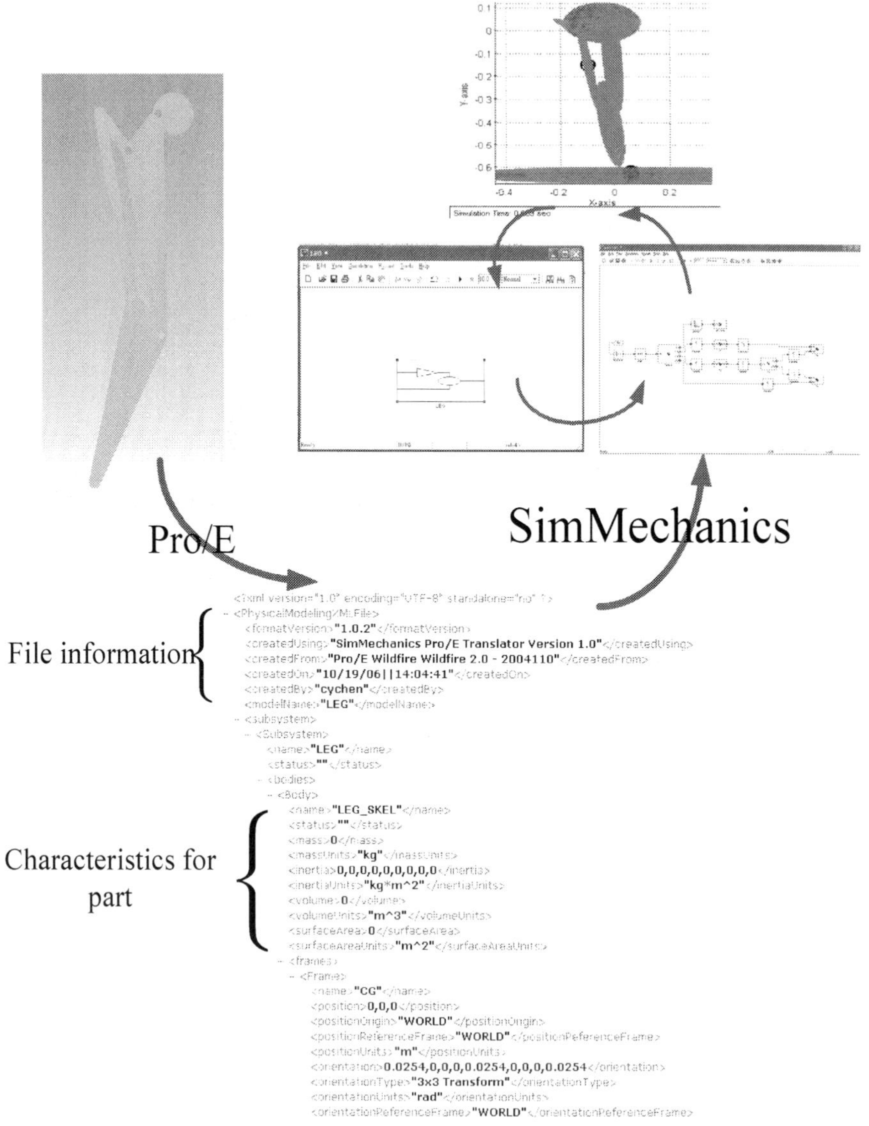

Fig. 9. (a) Multi-domain transforms

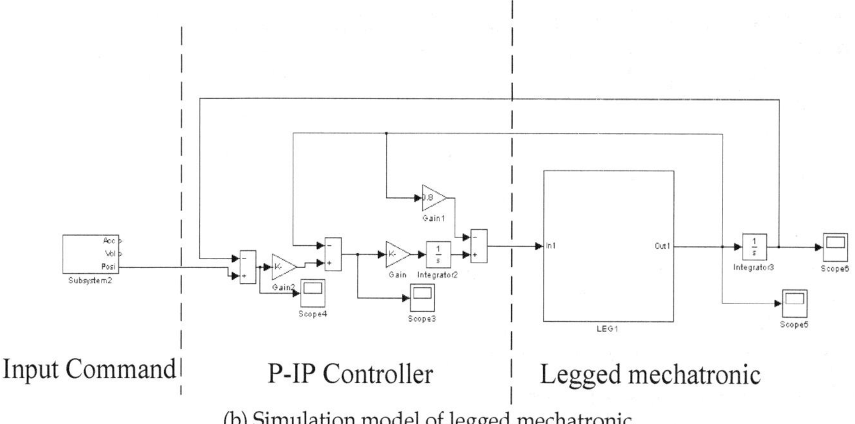

(b) Simulation model of legged mechatronic.

Fig. 9. Multi-domain graphical model.

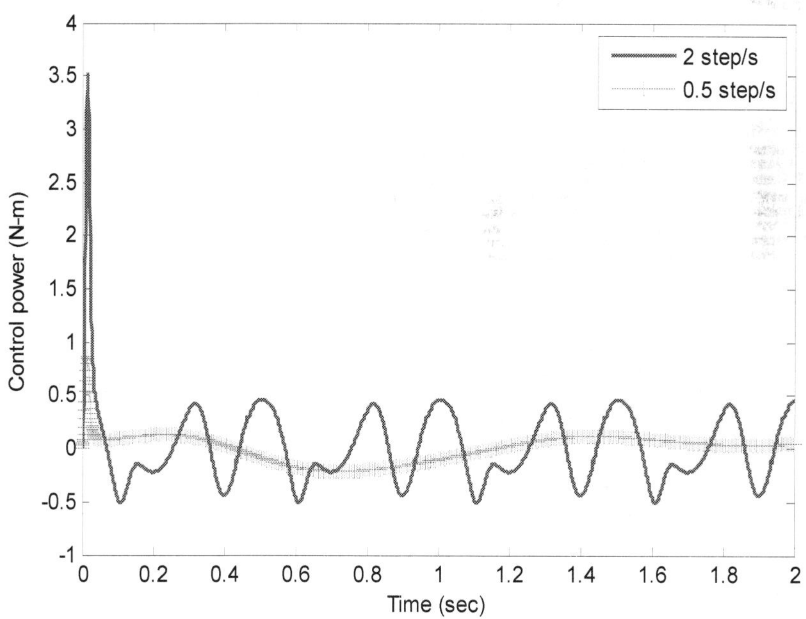

Fig. 10. (a)

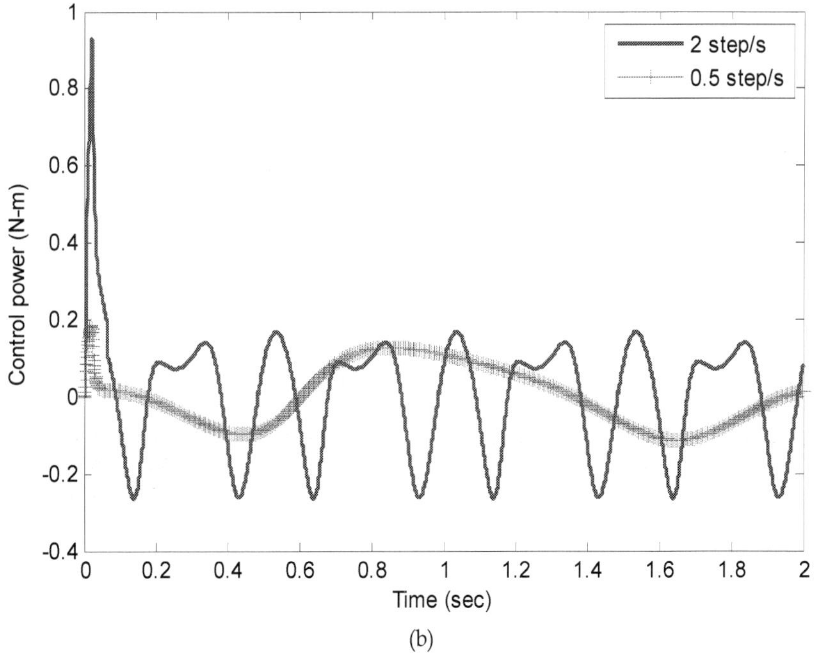

(b)

Fig. 10. Control power for DTC and DFC methods.

5. Rapid control prototyping

As control systems become increasingly complex with the development of control algorithms and controller designing techniques, manually interpreting and designing the control system using differential equations or numerical formulas is time-consuming and difficult. Additionally, various user-friendly graphs and interfaces are necessary as well as complex computations, and, moreover, because repetitive operations on the same work is mandatory when designing a control system, conventional handwork programming is not an easy job and is inefficient when faced with increased pressure for reducing product time-to-market.

Rather than conventional low-level programming languages, graphical model-based programming has been used increasingly for real-time simulation and hardware-in-the-loop (HIL) applications to obtain rapid prototyping of various electrical and mechanical systems. Compared with conventional low-level handwork programming, the most important feature of state-of-the-art control applications is the function that generates program codes automatically through some user-friendly graphic modules to decrease the time required for system development.

As mentioned, "Matlab/Simulink" software is a design and simulation tool used most in the control field. This software allows users to create models for dynamic systems simply by connecting blocks from given libraries, and also includes a library called SimMechanics, which simulates rigid body dynamics using a 3D graphical model. Some blocks of

Matlab/Simulink implement linear systems given as transfer functions or state-space realizations both in continuous and discrete time. When a simulate is complete, the Real-Time Workshop (RTW) toolbox generates C-code from the model without a need for programming knowledge. Therefore, rapid controller prototyping techniques facilitate implementation and validation of control strategies during the development process: users can work within the same environment from structure requirement analysis to the controller design and implementation phases. Based on this software, three implementation types are supported by the RTW toolbox, namely, Real-Time Windows Target, xPC target, and Real-Time Embedded Target. For the first two techniques, the target real-time devices are based on PC. Therefore, real-time performance or space must be considered in detail [20].

As previously stated, Figure 11 presents the "ICP_i8438" module, which is based on a micro-chip and provides some add-on modules such as analogy output (I-8024) and encoder feedback (I-8090) [21]. According to this model and legged mechatronic system (Fig. 12(a)), the simulation and experimental results are shows in Fig. 12(b).

As these results, the constant friction torque from each joins was assumed at 0.3 N-m. The experimental and simulation results are very close; however, one obvious problem with this result is that dynamic friction during the acceleration phase was not considered. Restated, integration of a graphical-based model and equation-based model to simulate a mechatronic system can easily obtain, predict and modify system model parameters to achieve the goal for a real system.

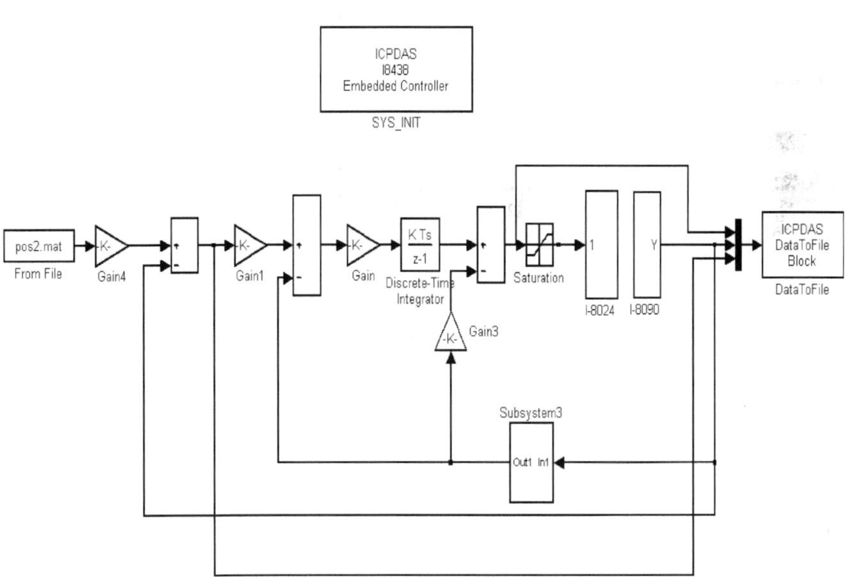

Fig. 11. Model for HIL.

(a) Legged mechatronic system

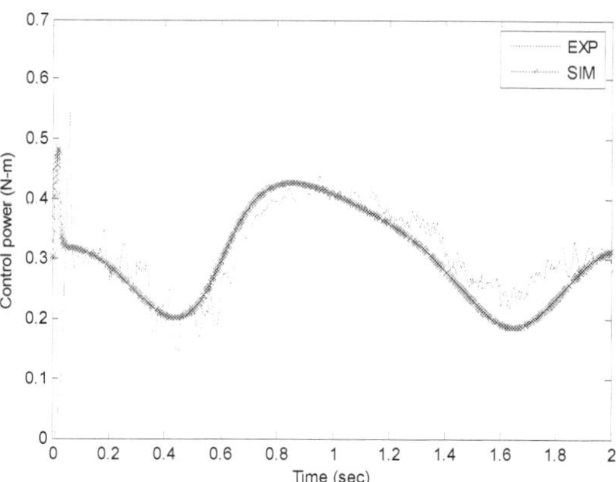

(b) Results for experimentation and simulation

Fig. 12. Legged mechatronic.

6. Conclusion

An integrated design concept DFC and rapid implementation CARSI for a walking machine are proposed in this paper. The DFC was utilized to design the mechanical structure of a mechatronic leg system by fully exploring the physical characteristics of the overall system while considering controller design and execution of control actions with the least significant hardware restriction. Restated, DFC not only helped the mechatronic system satisfy low driving power, its also helped easy to control the system. Additionally, the CARSI approach achieved structural design, controller design and system implementation simultaneously in the same design environment to reduce development time for the mechatronic leg system.

7. References

[1] A.L. Hale, R.J. Lisowski, and W.E. Dahl, "Optimal simultaneous structural and control design of maneuvering flexible spacecraft," *J. of Guidance Controller and Dynamics*, Vol. 8, No. 1, pp. 86-93, 1985.

[2] D.S. Bodden and J.L. Junkins, "Eigenvalue optimization algorithms for structure/controller design Iterations," *J. of Guidance Controller and Dynamics*, Vol. 8, No. 6, pp. 697-706, 1985.

[3] Messac, "Control-structure integrated design with closed-form design metrics using Physical Programming," *AIAA Journal*, Vol. 36, No. 5, pp. 855-864, 1998.

[4] J.H. Park and H. Asada, "Concurrent design optimization of mechanical structure and control for high speed robots," *ASME Journal of Dynamic System, Measurement, and Control*, Vol. 116, pp. 344-356, Sep. 1994.

[5] F. X. Wu, W. J. Zhang, Q. Li, and P. R. Ouyang. "Integrated design and PD control of High Speed Closed-loop Mechanisms," ASME Journal of Dynamics System, Measurement and Control, Vol. 124, No.4, pp. 522-527, 2002

[6] Q. Li, and F. X. Wu, "Control Performance Improvement of a Parallel Robot via the Design for Control Approach," *Mechatronics*, vol.14, no.8, pp.947-964, 2004

[7] A.C. Pil and H. Asada, "Integrated structure/control design of mechatronic system using a recursive experimental optimization method," *IEEE/ASME Transactions on Mechatronics*, Vol. 1, No. 3, pp. 191-203, 1996.

[8] K. Fu and J. K. Mills, "A Convex Approach Solving Simultaneous Mechanical Structure and Control System Design Problems With Multiple Closed-loop Performance Specifications," *ASME Journal of Dynamic Systems, Measurement, and Control*, Vol. 127, pp.57-68, 2005.

[9] Manfred Glesner, Andreas Kirschbaum, Frank-Michael Renner, and Burkart Voss, "State-of-the-art in rapid prototyping for mechatronic systems," *Mechatronics*, Vol. 12, No. 8, pp. 987-998, 2002

[10] Roberto Bucher and Silvano Balemi, "Rapid controller prototyping with Matlab/Simulink and Linux," *Control Eng Prac*, Vol. 14, No. 2, pp. 185-192, 2006.

[11] M. Deppe, M. Zanella, M. Robrecht and W. Hardt, "Rapid prototyping of real-time control laws for complex mechatronic systems: a case study," *Journal of Systems and Software*, Vol. 70, No.3, pp. 263-274, 2004.

[12] Chattopadhyay and N. Pagaldipti, "A multidisciplinary optimization using semi-analytical sensitivity analysis procedure and multilevel decomposition," *Computers Math. Applic.*, Vol. 29, No. 7, pp. 55-66, 1995.

[13] Manuel, Pablo Gonzalez de Santos. Climbing and Walking Robots. Springer Berlin Heidelberg, Germany, 2005.

[14] M. O. Tokhi , M. A. Hossain and G. S.Virk "Climbing and Walking Robots," Springer Berlin Heidelberg, Germany, 2006.

[15] Y. J. Chen, "Mechanism Design of An 8-link Type Biped Walking Machine with Auxiliary Wheels," Master thesis, National Cheng Kung Univ., 2004

[16] W. B. Shieh, L. W. Tsai and S. Azarm, "Design and Optimization of a One-Degree-of Freedom Six-Bar Leg Mechanism for a Walking Machine," *Journal of Robotic System*, Vol. 14, No.12, pp. 871-880, 1997.

[17] E. Ottaviano, C. Lanni and M. Ceccarelli, "Numerical and Experimental Analysis of a Pantograph-Leg with a Fully-Rorative Actuating Mechanism," *Proc. of the 11th World Congress in Mechanism and Machine Science*, Tianjin, China, 2004

[18] H. S. Yan and R. C. Soong, "Kinematic and dynamic design of four-bar linkages by links counterweighing with variable input speed," *Mech Mach Theory*, Vol. 36, No.9, pp. 1051-1071, 2001.

[19] J. Yang, S. Han, J. Cho, B. Kim, H. Y. Lee, "An XML-based macro data representation for a parametric CAD model exchange," *Computer-Aided Design and Applications*, Vol. 1, No. 1-4, pp. 153-162, 2004.

[20] C. Sun, "The Integration of MATLAB and Embedded Controller for Control Application," Master thesis, National Sun Yat-Sen Univ., 2004

[21] ICP DAS Co. User manual of I8438/8838 Matlab embedded control.

[22] K. Craig, F. Stolfi, "Teaching control system design through mechatronics: academic and industrial perspectives." Mechatronics, Vol 12, No. 2, pp. 371-381, 2002.

[23] Q. Li, W.J. Zhang and L. Chen, "Design for control-a concurrent engineering approach for mechatronic systems design," *Mechatronics, IEEE/ASME Transactions on Mechatronics*, Vol. 6 , No. 2, pp. 161-169, 2001.

[24] J, van Amerongen, "Mechatronic design", *Mechatronics*, Vol. 13, No. 10, pp. 1045-1066, 2003.

[25] K. Craig, "The role of computers in mechatronics", *Computing in Science & Engineering*, Vol. 5, No. 2, pp. 80-85, 2003.

Part 2

Robotics and Vision

6

On the Design of Underactuated Finger Mechanisms for Robotic Hands

Pierluigi Rea
DiMSAT, University of Cassino
Italy

1. Introduction

The mechatronic design of robotic hands is a very complex task, which involves different aspects of mechanics, actuation, and control. In most of cases inspiration is taken by the human hand, which is able to grasp and manipulate objects with different sizes and shapes, but its functionality and versatility are very difficult to mimic. Human hand strength and dexterity involve a complex geometry of cantilevered joints, ligaments, and musculotendinous elements that must be analyzed as a coordinated entity. Furthermore, actuation redundancy of muscles generates forces across joints and tissues, perception ability and intricate mechanics complicate its dynamic and functional analyses.

By considering these factors it is evident that the design of highly adaptable, sensor-based robotic hands is still a quite challenge objective giving in a number of cases devices that are still confined to the research laboratory.

There have been a number of robotic hand implementations that can be found in literature. A selection of leading hand designs reported here is limited in scope, addressing mechanical architecture, not control or sensing schemes. Moreover, because this work is concentrated to finger synthesis and design, the thumb description is excluded, as well as two-fingered constructions, because most of them were designed to work as grippers and would not integrate in the frame of multi-finger configuration.

Significant tendon operated hands are the Stanford/JPL hand and the Utah/MIT hand. In particular, the first one has three 3-DOF fingers, each of them has a double-jointed head knuckle providing 90° of pitch and jaw and another distal knuckle with a range of ±135°. The Utah/MIT dextrous hand has three fingers with 4-DOFs, each digit of this hand has a non anthropomorphic design of the head knuckle excluding circumduction. The inclusion of three fingers minimizes reliance on friction and adds redundant support to manipulations tasks. Each N-DOF finger is controlled by $2\text{-}N$ independent actuators and tension cables. Although these two prototypes exhibit a good overall behaviour, they suffer of limited power transmission capability.

The prototype of the DLR hand possesses special designed actuators and sensors integrated in the hand's palm and fingers. This prototype has four fingers with 3-DOFs each, a 2-DOFs base joint gives ± 45° of flexion and ±30° of abduction/adduction, and 1-DOF knuckle with 135° of flexion. The distal joint, which is passively driven, is capable of flexing 110°.

A prototype of an anthropomorphic mechanical hand with pneumatic actuation has been developed at Polytechnic of Turin having 4 fingers with 1-DOF each and it is controlled through PWM modulated digital valves.

Following this latter basic idea, several articulated finger mechanisms with only 1-DOF were designed and built at the University of Cassino and some prototypes allowing to carry out suitable grasping tests of different objects were developed.

More recently, the concept of the underactuation was introduced and used for the design of articulated finger mechanisms at the Laval University of Québec.

Underactuation concept deals with the possibility of a mechanical system to be designed having less control inputs than DOFs. Thus, underactuated robotic hands can be considered as a good compromise between manipulation flexibility and reduced complexity for the control and they can be attractive for a large number of application, both industrial and non conventional ones.

2. The underactuation concept

Since the last decades an increasing interest has been focused on the design and control of underactuated mechanical systems, which can be defined as systems whose number of control inputs (i.e. active joints) is smaller than their DOFs. This class of mechanical systems can be found in real life; examples of such systems include, but not limited to, surface vessels, spacecraft, underwater vehicles, helicopters, road vehicles, and robots.

The underactuation property may arise from one of the following reasons:

- the dynamics of the system (e.g. aircrafts, spacecrafts, helicopters, underwater vehicles);
- needs for cost reduction or practical purposes (e.g. satellites);
- actuator failure (e.g. in surface vessel or aircraft).

Furthermore, underactuation can be also imposed artificially to get a complex low-order nonlinear systems for gaining an insight in the control theory and developing new strategies. However, the benefits of underactuation can be extended beyond a simple reduction of mechanical complexity, in particular for devices in which the distribution of wrenches is of fundamental importance. An example is the automobile differential, in which an underactuated mechanism is commonly used to distribute the engine power to two wheels. The differential incorporates an additional DOF to balance the torque delivered to each wheel. The differential fundamentally operates on wheel torques instead of rotations; aided by passive mechanisms, the wheels can rotate along complex relative trajectories, maintaining traction on the ground without closed loop active control.

Some examples found in Robotics can be considered as underactuated systems such as: legged robots, underwater and flying robots, and grasping and manipulation robots.

In particular, underactuated robotic hands are the intermediate solution between robotic hands for manipulation, which have the advantages of being versatile, guarantee a stable grasp, but they are expensive, complex to control and with many actuators; and robotic grippers, whose advantages are simplified control, few actuators, but they have the drawbacks of being task specific, and perform an unstable grasp.

In an underactuated mechanism actuators are replaced by passive elastic elements (e.g. springs) or limit switches. These elements are small, lightweight and allow a reduction in the number of actuators. They may be considered as passive elements that increase the adaptability of the mechanism to shape of the grasped object, but can not and should not be handled by the control system.

The correct choice of arrangement and the functional characteristics of the elastic or mechanical limit (mechanical stop) ensures the proper execution of the grasping sequence. In a generic sequence for the grasping action, with an object with regular shape and in a fixed position, one can clearly distinguish the different phases, as shown in Fig. 1.

In Fig.1a the finger is in its initial configuration and no external forces are acting. In Fig.1b the proximal phalanx is in contact with the object. In the Fig.1c the middle phalanx, after a relative rotation respect to the proximal phalanx, starts the contact with the object. In this configuration, the first two phalanges can not move, because of the object itself. In Fig.1d, finally, the finger has completed the adaptation to the object, and all the three phalanges are in contact with it. A similar sequence can be described for an irregularly shaped object, as shown in Fig.2, in which it is worth to note the adaptation of the finger to the irregular object shape.

An underactuated mechanism allows the grasping of objects in a more natural and more similar to the movement obtained by the human hand. The geometric configuration of the finger is automatically determined by external constraints related with the shape of the object and does not require coordinated activities of several phalanges. It is important to note that the sequences shown in Figs.1 and 2 can be obtained with a continuous motion given by a single actuator.

Few underactuated finger mechanisms for robotic hands have been proposed in the literature. Some of them are based on linkages, while others are based on tendon-actuated mechanisms. Tendon systems are generally limited to rather small grasping forces and they lead to friction and elasticity. Hence, for applications in which large grasping forces are

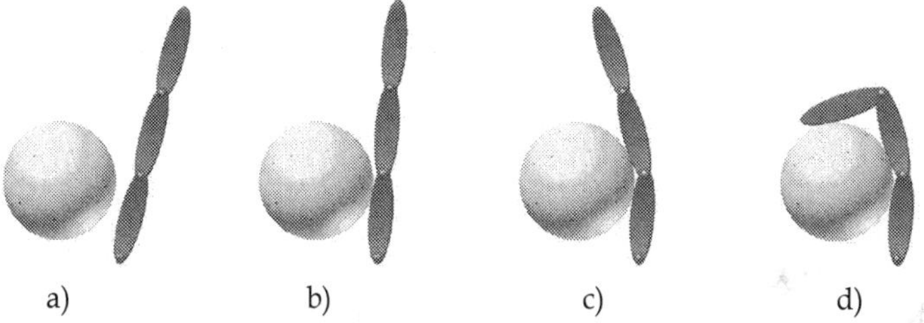

a)   b)   c)   d)

Fig. 1. A sequence for grasping a regularly shaped object: a) starting phase; b) first phalange is in its final configuration; c) second phalange is in its final configuration; d) third phalange is in its final configuration.

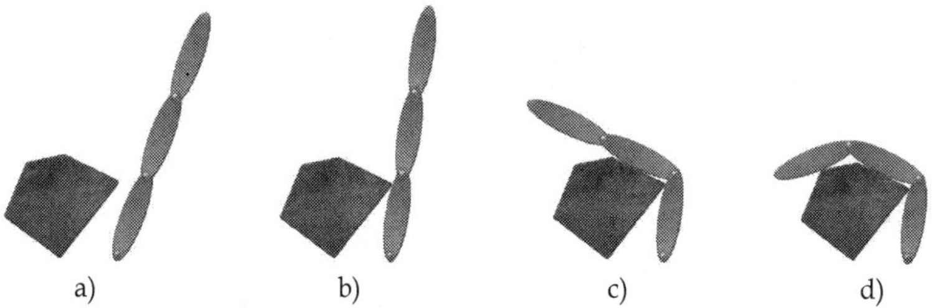

a)   b)   c)   d)

Fig. 2. A sequence for grasping an irregularly shaped object: a) starting phase; b) first phalange is in its final configuration; c) second phalange is in its final configuration; d) third phalange is in its final configuration.

required, linkage mechanisms are usually preferred and this Chapter is focused to the study of the latter type of mechanisms.

An example of underactuation based on cable transmission is shown in Fig.3a, it consists of a cable system, which properly tensioned, act in such a way as to close the fingers and grasp the object.

The underactuation based on link transmission, or linkages, consists of a mechanism with multiple DOFs in which an appropriate use of passive joints enables to completely envelop the object, so as to ensure a stable grasp. An example of this system is shown in Fig.3.b. This type solution for robotic hands has been developed for industrial or space applications with the aim to increase functionality without overly complicating the complexity of the mechanism, and ensuring a good adaptability to the object in grasp.

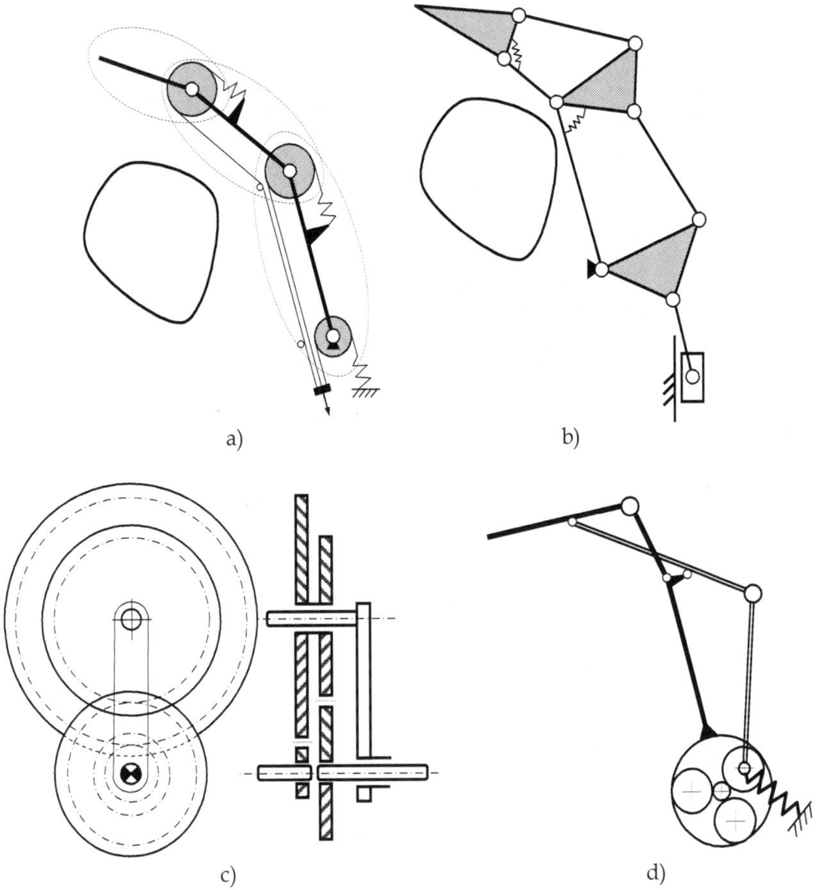

Fig. 3. Examples of underactuation systems: a) tendon-actuated mechanism; b) linkage mechanism; c) differential mechanism; d) hybrid mechanism.

A differential mechanism, shown in Fig. 3c, is a device, usually but not necessarily used for gears, capable of transmitting torque and rotation through three shafts, almost always used in one of two ways: in one way, it receives one input and provides two outputs, this is found in most automobiles, and in the other way, it combines two inputs to create an output that is the sum, difference, or average, of the inputs. These differential mechanisms have unique features like the ability to control many DOFs with a single actuator, mechanical stops or elastic limits. The differential gear, commonly used in cars, distributes the torque from the engine on two-wheel drive according to the torque acting on the wheels. Applying this solution to robotic hands, the actuation can be distributed to the joints according to the reaction forces acting to each phalanx during its operation.

Hybrid solutions have been also developed and make use of planetary gears and linkages, together with mechanical stops or elastic elements. An example is shown in Fig. 3d.

3. Design of underactuated finger mechanism

An anthropomorphic robotic finger usually consists of 2-3 hinge-like joints that articulates the phalanges. In addition to the pitch enabled by a pivoting joint, the head knuckle, sometimes also provides yaw movement. Usually, the condyloid nature of the human metacarpal-phalangeal joint is often separated into two rotary joints or, as in the case under-study, simplified as just one revolute joint.

Maintaining size and shape of the robot hand consistent to the human counterpart is to facilitate automatic grasp and sensible use of conventional tools designed for human finger placement. This holds true for many manipulative applications, especially in prosthesis and tele-manipulation where accuracy of a human hand model enables more intuitive control to the slave. Regarding to the actuation system in most of cases adopted solutions do not attempt to mimic human capabilities, but assume some of the pertinent characteristics of the force generation, since complex functionality of tendons and muscles that have to be replaced and somehow simplified by linear or revolute actuators and rotary joints.

The design of a finger mechanism proposed here uses the concept of underactuation applied to mechanical hands. Specifically, underactuation allows the use of $n - m$ actuators to control n-DOFs, where m passive elastic elements replace actuators, as shown in Fig. 4. Thus, the concept of underactuation is used to design a suitable finger mechanism for mechanical hands, which can automatically envelop objects with different sizes and shapes through simple stable grasping sequences, and do not require an active coordination of the phalanges. Referring to Figs. 4 and 5, the underactuated finger mechanism of Ca.U.M.Ha. (Cassino-Underactuated-Multifinger-Hand) is composed by three links m_j for j = 1, 2, 3, which correspond to the proximal, median and distal phalanges, respectively. Dimensions of the simplified sketch reported in Fig.4 have been chosen according to the overall characteristics of the human finger given in Table 1. In particular, in Fig. 4, θ_{iM} are the maximal angles of rotation, and torsion springs are denoted by S_1 and S_2. In the kinematic scheme of Fig.5, two four-bar linkages A, B, C, D and B, E, F, G are connected in series through the rigid body B, C, G, for transmitting the motion to the median and distal phalanges, respectively, where the rigid body A, D, P represents the distal phalange. Likewise to the human finger, links m_j (j = 1, 2, 3) are provided of suitable mechanical stoppers in order to avoid the hyper-extension and hyper-flexion of the finger mechanism. Both revolute joints in A and B are provided of torsion springs in order to obtain a statically determined system in each configuration of the finger mechanism.

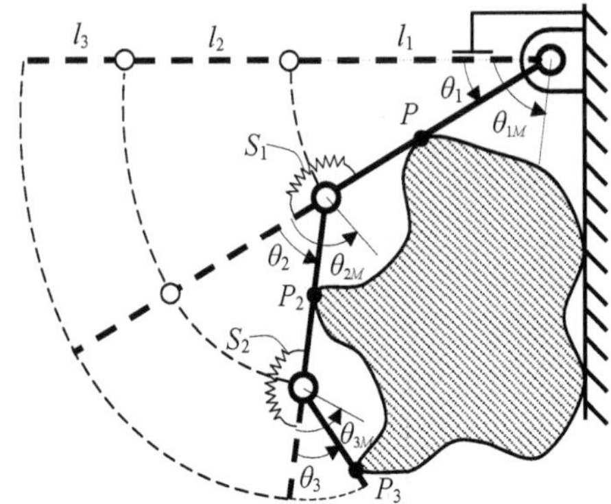

Fig. 4. Simplified sketch of underactuated finger mechanism.

Phalanx	Length	Angle
m_1	l_1 = 43 mm	θ_{1M} = 83°
m_2	l_2 = 25 mm	θ_{2M} = 105°
m_3	l_3 = 23 mm	θ_{3M} = 78°

Table 1. Characteristics of an index human finger.

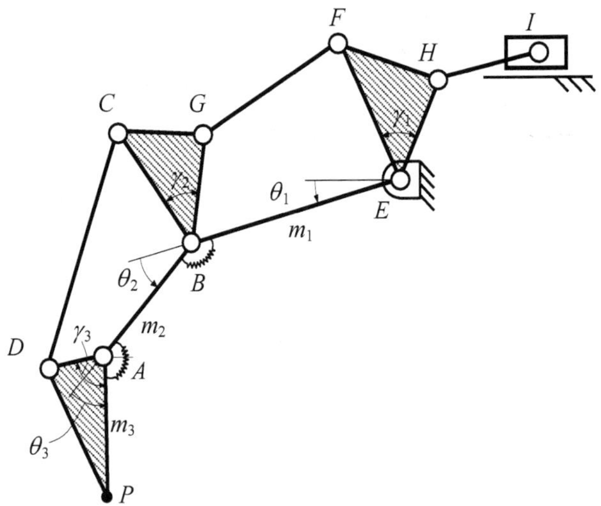

Fig. 5. Kinematic sketch of the underactuated finger mechanism.

3.1 Optimal kinematic synthesis

The optimal dimensional synthesis of the function-generating linkage shown in Fig. 5, which is used as transmission system from the pneumatic cylinder to the three phalanxes of the proposed underactuated finger mechanism, is formulated by using the Freudenstein's equations and the transmission defect, as index of merit of the force transmission. The three linkages connected in series are synthesized as in the following by starting from the four-bar linkage, which moves the third phalanx.

3.1.1 Synthesis of the four bar linkage A, B, C ,D

By considering the four-bar linkage A, B, C , D in Fig. 5, one has to refer to Fig.6 and the Freudenstein's equations can be expressed in the form

$$R_1 \cos \varepsilon_i - R_2 \cos \rho_i + R_3 = \cos(\varepsilon_i - \rho_i) \qquad i = 1,\ 2,\ 3 \tag{1}$$

with

$$R_1 = l_2 / a; \ R_2 = l_2 / c; \ R_3 = (a^2 - b^2 + c^2 + l_2^2)/2ac \tag{2}$$

where l_2 is the length of the second phalanx, a, b and c are the lengths of the links AD, DC and CB respectively, and ε_i and ρ_i for i = 1, 2, 3 are the input and output angles of the four-bar linkage $ABCD$.

Equations (1) can be solved when three positions 1), 2) and 3) of both links BC and AD are given through the pairs of angles (ε_i, ρ_i) for i = 1, 2, 3. According to a suitable mechanical design of the finger, (zoomed view reported in Fig.7) some design parameters are assumed, such as α = 50° for the link AD, γ = 40° and β_1 = 25° for the link BC, the pairs of angles $(\varepsilon_1$ = 115°, ρ_1 = 130°) and $(\varepsilon_3$ = 140°, ρ_3 = 208°) are obtained for the starting 1) and final 3) configurations respectively. Angle ρ_3 is given by the sum of ρ_1 and θ_{3M}. Since only two of the three pairs of angles required by the Freudenstein's equations are assigned as design specification of the function-generating four-bar linkage $ABCD$, an optimization procedure in terms of force transmission has been developed by assuming (ε_2, ρ_2) as starting values of the optimization, which correspond to both middle positions between 1) and 3) of links BC and AD respectively.

The transmission quality of the four-bar linkage is defined as the integral of the square of the cosine of the transmission angle. The complement of this quantity is defined "transmission defect" by taking the form

$$z' = \frac{1}{\varepsilon_3 - \varepsilon_1} \int_{\varepsilon_1}^{\varepsilon_3} \cos^2 \mu_1 \, d\varepsilon \tag{3}$$

where the transmission angle μ_1 is expressed as

$$\mu_1 = \cos^{-1}\left(\frac{l_2^2 + c^2 - a^2 - b^2 - 2 l_2 c \cos(\pi - \varepsilon)}{2\,ab} \right) \tag{4}$$

The optimal values of the pair of angles (ε_2, ρ_2) are obtained through the optimization of the transmission defect z'. In particular, the outcome of the computation has given $(\varepsilon_2$ = 132.5°, ρ_2 = 180.1°) and consequently, a = 22.6 mm, b = 58.3 mm and c = 70.9 mm have been obtained from the Eqs.(1) and (2).

It is worth to note that, as reported from Fig.8a to Fig.8c, these plots give many design solutions, the choice can be related to the specific application and design requirements. In the case under-study parameters ε_2 and ρ_2 have been obtained in order to have the maximum of the mean values for the transmission angle. The transmission angle μ_1 versus the input angle ε for the synthesized mechanism is shown in Fig.8d.

Figure 8 , shows a parametric study of the a, b, c, parameters as function of ε_2 and η_2. The colour scale represents the relative link length. For each plot the circle represents the choice that has been made for ε_2 and ρ_2, by assuming the length a = 23 mm, for the case under-study.

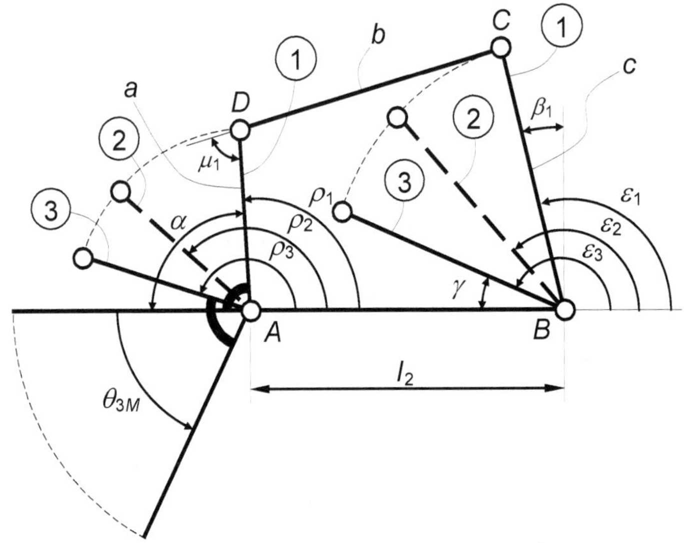

Fig. 6. Sketch for the kinematic synthesis of the four bar linkage *ABCD*, shown in Fig. 5.

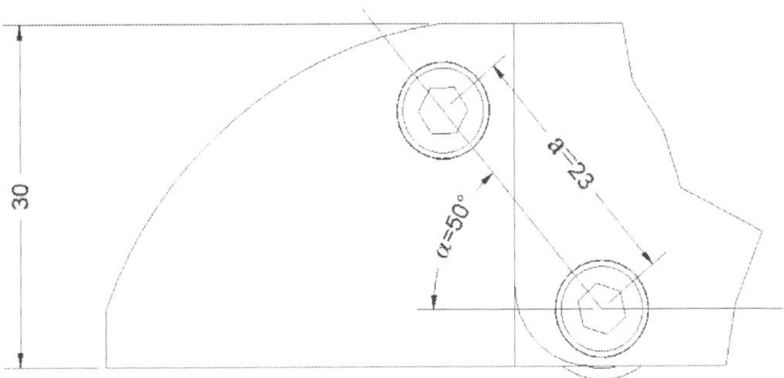

Fig. 7. Mechanical design of a particular used to define the angle α and the link length a of *A*, *B*, *C*, *D*, in Fig. 6.

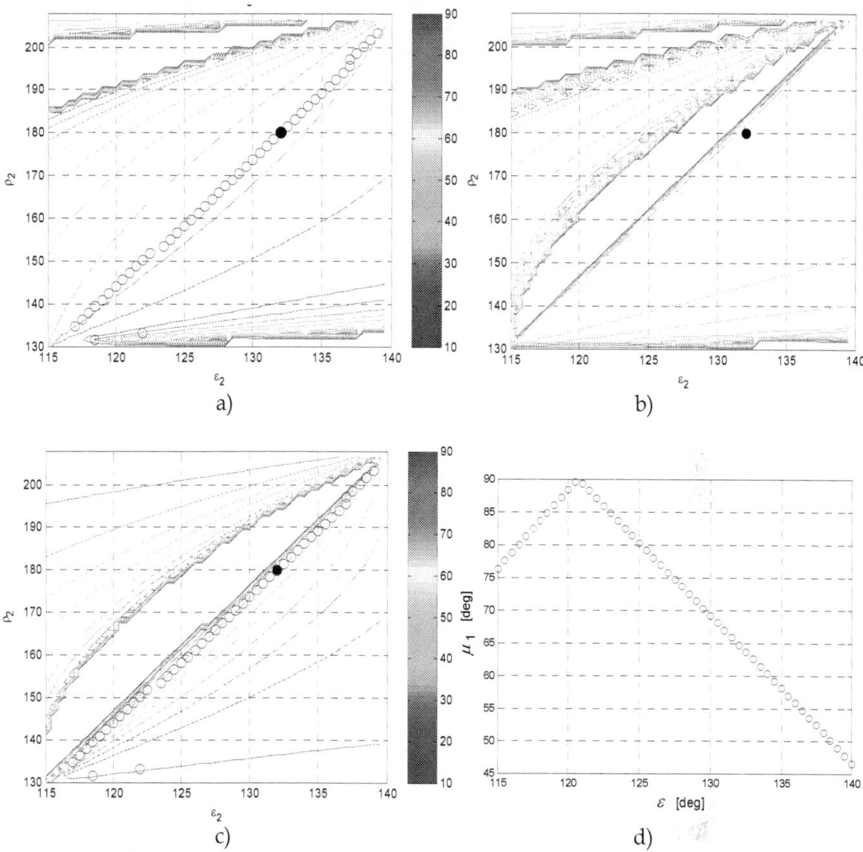

Fig. 8. Map of the link length versus the angles ε_2 and ρ_2.; a) link AD, b) link DC; c) link BC, d) Transmission angle μ_1 versus angle ε for the moving link c.

3.1.2 Synthesis of the four-bar linkage B, E, F, G

The same method has been applied to the synthesis of the function-generating four-bar linkage $BEFG$. In fact, referring to Fig.9, the Freudenstein's equations can be expressed in the form

$$R_1 \cos\psi_i - R_2 \cos\varphi_i + R_3 = \cos(\psi_i - \varphi_i) \qquad i = 1,\ 2,\ 3 \tag{5}$$

with

$$R_1 = l_1 / d; \ R_2 = l_1 / f; \ R_3 = (d^2 - e^2 + f^2 + l_1^2)/2df \tag{6}$$

where l_1 is the length of the first phalanx, d, e and f are lengths of the links BG, GF and FE respectively, and ψ_i and φ_i for $i = 1, 2, 3$ are the input and output angles of the four-bar linkage $BEFG$.

Likewise to the four-bar linkage $ABCD$, Eqs.(5) can be solved when three positions 1), 2) and 3) of both links EF and BG are given through the pairs of angles (ψ_i, φ_i) for i = 1, 2, 3. In particular, according to a suitable mechanical design of the finger, the design parameters γ = 40°, β_2 = 30° and δ = 10° are assumed empirically. Consequently, the pairs of angles (ψ_1 = 80°, φ_1 = 60°) and (ψ_3 = 140°, φ_3 = 190°) are obtained for the starting 1) and final 3) positions of both links EF and BG.

Since only two of the three pairs of angles required by the Freudenstein's equations are assigned as design specification of the function-generating four-bar linkage $BEFG$, an optimization procedure in terms of force transmission has been carried out by assuming (ψ_2, φ_2) as starting values of the optimization the middle positions between 1) and 3) of links EF and BG respectively. The transmission defect z' of the function-generating four-bar linkage $BEFG$ takes the form

$$z' = \frac{1}{\psi_3 - \psi_1} \int_{\psi_1}^{\psi_3} \cos^2\mu_2 \, d\psi \tag{7}$$

where the transmission angle μ_2 is expressed as

$$\mu_2 = \cos^{-1}\left(\frac{l_1^2 + f^2 - d^2 - e^2 - 2 l_1 f \cos(\pi - \psi)}{2\,de} \right) \tag{8}$$

The optimal values of the pair of angles (ψ_2, φ_2) are obtained and the output of the computation gives (ψ_2 = 115.5°, φ_2 = 133.7°). Consequently, d = 53.4 mm, e = 96.3 mm and f = 104.9 mm have been obtained from Eqs.(5) and (6). Figure 10 , shows a parametric study of the $d, e, f,$ parameters as a function of ψ_2 and φ_2. The colour scale represents the relative link length. For each plot the circle represents the choice that has been made for ψ_2 and φ_2, for the case under study. The diagram of the transmission angle μ_2 versus the input angle ψ of the moving link EF of the synthesized mechanism $BEFG$ is shown in Fig. 10d.

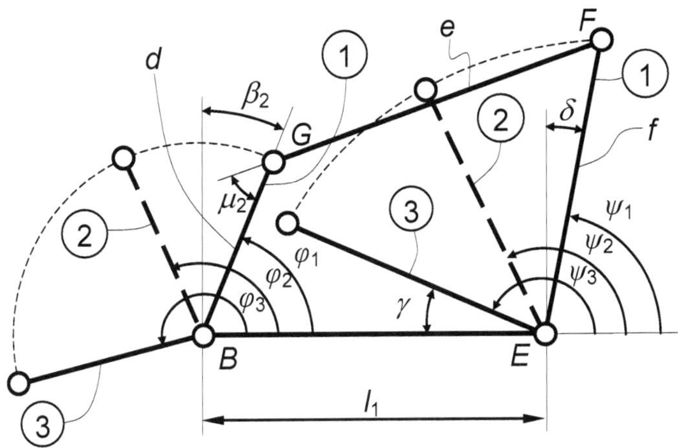

Fig. 9. Sketch for the kinematic synthesis of the four-bar linkage $BEFG$.

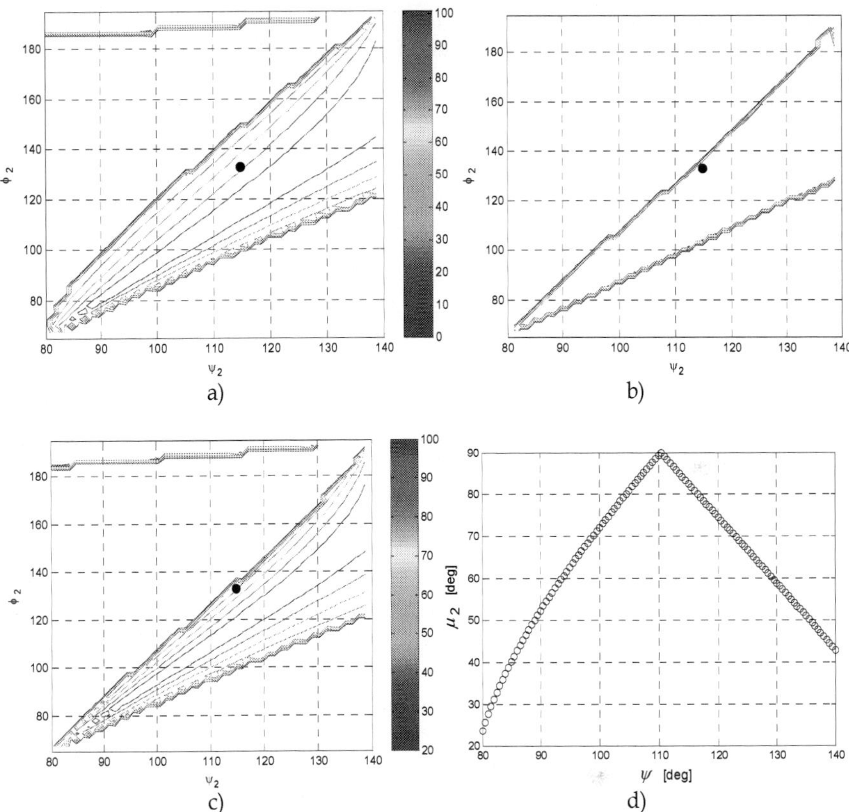

Fig. 10. Map of the link length versus the angles ψ_2 and φ_2; a) link BG, b) link GF, c) link EF, d) transmission angle μ_2 versus angle ψ of the moving link EF.

3.1.3 Synthesis of the slider-crank mechanism EHI

Likewise to both four-bar linkages $ABCD$ and $BEFG$, the offset slider-crank mechanism EHI of Fig. 11 is synthesized by using the Freudenstein's equations, which takes the form

$$R_1 (s_1 - x_i) \cos \lambda_i + R_2 \sin \lambda_i - R_3 = (s_1 - x_i)^2 \qquad i = 1, 2, 3 \tag{9}$$

with

$$\begin{aligned} R_1 &= 2g; \\ R_2 &= 2g\,o_f; \\ R_3 &= g^2 + o_f^2 - h^2 \end{aligned} \tag{10}$$

where o_f is the offset, g and h are the lengths of the links EH and HI respectively, and x_i and λ_i for $i = 1, 2, 3$ are the input displacement of the piston and the output rotation angle of the link *EH* of the slider-crank mechanism *EHI*. Equations (9) can be solved when three positions 1), 2) and 3) of both piston and link *EH* are given through the pairs of parameters (x_i, λ_i) for $i = 1, 2, 3$. In particular, according to a suitable mechanical design of the finger, the design parameters $(x_1 = 0$ mm, $\lambda_1 = 37°)$ and $(x_3 = 75$ mm, $\lambda_3 = 180°)$ are assumed empirically for the starting 1) and final 3) positions of both piston and link *EH*. The optimization procedure in terms of force transmission has been carried out by assuming as starting values of the optimization the middle position between 1) and 3) of the piston and link *EH* respectively. The transmission defect z' of the function-generating slider-crank mechanism *EHI* takes the form

$$z' = \frac{1}{x_3 - x_1} \int_{x_1}^{x_3} \cos^2 \mu_3 \, dx \tag{11}$$

where the transmission angle μ_3 is expressed as

$$\mu_3 = \cos^{-1}\left(\frac{(s_1 - x)^2 + o_f^2 - g^2 - h^2}{2gh} \right) \tag{12}$$

The optimal values of the pair of parameters (x_2, λ_2) are obtained, and the outcome of the computation has given $(x_2 = 47.5$ mm, $\lambda_2 = 126.9°)$. Consequently, $o_f = 43.4$ mm, $g = 35.7$ mm and $h = 74.7$ mm have been obtained from the Eqs. (9) and (10).

Figures 12a, 12b and 12c, show a parametric study of the parameters g, h and o_f, as a function of λ_2 and s_2. The colour scale represents the relative link length and for each plot the marked circle represents the choice that has been made for values λ_2 and s_2. The diagram of the transmission angle μ_3 versus the input displacement x of the moving piston of the synthesized slider-crank mechanism *EHI* is shown in Fig. 12d.

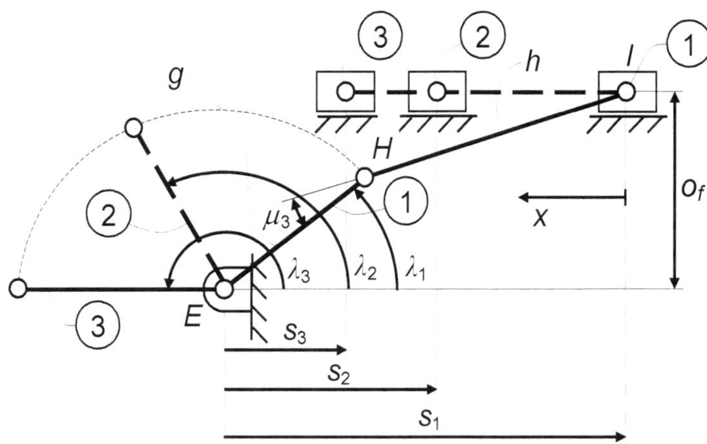

Fig. 11. Kinematic scheme of the offset slider-crank mechanism *EHI*.

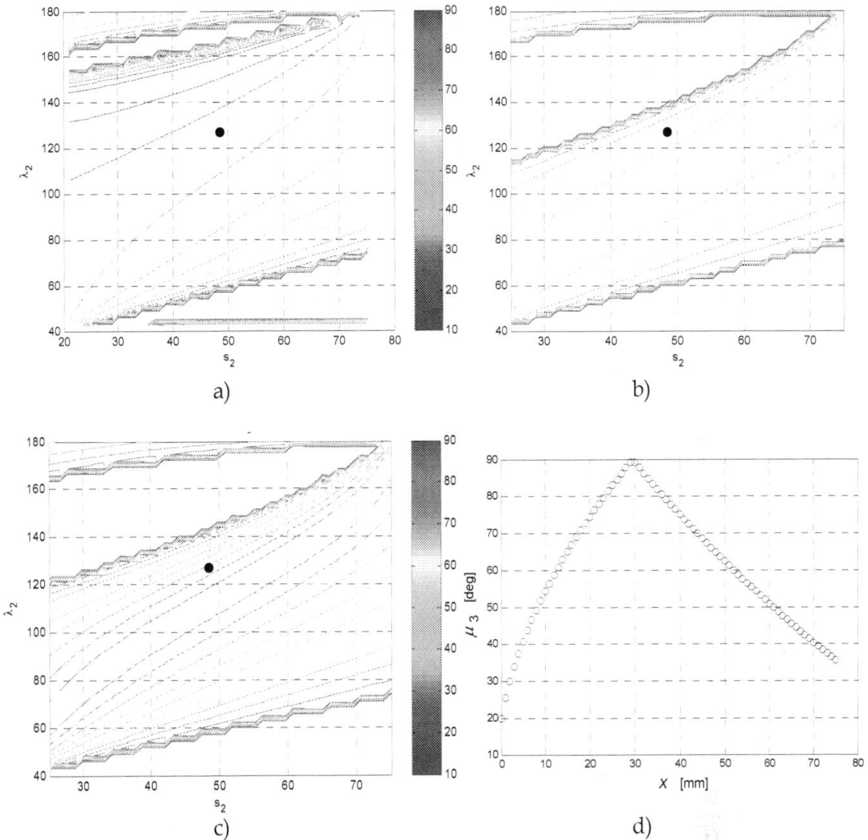

Fig. 12. Map of the link length versus angles λ_2 and s_2; a) link EI, b) link HI, c) link EH, d) transmission angle μ_3 versus distance x of the moving link.

3.2 Mechanical design

Figure 13 shows a drawing front view of the designed underactuated finger mechanism. In particular, EHI indicates the slider-crank mechanism, $ABCD$ indicates the first four-bar linkage, and $DEFG$ indicates the second four-bar linkage. In order to obtain the underactuated finger mechanism, two torsion springs (S_1 and S_2) have been used at joints A and B and indicated with 1 and 2, respectively.

Aluminium has been selected for its characteristics of lightness and low-cost. It has the disadvantage of low hardness, therefore for the manufacturing of the revolute joints, ferrules have been considered. In particular, in Fig. 13, it is possible to note that the finger mechanism, which allows the finger motion, is always on the upper side of the phalanges. This is to avoid mechanical interference between the object in grasp and the links' mechanism. Furthermore, the finger is asymmetric, this is due to the fact that is necessary to have a suitable side to mount the torsion spring.

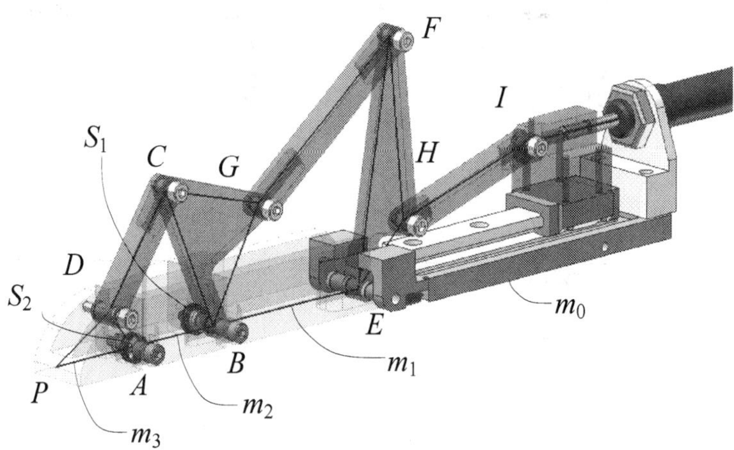

Fig. 13. Mechanical design of the underactuated finger.

Each phalange has a flat surface to interact with the object to be grasped. This is to further consider force sensors to develop a suitable force control of the robotic hand prototype, as reported in Section 4.

The common operation of the four underactuated fingers gives an additional auto-adaptability of the Ca.U.M.Ha. robotic hand, because each finger can reach a different closure configuration according to the shape and size of the object to grasp. This behaviour is due to the uniform distribution of the air pressure inside the pneumatic tank and pushing chambers, as it will be described below.

3.3 Actuation and control

The lay-out of the electro-pneumatic circuit of the proposed closed-loop pressure control system is sketched in Fig.14, where the pressure P_{OUT} in the rigid tank is controlled by means of two PWM modulated pneumatic digital valves V_1 and V_2, which are connected in supply, at the supply pressure P_S, and in exhaust, at the atmospheric pressure P_A, respectively.

Thus, both valves V_1 and V_2 approximate the behaviour of a three-way flow proportional valve, which allows the pressure regulation in the tank. These valves are controlled through the voltage control signals $V_{PWM\ 1}$ and $V_{PWM\ 2}$, which are modulated in PWM at 24 V, as it is required by the valves V_1 and V_2. These signals are given by a specific electronic board supplied at 24 V, which allows the generation of both signals $V_{PWM\ 1}$ and $V_{PWM\ 2}$ and the amplification at 24 V from the input signal V_{PWM} that lies within the range of [- 5; + 5] V. The PWM modulated control signal V_{PWM} is generated via software because of a suitable Lab-View program.

The feed-back signal $V_{F/B}$ is given by the pressure transducer Tp with static gain $K_T = 1$ V/bar, which is installed on the rigid tank directly.

Thus, a typical PID compensation of the ε error between the input electric signal V_{SET} and the feed-back electric signal $V_{F/B}$ is carried out through a PC controller, which is provided of the electronic board PCI 6052-E and a terminal block SCB-68.

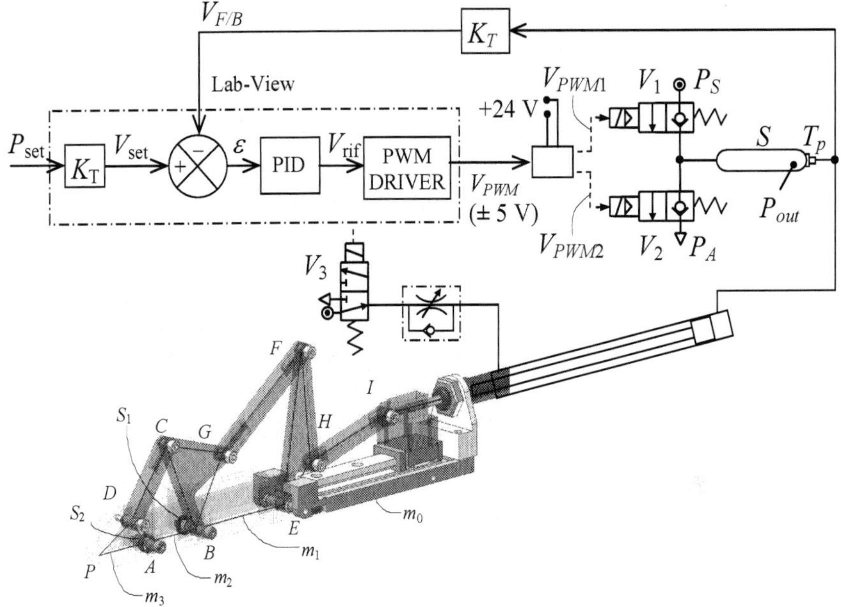

Fig. 14. Scheme for the pressure control of the robotic hand prototype finger.

3.3.1 Experimental test-bed

The closed-loop pressure control system and a test bed of Fig.15 have been designed and built according to the scheme of Fig.14. In particular, this test-bed is mainly composed by: 1) and 2), two 2/2 (two-way/two-position) pneumatic digital valves of type SMC VQ21A1-5Y0-C6-F-Q; 3) a tank of type Festo with a volume of 0.4 lt; 4) a pressure transducer of type GS Sensor XPM5-10G, connected to an electronic board of type PCI 6052-E with the terminal block SCB-68, which is connected to the PC in order to generate the control signal V_{PWM} ; 5) a specific electronic board to split and amplify at 24 V the control signals V_{PWM1} and V_{PWM2}.

The electronic circuit of Fig.15b splits and amplifies the modulated electric signal V_{PWM} that comes from the PWM driver into the signals V_{PWM1} and V_{PWM2}, which control the digital valves V_1 and V_2 respectively. This circuit is composed by a photodiode FD, three equal electric resistors R_1, a MOSFET M and a diode D. In fact, the working range of the electronic board NI DAQ AT MIO-16E-2 is amplified from [–5 / +5] V to the working range [0 / + 24] V of the digital valves because of the electric supply at 24 V DC. Moreover, this signal is decomposed and sent alternatively to V_1 and V_2 because of the effects of the MOSFET M.

A suitable software in the form of virtual instrument has been conceived and implemented by using the Lab-View software, as shown in Fig.16. This solutions gives the possibility of using the electronic board NI DAQ PCI-6052-E for driving the PWM modulated pneumatic digital valves and acquiring both voltage signals V_{SET} and $V_{F/B}$ of the proposed closed-loop pressure control system. Thus, the program can be considered as composed by three main blocks, where the first is for acquiring analogical signals through a suitable scan-rate, the second gives the PID compensation of the pressure error and the third one is for generating the PWM signal.

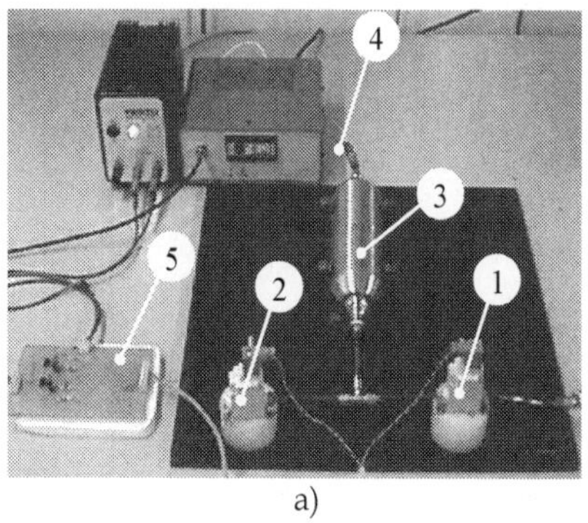

a)

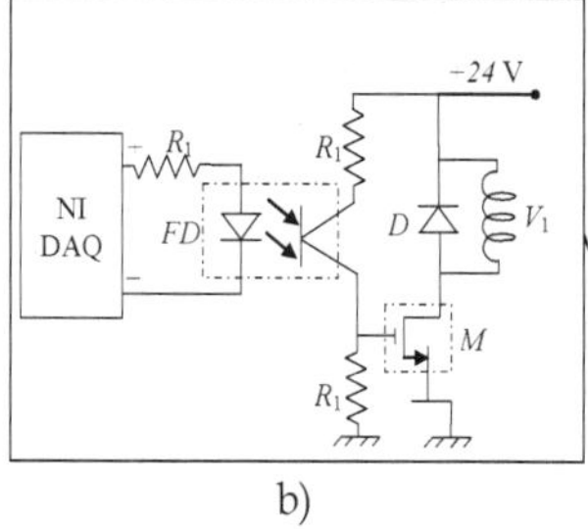

b)

Fig. 15. Test-bed a) of the proposed closed-loop pressure control system and b) a scheme of the electronic circuit.

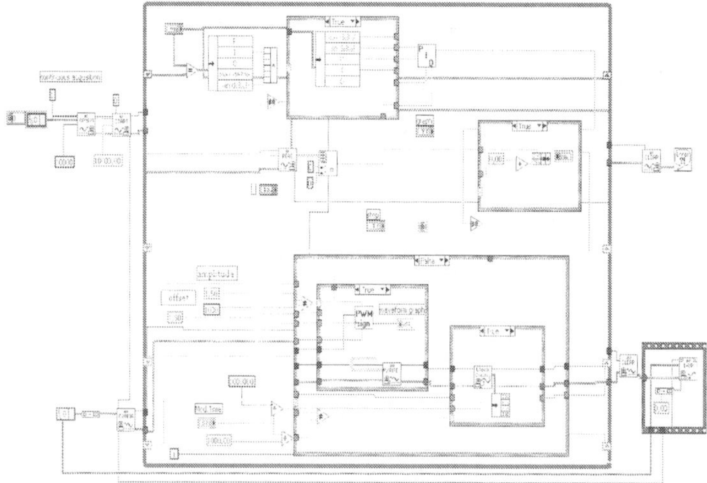

Fig. 16. Lab-View program for controlling the pressure in the tank through PWM modulated pneumatic digital valves.

3.3.2 Experimental results

The static and dynamic performances of the proposed closed-loop pressure control system have been analyzed by using the test-bed of Fig.15. Some experimental results in the time domain are reported in Fig.17 in order to show the effects of the proportional gain Kp of the PID compensator. In particular, the reference and output pressure signals P_{SET} and P_{OUT} are compared by increasing the values of the proportional gain Kp from 0.3 to 2.4, as shown in Figs.17a to 17d, respectively. Taking into account that the pressure transducer Tp is characterized by a static gain $K_T = 1$ V/bar, the pressure diagrams of P_{SET} and P_{OUT} show the same shape and values of the correspondent voltage diagrams V_{SET} and $V_{F/B}$, respectively. Moreover, the diagram of Fig.17c shows a good behaviour at high values of Kp, even if some instability of the system may appear, as shown in Fig.17d for $Kp = 2.4$. The experimental closed-loop frequency response of the proposed pressure control system has been carried out by using a Gain-Phase-Analyzer of type SI 1253. The Bode diagrams of Fig.18a and 18b have been obtained for the periods of the PWM modulation, $T = 50$ ms and $T = 100$ ms, respectively. Thus, the diagrams of the pressure signals P_{SET} and P_{OUT} versus time, which have been acquired through the Lab-View Data-Acquisition-System, are shown in continuous and dash-dot lines, respectively. In particular, Figs.19a and 19b show both frequency responses of Fig.18a and 18b in the time domain for a P_{SET} sinusoidal pressure signal with frequency $f = 0.1$ Hz, average value $Av = 3$ bar rel and amplitude $A = 2$ bar rel. Likewise to the diagrams of Fig.20 and still referring to the Bode diagrams of Fig.18, the frequency responses in the time domain for a P_{SET} with frequency $f = 1.5$ Hz are shown respectively in Fig.20a and 20b for $T = 50$ ms and $T = 100$ ms.

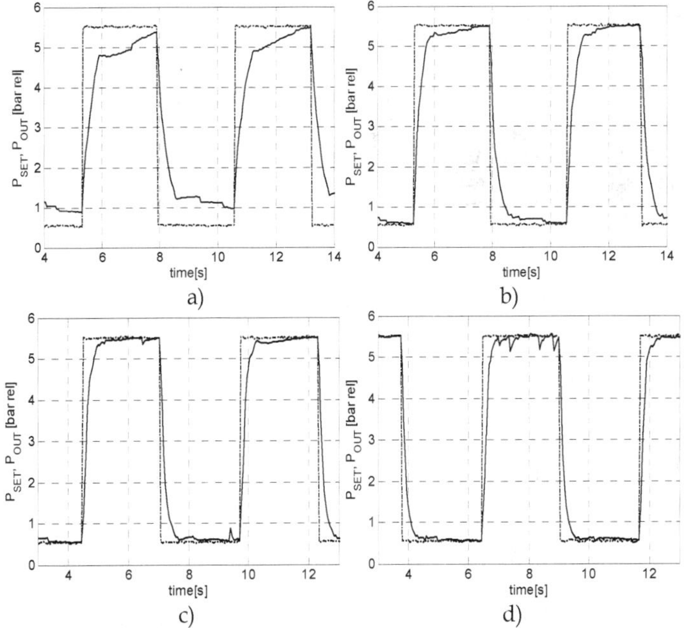

Fig. 17. Effects of the proportional gain: a) $Kp = 0.3$; b) $Kp = 0.9$; c) $Kp = 1.8$; d) $Kp = 2.4$.

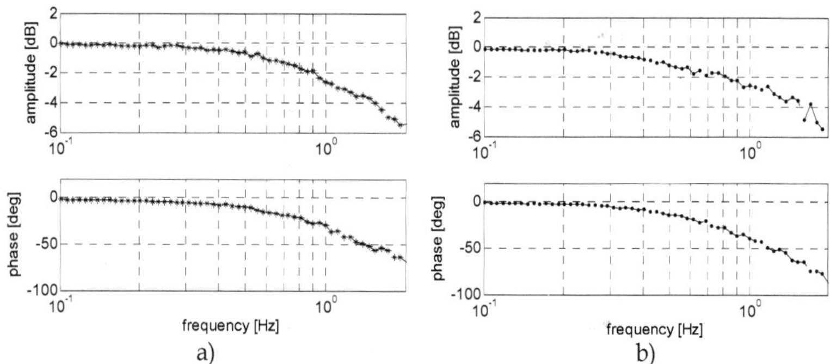

Fig. 18. Closed-loop frequency responses of the proposed pressure control system for different periods of the PWM modulation; a) T = 50 ms; b) T = 100 ms.

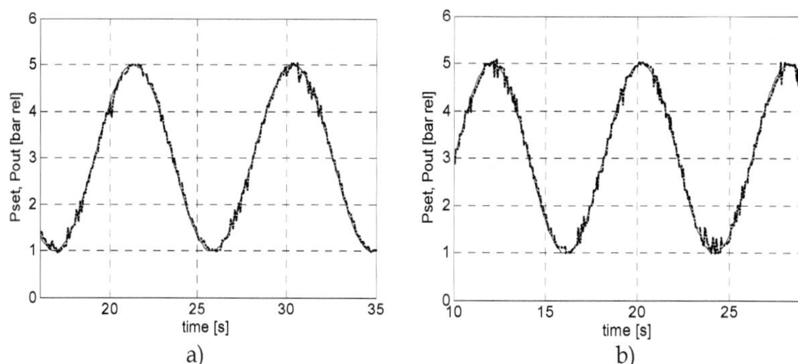

Fig. 19. Frequency responses in the time domain for a sinusoidal P_{SET} with f = 0.1 Hz, Av = 3 bar rel and A = 2 bar rel: a) T = 50 ms; b) T = 100 ms.

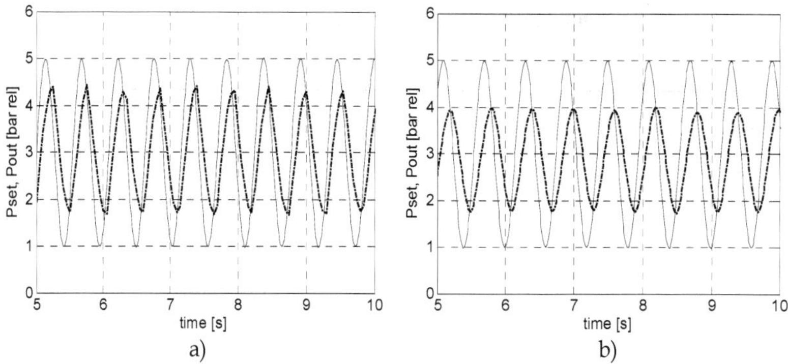

Fig. 20. Frequency responses in the time domain for a sinusoidal P_{SET} with f = 1.5 Hz, Av = 3 V and A = 2 V: a) T = 50 ms; b) T = 100 ms.

4. The CaUMHa underactuated robotic hand: overall design

According to the mechatronic design proposed and described in Sections II and III, a prototype of Ca.U.M.Ha. robotic hand has been built and tested by using the experimental test-bed of Fig. 21, which shows: 1) Ca.U.M.Ha. robotic hand prototype; 2) pneumatic cylinder; 3) PWM modulated pneumatic digital valves; 4) 3/2 pneumatic digital valve; 5) 5/2 pneumatic digital valve; 6) external block SCB-68; 7) electronic board to convert the signal V_{PWM} to $V_{PWM\,1}$ and $V_{PWM\,2}$; 8) electronic board to control the thumb of the robotic hand.

The mechanical parts of Ca.U.M.Ha., i.e. underactuated fingers along with their linkage systems, palm and thumb, have been manufactured in aluminum, while the tank is made by steel.

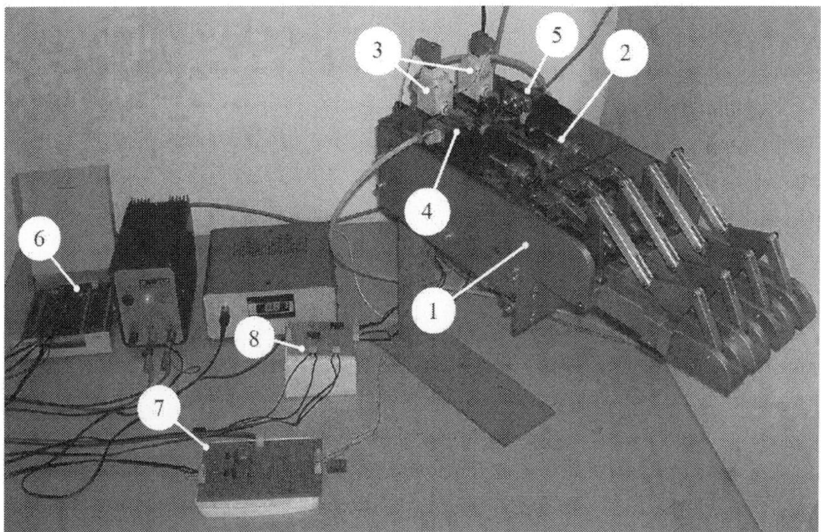

Fig. 21. Prototype and experimental test-bed of the Ca.U.M.Ha. robotic hand, 1) Ca.U.M.Ha. robotic hand; 2) double-acting pneumatic cylinder; 3) two PWM modulated pneumatic digital valves; 4) 3/2 pneumatic digital valve; 5) 5/2 pneumatic digital valve; 6) terminal block SCB-68; 7) electronic board to split and amplify at 24 V the control signals $V_{PWM\,1}$ and $V_{PWM\,2}$; 8) electronic board to split and amplify at 24 V both signals to control the thumb of the robotic hand.

5. Conclusions

In this Chapter the mechatronic design has been reported for the Ca.U.M.Ha. (Cassino-Underactuated-Multifinger-Hand) robotic hand. In particular, the underactuation concept is addressed by reporting several examples and kinematic synthesis and the mechatronic design have been developed for a finger mechanism of the robotic hand. As a result the Ca.U.M.Ha. robotic hand shows a robust and efficient design, which gives good flexibility and versatility in the grasping operation at low-cost. The kinematic synthesis and optimization of the underactuated finger mechanism of Ca.U.M.Ha. have been formulated and implemented. In particular, two function-generating four-bar linkages and one offset slider-crank mechanism have been synthesized by using the Freudenstein' equations and optimizing the force transmission, which can be considered as a critical issue because of the large rotation angles of the phalanxes. A closed-loop pressure control system through PWM modulated pneumatic digital valves has been designed and experimentally tested in order to determine and analyze its static and dynamic performances. The proposed and tested closed-loop control system is applied to the Ca.U.M.Ha. robotic hand in order to control the actuating force of the pneumatic cylinders of the articulated fingers. Consequently, a force control of the grasping force has been developed and tested according to a robust and low-cost design of the robotic hand.

6. References

Angeles J., Bernier A., (1987). The Global Least-Square Optimization of Function Generating Linkages. *Journal of Mechanisms, Transmissions and Automation in Design*, Vol.109, pp.204-209.

Arimoto S., (2003). Control for a family of nonlinear and under-actuated systems with DOF-redundancy and geometric constraints, *SICE Annual Conference in Fukui*, Fukui, pp.2205-2210.

Banks J.L. (2001). Design and control of an anthropomorphic robotic finger with multi-point tactile sensation MS Thesis, MIT, Cambridge.

Beccai L., Roccella S., Ascari L., Valdastri P., Sieber A., Carrozza M. C., Dario P., (2009). Development and experimental analysis of a soft compliant tactile microsensor for anthropomorphic artificial hand. *IEEE/ASME Transaction on Mechatronics*, Vol.13, no.2, pp.158-168.

Belforte G., Mauro S., Mattiazzo G.(2004). A method for increasing the dynamic performance of pneumatic servosystems with digital valves, *Mechatronics*, Vol.14 pp. 1105-1120.

Belforte, G.; Mattiazzo, G.; Mauro, S. & Cocito, C. (2001). A robotic system for apples harvesting. *10th RAAD International Workshop on Robotics in Alpe-Adria-Danube Region*, Vienna, paper: RD-103.

Bicchi A., (2000). Hands for Dexterous Manipulation and Robust Grasping: a Difficult Road Toward Simplicity. *IEEE/ASME Transaction on Robotics and Automation*, Vol.16, n.6, pp.652-662.

Birglen L., Gosselin C.M., (2003). On the Force Capability of Underactuated Fingers. *Proceedings of the 2003 IEEE International Conference on Robotics and Automation, ICRA 2003*, Taipei, pp.1139-1145.

Birglen L., Gosselin C.M., (2004). Kinetostatic analysis of underactuated fingers, *IEEE Transactions on Robotics and Automation*, Vol.20, n.2 pp. 211-221.

Birglen L., Gosselin C.M., (2005). Fuzzy Enhanced Control of an Underactuated Finger Using Tactile and Position Sensors. *Proceedings of the 2005 IEEE International Conference on Robotics and Automation, ICRA 2005*, Barcelona, pp.2320-2325.

Bulanon, D.M.; Kataoka, T.; Ota, Y. & Hiroma, T. (2001). A machine vision system for the apple harvesting robot. *Agricultural Engineering International: the CIGR J. of Scientific Research and Development*, Vol.3, paper: PM 01 006.

Butterfas J., Grebenstein M.,Liu H., Hirzinger G. (2001). DLR hand II: Next Generation of a Dextrous Robot Hand. *Proceedings of the 2001 IEEE International Conference on Robotics and Automation, ICRA 2001*, Seoul, pp.109-114.

Carrozza M.C., Massa B., Micera S., Lazzarini R., Zecca M. e Dario P. (2002). The Development of a Novel Prosthetic Hand Ongoing Research and Preliminary Results. *IEEE/ASME Transaction on Mechatronics*, Vol.7, n.2, pp.108-114.

Casolo F., Lorenzi V., (1990). Criteri di scelta e Ottimizzazione di Modelli per la Simulazione del Dito Umano. *X Congresso Nazionale dell'Associazione Italiana di Meccanica Teorica ed Applicata (AIMETA-90)*, Pisa, pp.415-420.

Chevallereau C., Grizzle J.W., Moog C.H., (2004). Non Linear Control of Mechanical Systems with One Degree of Underactuation. *Proceedings of the 2004 IEEE International Conference on Robotics and Automation, ICRA 2004*, New Orleans, pp.2222-2228.

Crisman J.D., (1996). Robot Arm End Effector, 1996. US Patent n. 5570920.

Figliolini G., Ceccarelli M. (2002) A novel articulated mechanism mimicking the motion of index fingers, *Robotica*, Vol.20, pp.13-22.

Figliolini G., Rea P.(2003) Ca.U.M.Ha. robotic hand for harvesting horticulture products, *XXX CIOSTA - CIGRV Congress on Management and Technology Applications to Empower and Agro-Food Systems*, Turin, pp.288-295.

Figliolini, F. & Rea, P. (2004). Actuation Force Control of Ca.U.M.Ha. Robotic Hand Through PWM Modulated Pneumatic Digital Valves. *3rd FPNI - PhD Symposium on Fluid Power*, Terrassa, pp. 149-156.

Figliolini, G. & Rea, P. (2005). Synthesis and optimization of an underactuated finger mechanism, *Proceedings of the 9th IFToMM Symposium on Theory of Machines and Mechanisms (SYROM'05)*, Bucharest, Vol.3, pp. 747-752.

Figliolini, G. & Rea, P. (2006). Overall design of Ca.U.M.Ha. robotic hand, *Robotica*, Vol.24 n.3, pp. 329-331.

Figliolini G. & Rea, P. (2007). Ca.U.M.Ha. Robotic Hand (Cassino-Underactuated-Multifinger-Hand). *Proceedings of the 2007 IEEE/ASME International Conference on Advanced Intelligent Mechatronics (AIM2007)*, Zurich, paper number 250.

Figliolini G., Rea P., Principe M.(2003). Mechatronic design of Ca.U.M.Ha (Cassino-Underactuated-Multifinger-Hand), *12th RAAD Workshop on Robotics in Alpe-Adria-Danube Region, Cassino*, paper: 026RAAD03.

Francisco J. V. C., (2000). Applying principles of robotics to understand the biomchanics, neuromuscolar control and clinical rehabilitation of human digits. *Proceedings of the 2000 IEEE International Conference on Robotics and Automation, ICRA 2000*, San Francisco, pp.270-275.

Freudenstein F.,(1955). Approximate synthesis of four-bar linkages, *Transactions of the ASME*, Vol. 77, pp.853-861.

Gokhale K., Kawamura A. and R.G. Hoft, (1987).Dead beat microprocessor control of PWM inverter for sinusoidal output waveform synthesis *IEEE/ASME Transaction on Industry application*, Vol.1A 23, no.5, pp.901-910.

Gosselin C., Angeles J., (1989). Optimization of Planar and Spherical Function Generators as Minimum-Defect Linkages. *Mechanism and Machine Theory*, Vol.24, pp.293-307.

Haulin E.N., Lakis A.A., Vinet R., (2001). Optimal Synthesis of a Planar Four-Link Mechanism used in a Hand Prosthesis, *Mechanism and Machine Theory*, Vol.36, pp.1203-1214.

Haulin E.N., Vinet R., (2003). Multiobjective Optimization of Hand Prothesis Mechanisms *Mechanism and Machine Theory*, Vol.38, pp.3-26.

Imai Y., Namiki A., Hashimoto K., Ishikawa M., (2004). Dynamic Active Catching Using a High-speed Multi⁻ngered Hand and a High-speed Vision System. *Proceedings of the 2004 IEEE International Conference on Robotics and Automation, ICRA*, New Orleans, pp.1849-1854.

Jiang L., Sun D. and Liu H., (2009). An inverse-kinematics table-based solution of a humanoid robot finger with nonlinearly coupled joints *IEEE/ASME Transaction on Mechatronics*, Vol.14, n.3, pp.273-281.

Jacobsen, S.C.; Wood, J.E.; Knutti, D.F. & Biggers, K.B. (1984). The Utah/MIT dexterous hand: work in progress, *The Int. J. of Robotics Research*, Vol.3, No.4, pp. 21-50.

Jobin J.P., Buddenberg H.S., Herder J.L., (2004). An Underactuated Prothesis Finger Mechanism with Rolling Joints. *Proceedings of DETC ASME, 28th Biennal Mechanisms Conference*, Salt Lake City,

Kim K., Edward Colgate J., Santos-Munné J.J., Makhlin A. and Peshkin M., (2010). On the design of miniature haptic devices for upper extremity prosthetics. *IEEE/ASME Transaction on Mechatronics*, Vol.15, n.1, pp.27-39.

Lalibertè, T., Gosselin, C.M. (1998). Simulation and design of underactuated mechanical hands. *Mechanism and Machine Theory*, Vol.33, No.1, pp.39-57.

Lalibertè T., Birglen L., Gosselin C.M. (2002). Underactuation in robotic grasping hands, *Japanese Journal of Machine Intelligence and Robotic Control*, Vol.4, n.3 pp.77-87.

Lee H.J., Yi B.J., Oh S.R., Suh I.H., (2001). Optimal Design and Development of a Five-Bar Finger with Redundant Actuation, *Mechatronics*, Vol. 11, pp.27-42.

Liu, H.; Butterfass, J.; Knoch, S.; Meisel, P. & Hirzinger, G. (1999). A new control strategy for DLR's multisensory articulated hand. *IEEE Control systems*, Vol.19, n.2, pp.47-54.

Liu G. and Li Z., (2004). Real time grasping-force optimization for multifingered manipulation: Theory and experiments *IEEE/ASME Transaction on Mechatronics*, Vol.9, n.1, pp.65-77.

Liu H., Meuse P., Hirzinger G., Jin M., Liu Y., and Xie Z. (2008).The modular multisensory DLR-HIT-hand: Hardware and software architecture *IEEE/ASME Transaction on Mechatronics*, Vol.13, n.4, pp.461-469.

Liu H., Wu K., Meusel P., Seitz N., Hirzinger G., Jin M.H., Liu Y.W., Fan S.W., Lan T. and Chen Z.P. (2008). Multisensory five-finger dexterous hand: The DLR/HIT hand II *IEEE/RSJ International Conference on Intelligent Robots and Systems*. Acropolis Convention Center Nice, Sept, 22-26.

Lotti F., Tiezzi P., Vassura G., Biagiotti L., Melchiorri C., Palli G., (2004). UBH 3: A Biologically Inspired Robotic Hand. *Proc. of IEEE International Conference on Intelligent Manipulation and Grasping (IMG 04)*, Genova, pp. 39-45.

Luo M., Mei T., Wang X., Yu Y., (2004). Grasp Characteristics of an Underactuated Robot Hand. *Proceedings of the 2004 IEEE International Conference on Robotics and Automation, ICRA 2004*, New Orleans, (USA), pp.2236-2241.

Mason M. T., Salisbury J.K., (1984). Robot hand and the mechanics of manipulation, Tokyo, pp.151-167.

Mason, M.T. & Salisbury, J.K. (1985). Robot Hand and the Mechanics of Manipulation, *MIT Press, Cambridge*, MA.

Montambault, S. & Gosselin, C.M. (2001). Analysis of underactuated mechanical grippers. *ASME J. of Mechanical Design*, Vol.123, n.3, pp. 39-57.

Namiki A., Imai Y., Ishikawa M., Kaneko M. (2003). Development of a High-Speed Multifingered Hand System and Its Application to Catching. *Proceedings of the 2003 IEEE/RSJ Intl. Conference on Intelligent Robots and Systems*, Las Vegas, pp.2666-2671.

Raparelli, T.; Mattiazzo, G. & Mauro, S., Velardocchia M. (2000). Design and development of a pneumatic anthropomorphic hand. *J. of Robotic Systems*, Vol.17, n.1, pp. 1-15.

Rash G.S., Belliappa P.P., Wachowiak M.P., Somia N.N., Gupta A., (1999). A Demon-stration of the Validity of a 3-D Video Motion Analysis Method for Mesuring Finger Flexion and Extension, *Journal of Biomechanics*, 32, pp.1337-1341.

Rea P., (2007). Mechatronic Design of underactuated robotic hand with pneumatic actuation, Ph.D. Dissertation, University of Cassino, Cassino. Available at: *http://www.scuoladottoratoingegneria.unicas.it/Tesi/Ciclo XIX/TesiRea.pdf*

Schultz S., Pylatiuk C., Bretthauer G., (2001). A New Ultralight Anthropomorphic Hand. *Proceedings of the 2001 IEEE International Conference on Robotics and Automation, ICRA 2001*, Seoul, (Korea), pp.2437-2441.

Schweizer A., Frank O., Ochsner P.E., Jacob H.A.C., (2003). Friction Between human Finger Tendons and Pulleys at High loads, *Journal of Biomechanics*, 36, pp.63-71.

Sorli M., Figliolini G., Pastorelli S., Modeling and experimental validation of a two-way pneumatic digital valve, *Bath Workshop on Power Transmission & Motion Control*, Eds. C.R. Barrows and K.A. Edge, Professional Engineering Publishing, London, 2003, pp.291-305.

Sorli M., Figliolini G. and Pastorelli S., (2004). Dynamic model and experimental investigation of a pneumatic proportional pressure valve. *IEEE/ASME Transaction on Mechatronics*, Vol.9, n.1, March 2004, pp.78-86.

Su F-C., Kuo L.C., Chiu H.Y., Chen-Sea M.J., (2003). Video-Computer Quantitative Evaluation of Thumb Function using Workspace of the Thumb. *Journal of Biomechanics*, Vol.36, pp.937-942.

Salisbury J. K., Craig J. J., (1981). Articulated Hands: Force Control and Kinematic Issues. *Joint Automatic Control Conference*, Charlottesville, pp.17-19.

Sutherland G., Roth B., (1973). A Transmission Index for Spatial Mechanisms. *ASME Journal of Engineering Industry*, pp.589-597.

Stellin G., Cappiello G., Roccella S., Carrozza M.C., Dario P., Metta G., Sandini G., Becchi F., (2006). Preliminary Design of an Anthropomorphic Dexterous Hand for a 2-Years-Old Humanoid: Towards Cognition. *The First IEEE/RAS-EMBS International Conference on Biomedical Robotics and Biomechatronics*, pp.290-295.

Taylor C.L., Schwartz R.J., (1955). The Anatomy and Mechanics of the Human Hand. *Artificial Limbs*, Vol.2 pp.22-35.

Van Varseveld R. B. and Bone G. M.,(1997). Accurate position control of a pneumatic actuator using On/Off solenoid valves IEEE/ASME Transaction on Mechatronics, Vol.2, n.3, pp.195-204.

Wenzeng Z., Chen Q., Sun Z., Khao D., (2004). Passive Adaptive Grasp Multi-Fingered Humanoid Robot Hand with High Under-actuated Function. *Proceedings of the 2004*

IEEE International Conference on Robotics and Automation, ICRA 2004, New Orleans, pp.2222-2228.

Yamano I., Maeno T., (2005). Five Fingered Robot Hand using Ultrasonic Motors and Elastic Elements. *Proceedings of the 2005 IEEE International Conference on Robotics and Automation,* pp.2673-2678.

Yuan X., Stemmler H. and I. Barbi (2001).Self-balancing of the clamping-capacitor-voltages in the multilevel capacitor-clamping-inverter under sub-harmonic PWM modulation *IEEE/ASME Transaction on Power Electronics,* Vol.16, n.2, pp.256-263.

Yun M.H., Eoh H.J., Cho J., (2002). A Two Dimensional Dynamic Finger Modeling for the Analysis of Repetitive Finger Flexion and Extension, *Journal of Industrial Ergonomics,* Vol.29, pp.231-248.

7

Robotic Grasping and Fine Manipulation Using Soft Fingertip

Akhtar Khurshid[1], Abdul Ghafoor[2] and M. Afzaal Malik[1]

[1]College of Electrical and Mechanical Engineering, Rawalpindi, National University of Sciences and Technology, H-12, Islamabad,
[2]School of Mechanical and Manufacturing Engineering, National University of Sciences and Technology, H-12, Islamabad,
Pakistan

1. Introduction

The ability to create stable, encompassing grasps with subsets of fingers is greatly increased by using soft fingertips that deform during contact and apply a larger space of frictional forces and moments than their rigid counterparts. This is true not only for human grasping, but also for robotic hands using fingertips made of soft materials.

The superiority of deformable human fingertips as compared to hard robot gripper fingers for grasping and manipulation has led to a number of investigations with robot hands employing elastomers or materials such as fluids or powders beneath a membrane at the fingertips.

When the fingers are soft, during holding and for manipulation of the object through precise dimensions, their property of softness maintains the area contact between, the fingertips and the manipulating object, which restraints the object and provides stability. In human finger there is a natural softness which is a combination of elasticity and damping. This combination of elasticity and damping is produced by nature due to flesh and blood beneath the skin. This keeps the contact firm and helps in holding the object firmly and stably.

2. Background

Over the past several decades, object manipulation and grasping by robot hands has been widely studied [1-7]. Multifingered-hand research has focused on grasping control [8] and visual and tactile control [9]. Grasp-less manipulation [10], i.e., manipulation without hand grasping and power grasping [11] using the palm of the robot hand have also been proposed. These studies assume that fingertips and manipulated objects are rigid, making point contact, and have analyzed manipulation as quasi-static. Such assumptions are useful for kinematic and static analysis of robot finger grasping and manipulation, but rarely apply in actual dynamic grasping and manipulation. Manipulation and grasping by soft fingertips contribute to grasping stability due to the area of contact and high friction involved.

One approach to investigate in this area is by first analyzing the stability of dynamic control of an object grasped between two soft fingertips through a soft interface using the viscoelastic material between the manipulating fingers and a manipulated object and then modeling it through bond graph method (BGM). The fingers are made viscoelastic by using springs and dampers. Detailed bond graph modeling of the contact phenomenon with two soft-finger contacts considered to be placed against each other on the opposite sides of the grasped object as is generally the case in a manufacturing environment is made. The viscoelastic behavior of the springs and dampers is exploited in order to achieve the stability in the soft-grasping which includes friction between the soft finger contact surfaces and the object. This work also analyses stability of dynamic control through a soft interface between a manipulating finger and a manipulated object. It is shown in this work that the system stability depends on the viscoelastic material properties of the soft interface. Method of root locus is used to analyze this phenomenon.

Ultimate objective of this work is to design and develop a robotic gripper which has soft fingers like human fingers. Soft fingers have ability to provide area contact which helps in dexterous grasping, stability and fine manipulation of the gripping object.

Robotics is gaining new and extensive application fields, becoming pervasive in the daily life. Manipulation skills at macro and micro scale are very important requirements for the emergent robot applications, both in industry (e.g. handling food, fabrics, leather) and in less structured domains (e.g. surgery, space, undersea). The manipulation and grasping devices and systems are a vital part of industrial, service and personal robotics for various applications and environments to advance manufacturing automation, to make safe hazardous operations and to enhance in different ways to the living standards.

The human hand which has the three most important functions: to explore, to restrain objects, and to manipulate objects with arbitrary shapes (relative to the wrist and to the palm) is used in a variety of ways [12]. The first function falls within the realm of haptics, an active research area in its own merits [13]. My work does not attempt an exhaustive coverage of this area. This work in robotic grasping is to understand and to emulate the other two functions. The task of manipulating objects with fingers (in contrast to manipulation with the robot arm) sometimes is called dexterous manipulation. This work will be fascinated with constructing mechanical analogues of human hands and will lead us to place all sorts of hopes and expectations in robot capabilities.

Probably the first occurrence of mechanical hands was in prosthetic devices to replace lost limbs. Almost without exception prosthetic hands have been designed to simply grip objects [14]. In order to investigate the mechanism and fundamentals of restraining and manipulating objects with human hands, later a variety of multifingered robot hands are developed, such as the Stanford/JPL hand [14], the Utah/MIT hand [15], and other hands. Compared to conventional parallel jaw grippers, multifingered robot hands have three potential advantages: (1) they have higher grip stability due to multi-contact points with the grasped object; (2) they can grasp objects with arbitrary shapes; (3) it is possible to impart various movements onto the grasped object. However, multifingered robot hands are still in their infancy. In order for the multifingered robot hands to possess the properties so that robots implement autonomously the tasks of grasping in industry, it is necessary to study the planning methods and fundamentals of robotic grasping as well.

The vast majority of robots in operation today consist of six-jointed "arms" with simple hands or "end effectors" for grasping objects. The applications of robotic manipulations range from pick and place operations, to moving cameras and other inspection equipment, to performing delicate assembly tasks. They are certainly a far cry from the wonderful fancy about the stuff of early science fiction, but are useful in such diverse arenas as welding, painting, transportation of materials, assembly of printed circuit boards, and repair and inspection in hazardous environments [14, 16].

The hand or end effector is the bridge between the manipulator (arm) and the environment. The traditional mechanical hands are simple, out of anthropomorphic intent. They include grippers (either two- or three-jaw), pincers, tongs, as well as some compliance devices. Most of these end effectors are designed on an ad hoc basis to perform specific tasks with specific tools. For example, they may have suction cups for lifting glass which are not suitable for machined parts, or jaws operated by compressed air for holding metallic parts but not suitable for handling fragile plastic parts. Further, a difficulty that is commonly encountered in applications of robotic manipulations is the clumsiness of a robot equipped only with these simple hands, which is embodied in lacking of dexterity because simple grippers enable the robot to hold parts securely but they cannot manipulate the grasped object, limited number of possible grasps resulting in the need to change end effectors frequently for different tasks, and lacking of fine force control which limits assembly tasks to the most rudimentary ones [16].

3. Gripper

Any mechanism which can grasp different objects is called as gripper. It is actually a subsystem of handling mechanism which provides a temporary contact with the object to be grasped. The Gripper ensures that the position and the orientation of the object that is grasped are constrained enough so that the process of carrying, joining etc is done efficiently. This term "gripper" is also used where no actual grasping, rather holding of the object for example in vacuum suction takes place. [17]

4. Classification

Grippers can be classified on the basis of various aspects ranging from type of grasping to number of fingers as discussed below:

4.1 Classification on basis of type of contact
There are three basic types of grippers on the basis of type of contact, shown in figure 1:
- Point Contact
- Line Contact
- Area Contact

4.1.1 Point contact
As the name indicates, point contact gripping takes place when the gripping fingers and the object to be grasped come in contact at some particular points. In this type of gripping there are at least three to four points of contact between the gripping fingers and the object to be grasped.

4.1.2 Line contact
In line contact the contact between the gripper jaw / finger takes place in the form of a line which is dependent on the shape of the object. In Line Contact one has to make sure that the hypothetical lines which are formed during contact are parallel or as close to parallel as possible otherwise proper grasping becomes far too difficult.

4.1.3 Area contact
Instead of points or lines, there is a whole surface area of the fingers that is coming in contact with the object. Generally in area contact, contact of two surface areas from opposite sides is enough to completely constrain the object.

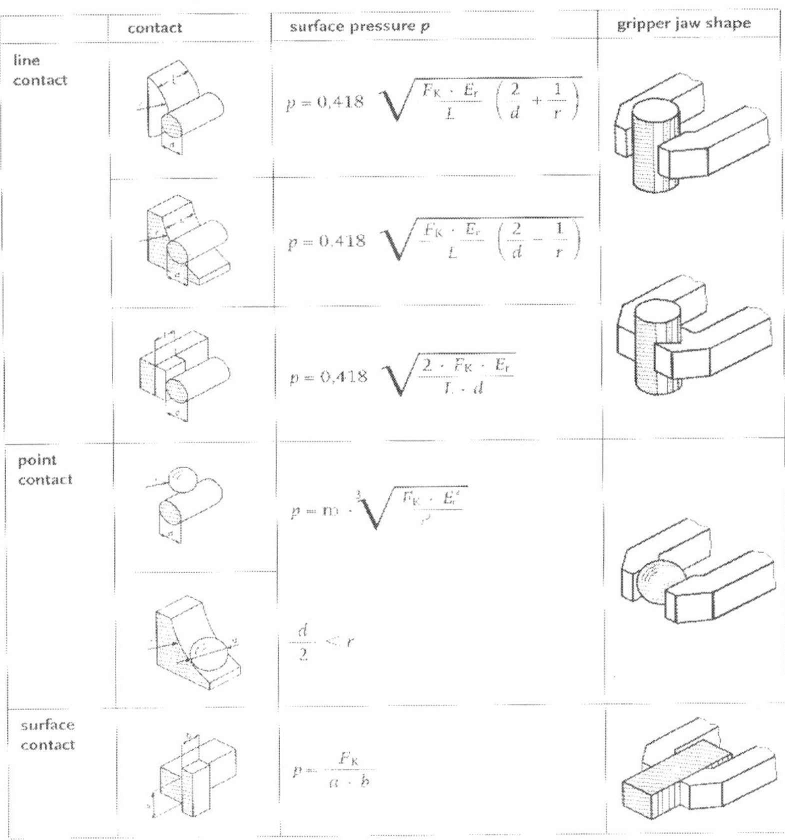

Fig. 1. Types of contacts, their pressure force and the general gripper jaw shape. [18]

Where:

F_k	=	Contact Force
E_r	=	$(2{*}E_t{*}E_s)/(E_t + E_s)$
E_t	=	Young's Modulus of object
E_s	=	Young's Modulus of gripper finger/Jaw.

4.2 Classification on basis of number of fingers
On the basis of number of fingers, grippers can be classified into two, three, four, and more number of fingers:

4.2.1 Two finger grippers

Two finger grippers only have two fingers by which they grasp the object. These types of grippers have generally area contact because they generally due to the shape of the fingers cannot give more than two points of contact and as discussed before, this is not enough to grasp the object firmly and to constrain its degrees of freedom.

4.2.2 Three finger grippers

Due to the fact that the gripper has three fingers, it can have both area and surface contact. But the usage of three fingers in grasping also increases the design complexity and the complexity of the control that must be developed for it.

4.2.3 Four finger grippers

Four Finger Grippers are sometimes a combination of two finger grippers and at other times a combination of independent fingers working together. These grippers are used in relatively high cost and precision demanding applications.

4.2.4 Five finger grippers

These grippers are developed purely on research basis to make the grasping dexterous (closer to human hand gripping approach). There are grippers with more than five fingers as well.

4.3 Classification on the basis of gripping method

There are basically four types of grippers on the basis of their gripping method. Table 1 shows the types of gripping methods along with the non-penetrating and the penetrating examples.

Gripping Method	Non Penetrating	Penetrating
Impactive	Clamping Jaws, Chucks, Collets	Pincers, Pinch Mechanisms
Ingressive	Brush elements, hooks, hook and loop	Needles ,pins , hackles
Contigutive	Chemical adhesion(glues) ,Surface Tension Forces	Thermal adhesion
Astrictive	Electrostatic adhesion	Magnetic grippers, vacuum suction

Table 1. Table of Gripping methods along with the non-penetrating and the penetrating examples. [19]

4.3.1 Impactive grippers

Impactive Grippers make the use of impact of jaws and the object to be grasped through the motion of solid jaws in order to produce the necessary grasping force.

4.3.2 Ingressive grippers

Ingressive grippers deform or penetrate the surface to some predefined depth (force shape mating).

4.3.3 Contigutive grippers

Contigutive type of gripping means that there is a direct contact to facilitate the gripping. For example: chemical adhesion.

4.3.4 Astrictive grippers

Astrictive grippers make use of the binding forces between the surfaces. For example: magnets or electrostatic adhesion. Some of these types of grippers do not even have physical contact with the grasped object.

4.3.5 Importance of gripper

The end effector that is usually used by the manipulator is the gripper. There are a lot of Grippers having different mechanisms and design that exist in the industry. Grippers essentially replace most of the work that is done by the human hand. If the gripping abilities of a mechanical five-finger "hand" are denoted as 100%, then a four-finger hand has 99% of its ability, a three-finger hand about 90%, and a two-finger hand 40%.

4.4 Taxonomy of common end effectors

Nowadays Robotic end effectors include everything from simple two-fingered grippers and vacuum attachments to elaborate multi fingered hands. Perhaps the best way to become familiar with end effector design issues is to first review the main end effector types.

Figure 2 is taxonomy of common end effectors. It is inspired by an analogous taxonomy of grasps that humans adopt when working with different kinds of objects and in tasks requiring different amounts of precision and strength .The left side includes "passive" grippers that can hold parts, but cannot manipulate them or actively control the grasp force. The right-hand side includes active servo grippers and dexterous robot hands found in research laboratories and teleoperated applications.

4.4.1 Passive end effectors

Most end effectors in use today are passive; they emulate the grasps without manipulating it in the fingers. However, a passive end effector may (and generally should) be equipped with sensors, and the information from these sensors may be used in controlling the robot arm. The left-most branch of the "passive" side of the taxonomy includes vacuum, electromagnetic, and Bernoulli-effect end effectors. Vacuum grippers, either singly or in combination, are perhaps the most commonly used gripping device in industry today. They are easily adapted to a wide variety of parts from surface mount microprocessor chips and other small items that require precise placement to large, bulky items such as automobile windshields and aircraft panels. These end effectors are classified as "no prehensile" because they neither enclose parts nor apply grasp forces across them. Consequently, they

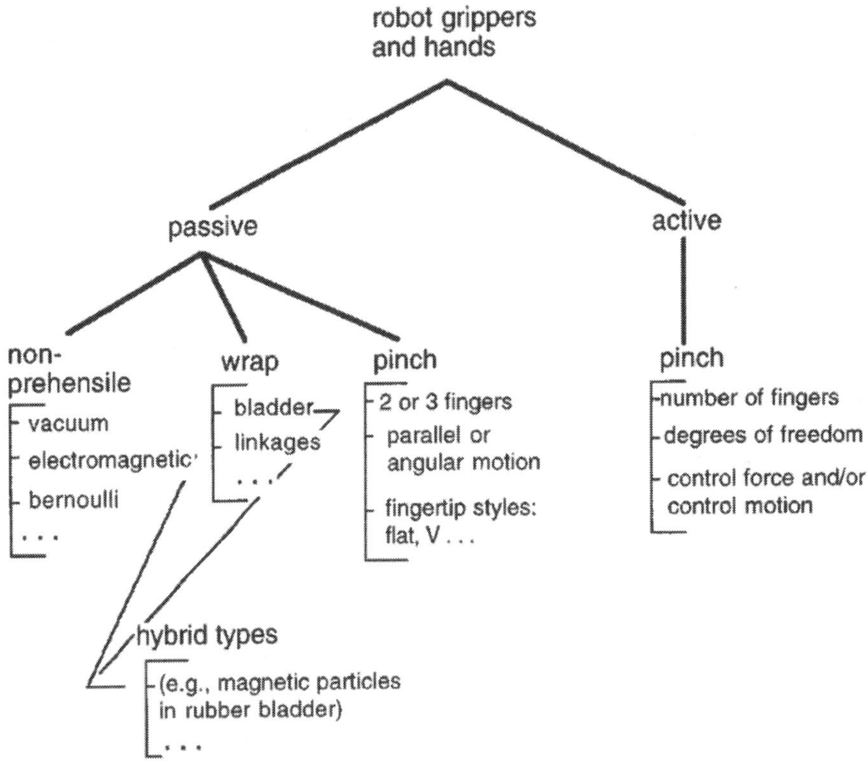

Fig. 2. A taxonomy of the basic end effector types. [20]

are ideal for handling large and delicate items such as glass panels. Unlike grippers with fingers, vacuum grippers do not tend to "center" or relocate parts as they pick them up. If difficulties are encountered with a vacuum gripper, it is helpful to remember that problem can be addressed in several ways, including increasing the suction cup area through larger cups or multiple cups, redesigning the parts to be grasped so that they present a smoother surface (perhaps by affixing smooth tape to a surface), and augmenting suction with grasping. The second branch of end effector taxonomy includes "wrap" grippers that hold a part in the same way that a person might hold a heavy hammer or a grapefruit. In such applications, humans use wrap grasps in which the fingers envelop a part, and maintain a nearly uniform pressure so that friction is used to maximum advantage.

Another approach to handling irregular or soft objects is to augment a vacuum or magnetic gripper with a bladder containing particles or a fluid. When handling ferrous parts, one can employ an electromagnet and iron particles underneath a membrane. Still another approach is to use fingertips filled with an electro rheological fluid that stiffens under the application of an electrostatic field. The middle branch of the end effector taxonomy includes common two-fingered grippers. These grippers employ a strong "pinch" force between two fingers, in the same way that a person might grasp a key when opening a lock. Most such grippers are sold without fingertips since they are the most product-specific part of the design. The fingertips are designed to match the size of components, the shape of components (e.g., flat or V-grooved for cylindrical parts), and the material (e.g., rubber or plastic to avoid damaging fragile objects). Note that since two-fingered end effectors typically use a single air cylinder or motor that operates both fingers in unison, they will tend to center parts that they grasp. This means that when they grasp constrained parts (e.g., pegs that have been set in holes or parts held in fixtures) some compliance must be added. [21]

4.4.2 Active end effectors and hands

The right-hand branch of the taxonomy comprises of servo grippers and dexterous multi fingered hands. Here the distinctions depend largely on the number of fingers and the number of joints or degrees of freedom per finger. For example, the comparatively simple two-fingered servo gripper is confined to "pinch" grasps, like commercial two-fingered grippers, shown in figure 3:. Servo-controlled end effectors provide advantages for fine-motion tasks. In comparison to a robot arm, the fingertips are small and light, which means that they can move quickly and precisely. The total range of motion is also small, which permits fine-resolution position and velocity measurements. When equipped with force sensors such as strain gages, the fingers can provide force sensing and control, typically with better accuracy than can be obtained with robot wrist or joint-mounted sensors. A servo gripper can also be programmed either to control the position of an unconstrained part or to accommodate to the position of a constrained part .The sensors of a servo-controlled end effector also provide useful information for robot programming. For example, position sensors can be used to measure the width of a grasped component, thereby providing a check that the correct component has been grasped. Similarly, force sensors are useful for weighing grasped objects and monitoring task-related forces which can help in checking the weights of the objects.

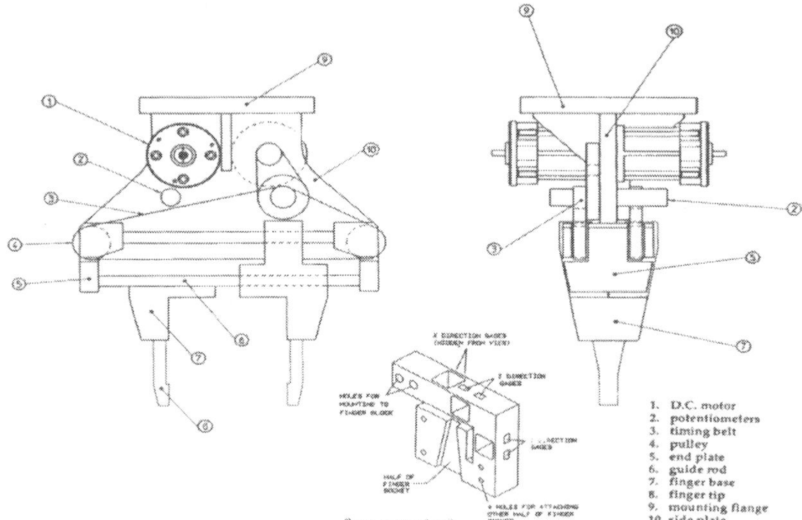

Fig. 3. A two-finger servo gripper with force sensing and changeable fingertips. [3]

For a wide range of applications requiring a combination of dexterity and versatility for grasping a wide range of objects, a dexterous multi fingered hand is the ultimate solution. A number of multi fingered hands have been described in the literature and commercial versions are available. Most of these hands are frankly anthropomorphic, although kinematic criteria such as workspace and grasp isotropy (basically a measure of how accurately motions and forces can be controlled in different directions) have also been used. Despite their practical advantages, dexterous hands have thus far been confined to a few research laboratories. One reason is that the design and control of such hands present numerous difficult tradeoffs among cost, size, power, flexibility and ease of control. For example, the desire to reduce the dimensions of the hand, while providing adequate power, leads to the use of cables that run through the wrist to drive the fingers. These cables bring attendant control problems due to elasticity and friction .A second reason for slow progress in applying dexterous hands to manipulation tasks is the formidable challenge of programming and controlling them. The equations associated with several fingertips sliding and rolling on a grasped object are complex, the problem amounts to coordinating several little robots at the end of a robot. In addition, the mechanics of the hand/object system are sensitive to variations in the contact conditions between the fingertips and object (e.g., variations in the object profile and local coefficient of friction). Moreover, during manipulation the fingers are continually making and breaking contact with the object, starting and stopping sliding, etc., with attendant changes in the dynamic and kinematic equations which must be accounted for in controlling the hand.

When the fingers are soft, during holding and for manipulation of the object through precise dimensions, their property of softness maintains the area contact between, the fingertips and the manipulating object, which restraints the object and provides stability. In human finger there is a natural softness which is a combination of elasticity and damping. This combination of elasticity and damping is produced by nature by putting flush and blood beneath the skin. When it holds the object due to the applied force reaction comes according to 3rd law of Newton and presses the skin and when it rotates the object for manipulating, skin comes back to its original state softly and the other portion compresses against the object. This keeps the contact firm and helps in holding the object firmly and stably. The setup shown in figure 4 is an example of such grasping. It is being modeled in figure 5.

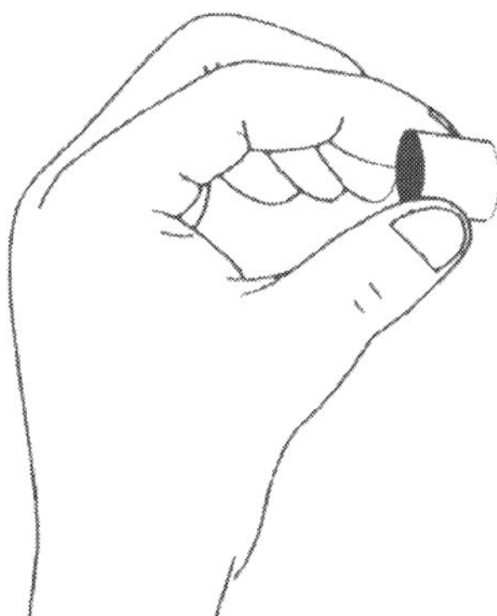

Fig. 4. A set up showing two soft fingers grasping a knob.

To analyze the stability of dynamic control through a *soft interface* that is the viscoelastic material between manipulating fingers and a manipulated object, we have to model dynamic control through the soft interface.

Thus we have to find out:

- How to make the robotic fingers soft?
- The method for measuring the soft fingers grasping force
- Their modeling and simulation
- Analyzing the stability effect due to soft fingers
- Role of springs and dampers in providing stability and accuracy in dexterous manipulation

We shall show that system stability depends on the viscoelasticity of the soft interface for feedback control. The relationship between material viscoelastic property and the settling time shall be analyzed by root locus. Stability analysis is done by using bondgraph methodology to precisely model the situation and then simulating based on Runge-Kutta method.

4.5 Work objective

During grasping the weight of the object being grasped is controlled from slipping downward with the friction between the grasping fingers and the object. This friction further depends upon the applied force. For securing the object from damaging, the contact fingers are made soft by introducing springs and dampers at contacts.

Human finger tips are fleshy, soft and deformable. They locally mold to the shape of a touched or grasped object due to their viscoelastic behavior, and for these reasons, are capable of extremely dexterous manipulation tasks. Viscoelastic materials have an interesting mix of material properties that exhibit viscous behavior (like the gradual deformation of molasses) as well as elasticity (like a rubber band that stretches instantaneously and quickly returns to its original state once a load is removed). The clearest way to visualize the behavior of a material containing both elastic and viscous components is to think of a spring (exerting forces to return to its unstressed state) in series with a dashpot (a damper that resists sudden motion, similar to the pneumatic cylinder that prevents a storm door from slamming shut).

Most robot fingers are crude and therefore rather limited in capability. This realization has led to the investigation of robotic manipulation with soft, human like fingers, for example, Sun and Howe [22], Trembley and Cutkosky[23], Howe and Cutkosky [24], Russel and Parkinsan[25], and Shinofa and Goldenburg[26] report on experiments in which either foam-backed or fluid-filled fingers successfully enhanced dexterous capability. Therefore, in this work, the robot fingers are made soft by introducing springs and dampers at contact. Thus by varying the damping and stiffness, control of the grasped object is achieved.

In the first step we put spring and damper to provide this viscoelastic effect in the grasping finger tips to make the grasping dexterous. Then we made its precise mathematical model by using BGM and put it in virtual environment. Then by using 20-Sim software we have studied by modeling and simulation different experiments in virtual environment the behavior of springs and dampers during grasping. Then we came to this conclusion that by changing spring stiffness and damping coefficient of damper the object can be grasped effectively. They influence the friction between the fingers and the object which plays a vital role in grasping stability. Figure 5 shows this setup.

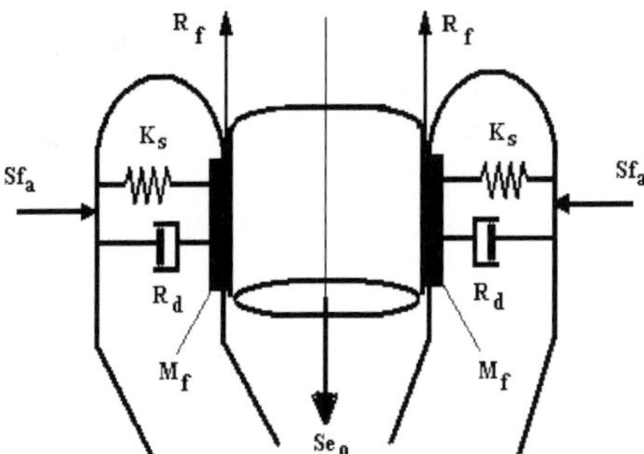

Fig. 5. Model of two soft fingers grasping the object.

Bond-graph modeling and simulation methodology is an attempt to explore the modeling intricacies encountered in the system, using an alternative but powerful modeling technique. The bond graph technique is a graphical representation of the power flow using bonds. It can be very helpful in the preparation of the equations of motion for digital simulation. This method is especially powerful when several physical domains have to be modeled within a system simultaneously. Although the bond graph is of relatively recent origin, it has been widely used in the simulation of system dynamics.

The bond graph is a graphic language for modeling dynamic systems. It displays, by letter elements and half arrows, the energy phenomena modeled in a system and the topology of the energy transfer between these phenomena. The bond graph representation of a system may be constructed in total abstraction from the mathematical model of the system. Even the individual phenomena may be graphically represented without considering the characteristic laws. This constitutes the physical level of the description contained in the bond graph representation.

A.Khurshid and M A Malik [27-29] have used bond graph techniques for modeling and simulation of different dynamic systems. Bond-graphs represent the dynamics of the system pictorially and are extensively used for the modeling of physical system dynamics in multiple energy domains, as discussed by A.Khurshid and M A Malik [30]. A bond-graph model is based on the interaction of power between the elements of the system. The cause-effect relationships help in deriving the system state equations. Further, the model yields insight into various aspects of the control of the system [31, 32].

The effects due to softness of the finger tips while manipulating an object and due to the friction at the finger contacts, and their internal damping and stiffness are modeled and successfully analyzed.

We have modeled dynamic control through a soft interface and formulate system dynamics through a soft interface represented by continuous-discrete time. Taking as an example force control based on the linear mass-damper-spring model.

4.6 Modeling the soft interface

Figure 5 shows a simplified model for a soft interface using linear mass-damper-spring components. The model describes soft robotic fingertips holding an object.

It consists of two fingers which are used to manipulate the objects as done by human fingers. The two fingers are made soft by introducing *linear mass, spring, and damper effects* in these. Force Sf_a is applied to both fingers for the grasping of the object. The weight of the object tries to slip it from the grasping of fingers, whereas the friction between the fingers contacts surfaces with the object balance it. Friction is represented as damping by R_f at the finger's contact surfaces with the object and two dampers are part of the fingers having damping R_d. The stiffness of the springs used in the fingers is K_s. The mass of the outer surface layers of the fingers in contact with the object is M_f. The mass of object is M_o and its weight is taken as Se_o. *We have assumed that the soft material shows linear viscoelasticity. Parameters R_f, K_s, and R_d are time-invariant and positive.*

4.7 Bond-graph model and state space equations

The physical system shown in figure 1 is converted into bond graph shown in figure 6. A bond graph model is a precise mathematical description of the physical system in the sense that it leads to the state space description of the system.

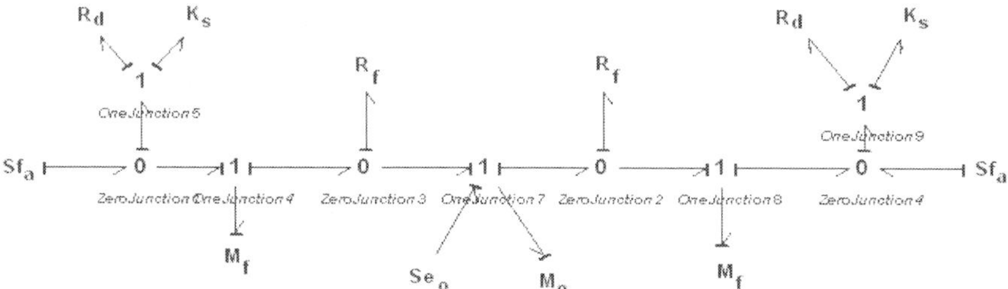

Fig. 6. The bond graph model of two fingers grasping an object as shown in fig-5.

The bond graph model shown in figure 6 leads to the following state space equations from 1-5 of the soft finger system:

$$p\dot{}_f = R_d*(Sf_a - p_f/M_f) + K_s*q_s + R_f*(- p_f/M_f - p_o/M_o) \tag{1}$$

$$p\dot{}_f = R_f*(- p_o/M_o - p_f/M_f) + K_s*q_s + R_d*(- p_f/M_f + Sf_a) \tag{2}$$

$$p\dot{}_o = R_f*(- p_f/M_f - p_o/M_o) + Se_o + R_f*(- p_o/M_o - p_f/M_f) \tag{3}$$

$$q\dot{}s = - p_f/M_f + Sf_a, \tag{4}$$

$$q\dot{}s = Sf_a - p_f/M_f \tag{5}$$

The description of different variables and parameters appearing in the above equations is given below.

$p\dot{}_f$ = force on finger by the object (N), $p\dot{}_o$ = force on object by the fingers (N)

$q\dot{}s$ = velocity of spring in finger (m/s), Sf_a = applied force on finger (m/s)

R_d = damping in soft finger (Ns/m), R_f = friction at soft finger contact (Ns/m)

M_f = mass of finger & M_o = mass of object (kg),

The state variables are q_s = displacement of spring (m), p_f = momentum on finger (kg m/s), p_o = momentum on object (kg m/s). Total five state variables, for both the fingers q_s and p_f are assumed same.

Se_o = weight of object (N), K_s = spring stiffness (N/m)

4.8 Simulation and results

For simulation of the state space equations of the physical system, we have used 20-sim computer software [33]. The results of grasping the object by soft contact fingers and their corresponding root locus, based on BGM modeling and simulation are shown in figure 7-9. The flow signal on each finger is 0.1 m/s.

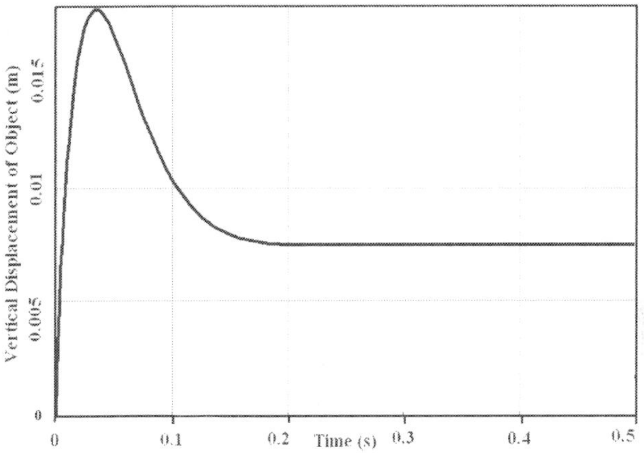

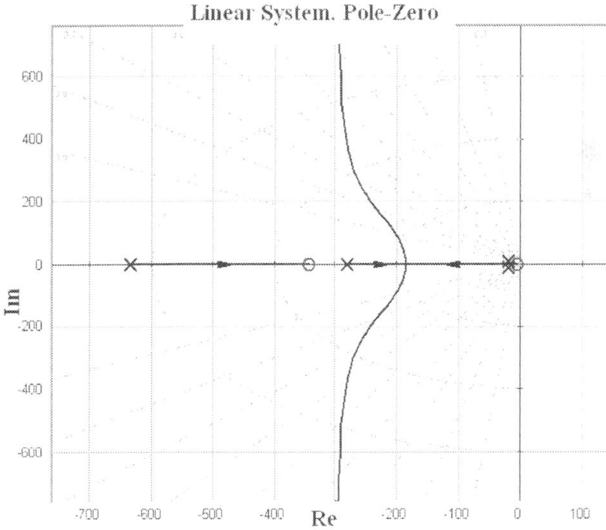

Fig. 7. The vertical displacement of object vs time adjusting stiffness and corresponding rootlocus of the dynamic system.

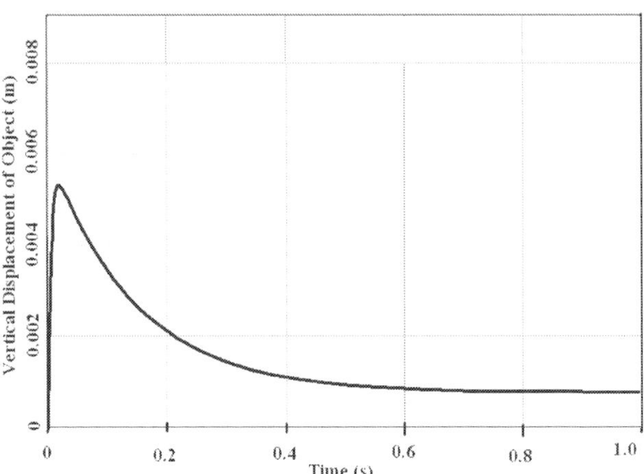

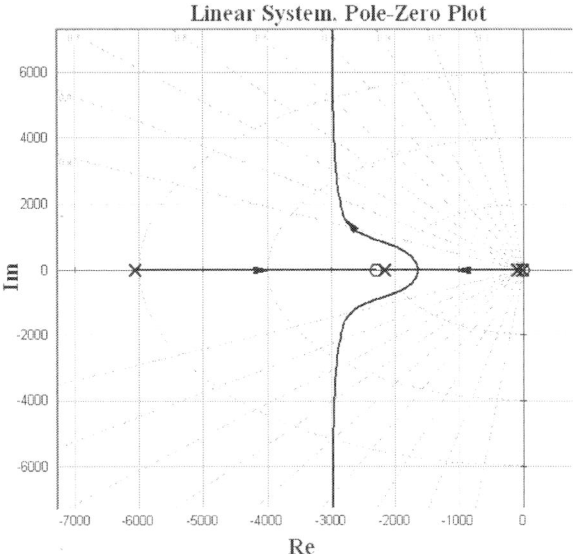

Fig. 8. The vertical displacement of object vs time adjusting finger damping and corresponding rootlocus of the dynamic system.

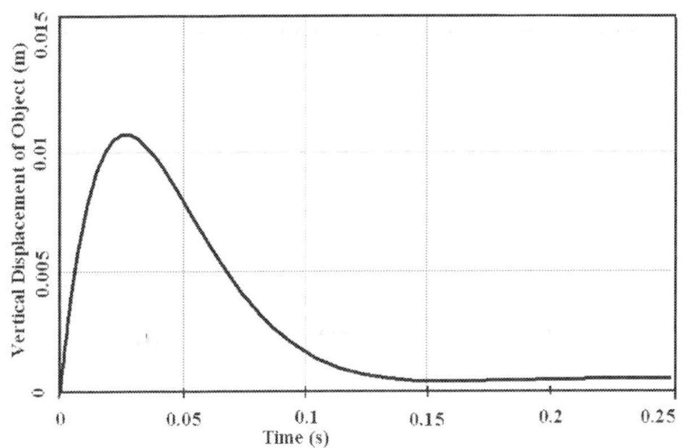

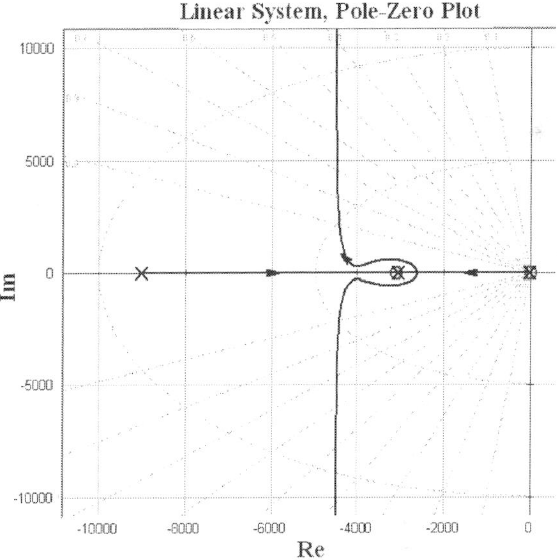

Fig. 9. Final adjusted vertical displacement 5of object vs time and corresponding rootlocus of the dynamic system.

5. Discussion

The objective of this work is to design and develop a robotic gripper which has soft fingers like human fingers. Soft fingers have ability to provide area contact which helps in dexterous grasping, stability and fine manipulation of the gripping object. This work is a step towards this final goal. We have carried out a detailed parametric study of the dynamic system and have observed the effects of changing material properties on the dynamics of the soft contact grasping system. In this work my objective is to optimize the values of spring stiffness and damping in the soft finger for an effective grasping. This has been achieved by making many simulated experiments.

The poles of the system have negative real parts (-0.9017, -0.3050, -16.59+23.3j, -16.59-23.3j) thus the exponential terms will eventually decay to zero. Since, for the springs and the dampers which specify the viscoelastic property of the soft contact fingers, the poles have negative real parts, the system is stable. *Table 2 to 5 show the consolidated results found from the simulated experiments shown in figures 7-9.* The left side curves present the response of the object vertical displacement with respect to time and the right side curves present the root locus for the corresponding system poles. Initially the system was settling down slow as the dominant poles are very close to the imaginary axis. Thus a zero is introduced to cancel the effect of dominant pole as seen by comparing the figures 7 and 9, and the root locus is pulled away from the imaginary axis to settle down the system quickly.

Sr #	$K_{s_1} = K_{s_2}$ [N/m]	$R_{d_1} = R_{d_2}$ [Ns/m]	$R_{f_1} = R_{f_2}$ [Ns/m]	Peak value [mm]	Peak Time [ms]	Steady State Displacement Value [mm]	Settling Time [s]
1	10	10	20	21.7	73.5	7.67	4.4
2	25	10	20	20.9	67.7	7.5	2.2
3	50	10	20	20.2	56.0	7.5	1.06
4	100	10	20	19.2	44.4	7.5	0.5
5	200	10	20	17.5	26.8	7.5	0.19

Table 2. Results of the simulated experiments by varying stiffness of springs and keeping damping and friction constant.

Sr #	$K_{s_1} = K_{s_2}$ [N/m]	$R_{d_1} = R_{d_2}$ [Ns/m]	$R_{f_1} = R_{f_2}$ [Ns/m]	Peak value [mm]	Peak Time [ms]	Steady State Displacement Value [mm]	Settling Time [s]
1	200	15	20	15.3	33.5	7.5	0.276
2	200	20	20	13.7	31.5	7.5	0.412
3	200	25	20	12.7	29.2	7.5	0.519
4	200	30	20	11.9	24.2	7.5	0.604

Table 3. Results of the simulated experiments by varying damping and keeping stiffness of springs and friction constant.

Sr #	$K_{s_1} = K_{s_2}$ [N/m]	$R_{d_1} = R_{d_2}$ [Ns/m]	$R_{f_1} = R_{f_2}$ [Ns/m]	Peak value [mm]	Peak Time [ms]	Steady State Displacement Value [mm]	Settling Time [s]
1	200	30	50	7.5	22.9	3.05	0.679
2	200	30	100	6.06	19.4	1.53	0.76
3	200	30	150	5.56	19.2	1.03	0.759
4	200	30	200	5.3	18.14	0.79	0.699

Table 4. Results of the simulated experiments by varying friction and keeping damping and stiffness of springs constant.

Sr #	$K_{s_1} = K_{s_2}$ [N/m]	$R_{d_1} = R_{d_2}$ [Ns/m]	$R_{f_1} = R_{f_2}$ [Ns/m]	Peak value [mm]	Peak Time [ms]	Steady State Displacement Value [mm]	Settling Time [s]
1	200	10	200	11.36	31.8	0.74	0.224
2	250	10	200	10.9	28.8	0.64	0.176
3	250	10	250	10.8	25.2	0.47	0.155
4	250	10	300	10.4	25.2	0.17	0.154

Table 5. Optimum results of the simulated experiments.

6. Conclusion

A new approach to design an effective soft contact grasping system is presented in this research work portion. The parametric study is made to evolve suitable values of material properties for an effective grasping. The bond graph modeling technique has been applied to obtain the precise mathematical model of the two soft contact robotic fingers. The two fingers are made soft by introducing linear mass, spring, and damper effects in them. The object is controlled by the friction between the fingers from slippage. It would have taken a lot more effort to get these results using traditional methods.

From the simulated results presented in Table 2 to 5, it is concluded that the friction, when increased between the contact surfaces, reduces the displacement of the object. Secondly the damping of the soft fingers when increased controls the peak value of displacement of object and also brings the stable value close to zero. Thirdly the stiffness of the spring effects the settling time of the object. Therefore, the damping of soft finger and the stiffness of the spring in the soft finger and the friction between the soft contact surfaces effects considerably in manipulation of the object. Combination of the stiffness and the damping is the viscoelastic property of the material. The flow signal is produced due to the applied forces on the fingers by some separate mechanism which is not the part of this work but may be designed or procured for experiments.

7. Acknowledgement

The author is indebted to College of E & ME, National University of Sciences and Technology, Rawalpindi, Pakistan for having made this research work possible.

8. References

[1] M. R. Cutkosky, "Robotic Grasping and Fine Manipulation," Kluwer Academic Publishers, 1985.

[2] M. Mason, and K. Salisbury, "Robot Hands and Mechanics of Manipulation,"MIT Press, 1986.

[3] R. Murray, Z. Li, and S. Sastry, "A mathematical introduction torobotic manipulation," CRC Press, 1999.

[4] T. Yoshikawa, and K. Nagai, "Manipulating and Grasping Forces in Multifingered Robot Hands," IEEE Tras. on Robotics and Automation, Vol.7-1, pp. 67-77, 1991.

[5] A. Namiki, and M. Ishikawa, "Optimal grasping using visual and tactile feedback," Proc. of IEEE Int. Conf. on Multisensor Fusion and Intelligent Systems, pp. 584-596, 1996.

[6] Y. Maeda, and T. Arai, "A Quantitative Stability Measure for Graspless Manipulation," Proc. of IEEE Int. Conf. on Robotics and Automation, pp. 2473-2478, 2002.

[7] A. Bicchi, "Force Distribution in Multiple Whole-Limb Manipulation," Proc. of IEEE Int. Conf. on Robotics and Automation, pp. 196-201, 1993.

[8] S. Arimoto, P. T. A. Nguyen, H. Y. Han, and Z. Doulgeri, "Dynamics and control of a set of dual fingers with soft tips," Robotica, Vol.18, No.1, pp. 71-80, 2000.

[9] S. Arimoto, Z. Doulgeri, P. T. A. Nguyen, and J. Fasoulas, "Stable pinching by pair of robot fingers with soft tips under the effect of gravity," Robotica, Vol.20, No.1, pp. 1-11, 2002.

[10] S. D. Eppinger, and W. P. Seering, "Three Dynamics Problems in Robot Force Control," Proc. of IEEE Int. Conf. on Robotics and Automation, pp. 392-397, 1989.

[11] K. B. Shimoga, and A. A. Goldenberg, "Soft Robotic Fingers: Part I. A Comparison of Construction Materials," International Journal of Robotics Research, pp. 320-334, 1996.

[12] K. B. Shimoga, and A. A. Goldenberg, "Soft Robotic Fingers: Part II. Modeling and Impedance Regulation," International Journal of Robotics Research, pp. 335-350, 1996.

[13] E.N.Ohwovoriole *"Kinematics and Friction in Grasping by Robotic Hands"* 398/ Vol. 109, Sep 1987, ASME Transactions.

[14] *Lakshminarayana, K., "Mechanics of Form Closure", 1978, ASME 78-DET-32.*

[15] *Trinkle, J.C, Abel, J.M. and Paul, R. P., 1988, "An Investigation of Enveloping Grasping in the Plane", International Journal of Robotics Research, vol. 3 no. pp. 33-55.*

[16] *Trinkle, J.C., 'On the Stability and Instantaneous Velocity of Grasped Frictionless Objects', IEEE J. Robotics and Automation, vol. 8, no. 5, 1992, pp. 560-572.*

[17] Robot Grippers by Gareth J. Monkman, Stefan hesse, Ralf Strinmann, Henrick Schunk. Edited, designed and published by Wiley-vch, pp 2.

[18] Robot Grippers by Gareth J. Monkman, Stefan hesse, Ralf Strinmann, Henrick Schunk. Edited, designed and published by Wiley-vch, pp 24.

[19] Robot Grippers by Gareth J. Monkman, Stefan hesse, Ralf Strinmann, Henrick Schunk. Edited, designed and published by Wiley-vch, pp 19.

[20] Mechanical engineering handbook By Lewis F.L. CRC Press LLC, 1999; page 14-24

[21] Journal of the Brazilian Society of Mechanical Sciences and Engineering version ISSN 1678-587 J. Braz. Soc. Mech. Sci. & Eng. vol.31 no.4

[22] J.S. Son, E.A. Monteverde, and R.D. Howe, "A Tactile Sensor for Localizing Transient Events in extrapolated from our findings for the two-dimensional problem. We note that for the case of fingertips with two Manipulation," *Proceedings of the 1994 IEEE International. Conference on Robotics and Automation* pp. 471-476, San Diego, May 1994.

[23] M. Tremblay and M.R. Cutkosky, "Estimating friction using incipient slip sensing during a manipulation will cause contact trajectories to deviate from the expected paths. This effect is illustrated in Figure 4, for the case of task," *Proceedings of the 1993 IEEE International Conference on Robotics and Automation* , pp. 429-434, Atlanta, Georgia, May 1993.

[24] R.D. Howe and M.R. Cutkosky, "Sensing skin acceleration for texture and slip perception," rigid or undeformed fingertip and for the case of a deformed fingertip for which rolling velocities are *Proceedings of the 1989 IEEE International Conference on Robotics and Automation*, pp. 145-150, Scottsdale, Arizona, May 1989.

[25] R.A. Russell, S. Parkinson, "Sensing Surface Shape by Touch," that the deformed fingertip follows a trajectory that diverges from the trajectory predicted by rigid body *Proceedings of the 1993 IEEE International Conference on Robotics and Automation* , pp. 423-428, Atlanta, Georgia, May 1993.

[26] K.B. Shimoga and A.A. Goldenberg, "Soft Materials for Robotic Fingers," *Proceedings of the 1992 IEEE International Conference on Robotics and Automation* pp. 1300-1305, Nice, France, May 1992.

[27] A Khurshid and M A Malik, *"Modeling and Simulation of an automotive system by using Bond Graphs"* 10th International Symposium on Advanced Materials ISAM 2007 Islamabad, Pakistan.

[28] A Khurshid and M A Malik, "Bond Graph Modeling and Simulation of Impact Dynamics of a Car Crash" 5th International Bhurban Conference On Applied Sciences And Technology 5th IBCAST-2007, Islamabad, Pakistan.

[29] A Khurshid and M A Malik, "Modeling and Simulation of a Quarter Car Suspension System using Bond Graphs" 9th International Symposium on Advanced Materials ISAM 2005, Islamabad, Pakistan.

[30] A Khurshid and M A Malik, "Bond Graph Modeling and Simulation of Mechatronic Systems" International Multi-topic Conference 2003, INMIC 2003, In association with IEEE, Islamabad, Pakistan.

[31] A. Mukherjee, R. Karmakar, Modeling and simulation of engineering systems through bond graphs, Narosa Publishing House, New Delhi, 2000.

[32] D. C. Karnopp, D. L. Margolis, and R. C. Rosenberg, System Dynamics: Modeling and simulation of mechatronic systems, third edition, Wiley-Interscience, 2000.

[33] 20-sim Control Laboratory, University of Twente Controllab Products B.V. Drienerlolaan 5 EL-CE, 7522 NB Enschede the Netherlands. 2003.

8

Recognition of Finger Motions for Myoelectric Prosthetic Hand via Surface EMG

Chiharu Ishii
Hosei University
Japan

1. Introduction

Recently, myoelectric prosthetic arms/hands, in which arm/hand gesture is distinguished by the identification of the surface electromyogram (SEMG) and the artificial arms/hands are controlled based on the result of the identification, have been studied (Weir, 2003). The SEMG has attracted an attention of researchers as an interface signal of an electric actuated arm for many years, and many of studies on the identification of the SEMG signal have been executed. Nowadays, it can be said that the SEMG is the most powerful source of control signal to develop the myoelectric prosthetic arms/hands.

From the 1970s to the 1980s, elementary pattern recognition technique such as linear discriminant analysis, was used for the identification of the SEMG signals in (Graupe et al., 1978) and (Lee et al., 1984). In the 1990s, research on learning of a nonlinear map between the SEMG pattern and arm/hand gesture using a neural network has been performed in (Hudgins et al., 1993). Four kinds of motions of the forearm were distinguished by combining Hopfield-type neural network and back propagation neural network in (Kelly et al., 1990).

The amplitude and the frequency band are typical information extracted from the SEMG signal, which can be used for the identification of arm/hand gesture. (Ito et al., 1992) presumed muscle tension from the EMG signal, and tried to control the forearm type myoelectric prosthetic arm driven by ultrasonic motor. (Farry et al., 1996) has proposed a technique of teleoperating the robot hand through the identification of frequency spectrum pattern of the SEMG signal.

At present, however, most of the myoelectric prosthetic arms/hands can only realize some limited motions such as palmar seizure, flexion-extension of a wrist, and inward-outward rotation of a wrist. To the best of our knowledge, myoelectric prosthetic hands which can distinguish motions of plural fingers and can independently actuate each finger have not been developed yet, since recognition of independent motions of plural fingers through the SEMG is fairly difficult.

Probably, a present cutting edge practical myoelectric prosthetic hand is the "i-LIMB Hand" produced by Touch Bionics Inc.. However, myoelectric prosthetic hands which imitate the hand of human, such as the "i-LIMB Hand", are quite expensive, since they require accurate measurement of SEMG signal and use many actuators to drive finger joints. Therefore, improvement of operativity of the myoelectric prosthetic arms/hands and simplification of structure of the artificial arms/hands to lower the price are in demand.

The purpose of this study is to develop a myoelectric prosthetic hand which can independently actuate each finger and can realize fundamental motions, such as holding and grasping, required in daily life. In order to make it budget price, an underactuated robotic hand structure which realizes flexion and extension of fingers by tendon mechanism, is introduced. In addition, the "fit grasp mechanism" in which the fingers can fit the shape of the object when the fingers grasp the object, is proposed. The "fit grasp mechanism" makes it possible for the robotic hand to grasp a small object, a cylindrical object, a distorted object, etc.. In this study, a robotic hand with the thumb and the index finger was designed and built as a prototype.

As for the identification of independent motion of each finger, using the neural network, an identifier which distinguishes four finger motions, namely flexion and extension of the thumb and the index finger in respective metacarpophalangeal (MP) joint, is constructed. Four patterns of neural network based identifiers are proposed and the recognition rates of each identifier are compared through simulations and experiments. The online control experiment of the built robot hand was conducted using the identifier which showed the best recognition rate.

2. Robot hand

In this section, details of the robot hand for myoelectric prosthetic hand are explained. Overview of the built underactuated robot hand with two fingers, namely the thumb and the index finger, is shown in Fig.1.

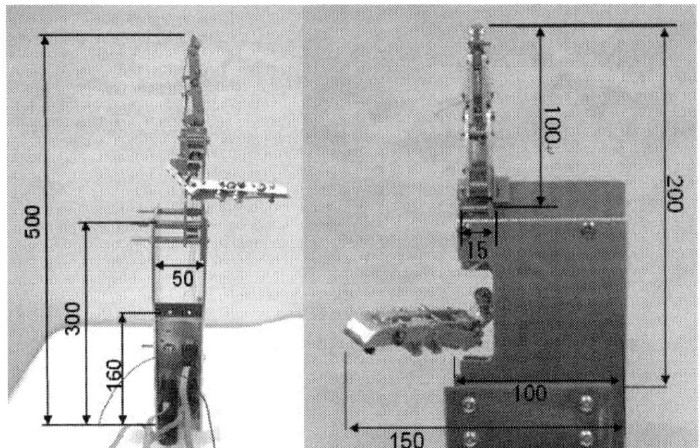

Fig. 1. Overview of robot hand.

2.1 Specifications
The primary specifications of the robot hand are shown as follows.
1. Entire hand: 500mm total length, and 50mm thickness
2. Palm: 100mm length, 110mm width, and 20mm thickness
3. Finger: 100mm length, 15mm width, and 10mm thickness
4. Pinching force when MP joint is driven: 3N

2.2 Mechanism of finger

As shown in Fig.2, imitating the human's frame structure, the robot hand has finger mechanism which consists of three joints, namely distal interphalangeal joint (DIP: the first joint), proximal interphalangeal joint (PIP: the second joint), and metacarpophalangeal joint (MP: the third joint). The fingers are driven by the wire actuation system like human's tendon mechanism. When the wire connected with each joint is pulled by driving force of the actuator, the finger bends. While, when the tension of the wire is loosed, the finger extends due to the elastic force of the rubber. This makes it possible to omit actuators used to extend the finger. The built robot hand can realize fundamental operation required in daily life, such as holding and grasping.

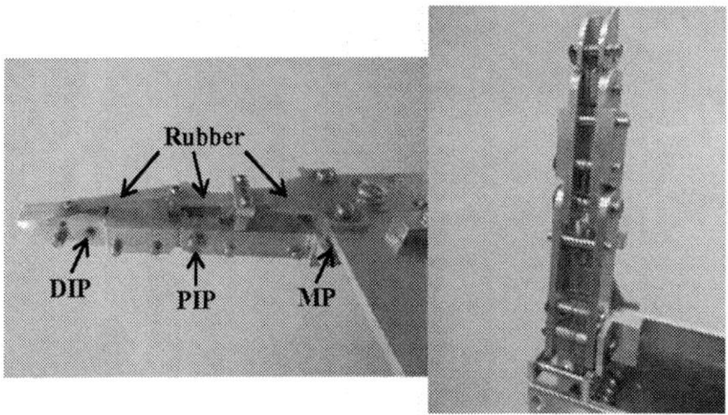

Fig. 2. Mechanism of finger.

2.3 Fit grasp mechanism

In general, when human holds the object, the fingers flexibly fit the shape of the object so that the object can be wrapped in. We call this motion "fit grasp motion". As shown in Fig.3, the finger of the robot hand has two kinds of wires which perform interlocked motion in DIP and PIP joints and motion in MP joint respectively. Therefore, the interlocked bending in DIP and PIP joints and the bending in MP joint can be performed independently.

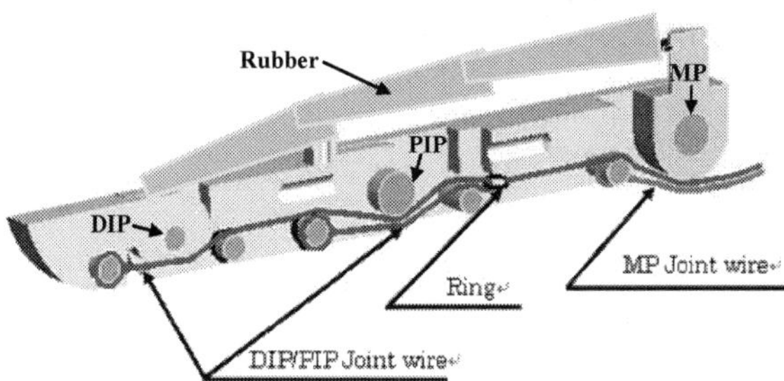

Fig. 3. Arrangement of wires.

In addition, as shown in Fig.3, the ring is attached to the wire between DIP joint and PIP joint, and the interlocked motion of DIP and PIP joints is achieved by pulling the ring by other wire connected to the ring. This mechanism allows to realize "fit grasp motion". We call this mechanism "fit grasp mechanism." Details of the "fit grasp motion" are illustrated in Fig.4.

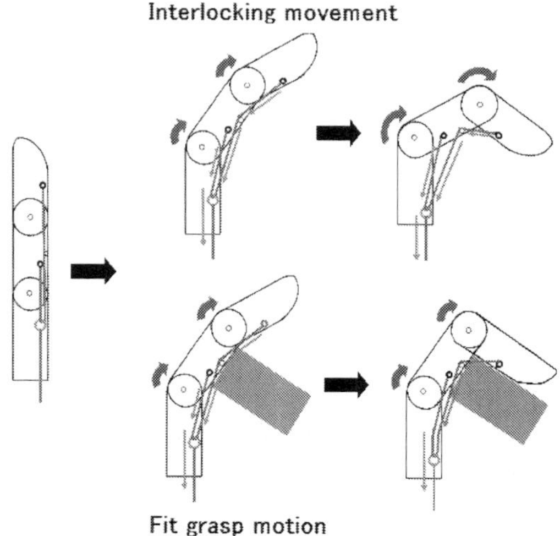

Fig. 4. Bending motion by fit grasp mechanism.

In the case where there is no object to hold, when the wire is pulled by the actuator, DIP and PIP joints bend at the almost same angle (Fig.4 upper). On the other hand, in the case where there is object to hold, when the object contacts the finger, only one side of the wire is pulled since the wire between DIP joint and PIP joint can slide inside of the ring. As a result, DIP joint can bend in accordance with the shape of the object (Fig.4 lower). Thus, "fit grasp motion" is achieved. The "fit grasp mechanism" makes it possible for the robotic hand to grasp a small object, a cylindrical object, a distorted object, etc..

3. Measurement and signal processing of SEMG

In this section, measurement and signal processing of the SEMG are described.

3.1 Measurement positions of SEMG

The built robot hand for myoelectric prosthetic hand has thumb and index finger to operate, and the thumb and the index finger are operated independently. Various motions of each finger can be considered, however in this study, flexion and extension of the thumb and the index finger in MP joint are focused on. Namely, flexion and extension in interlocked DIP and PIP joints are not considered here. Inward rotation and outward rotation of each finger are also not taken into consideration.

The measurement positions of SEMG are shown in Fig.5. Those are the following three positions; the vicinity of a musculus flexor carpi radialis / a musculus flexor digitorum superficialis (ch1), the vicinity of a musculus flexor digitorum profundus (ch2), and the vicinity of a musculus extensor digitorum (ch3). The former two musculuses are used for flexion of each finger and the latter musculus is used for extension of each finger.

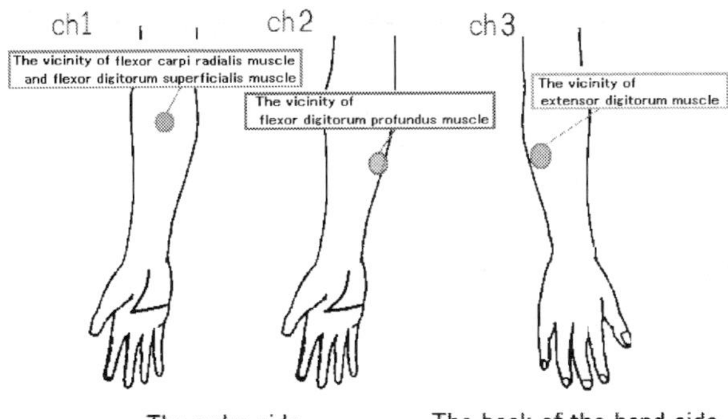

Fig. 5. Measurement positions of SEMG

3.2 Signal processing

One finger motion is performed in approximately 0.5 second, and the SEMG signal is measured by 1kHz of sampling frequencies. Fast Fourier Transform (FFT) is performed to the measured SEMG signals, and spectral analysis is conducted. The number of samples for FFT was set as 256. When performing FFT, the humming window function was utilized to the processing signals.

However, influence of the alternate current (AC) power source, which is regarded as an external noise, appears in the amplitude value of the SEMG by which FFT processing was carried out. This AC power source noise appears at odd times frequencies of the fundamental frequency. Since the area where this experiment was conducted is East Japan, as shown in Fig.6, the influence of the AC power source noise appears at 50Hz and 150Hz.

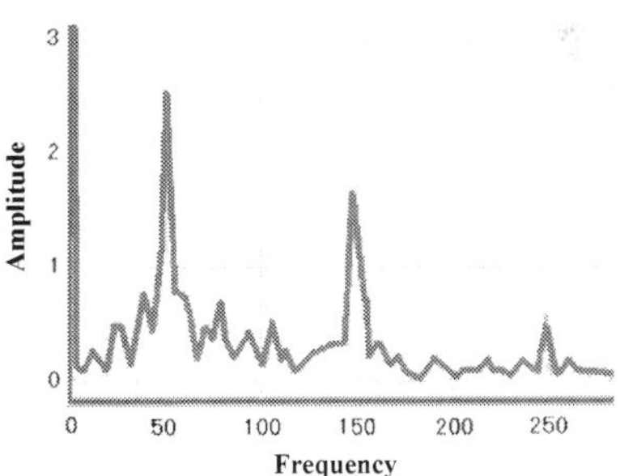

Fig. 6. Spectrum of SEMG signal

Since it is considered that least influence of the AC power source noise is at 100Hz, the amplitude value at 100Hz is used for recognition of the finger motions.

Three-dimensional graph of the amplitude values at 100Hz of each motion in MP joint is shown in Fig.7, in which each measurement position, namely each electrode, is taken as an axis of the coordinates. In addition, the distribution in Fig.7 was divided into the distribution along the thumb and the index finger respectively, which are shown in Fig.8 and Fig.9.

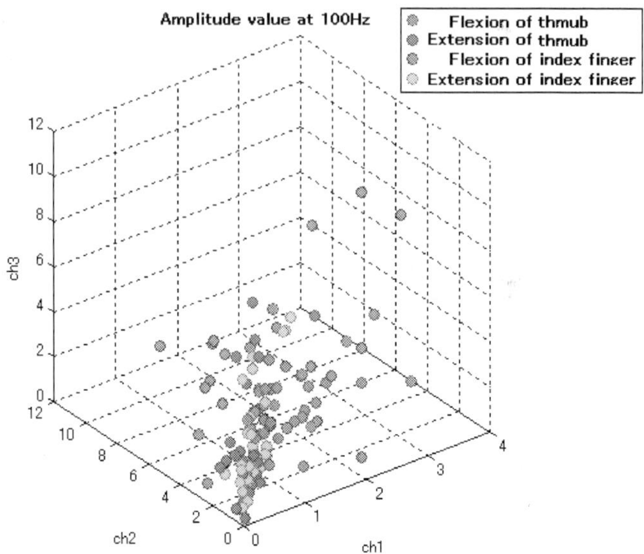

Fig. 7. Distribution of amplitude values at 100 Hz.

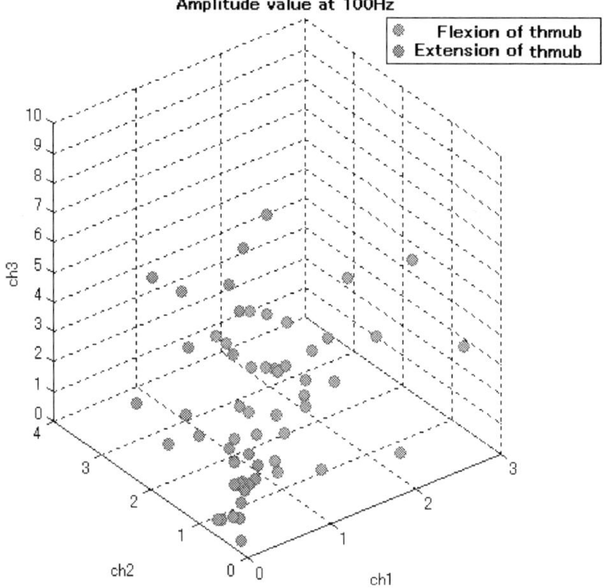

Fig. 8. Distribution of amplitude values at 100 Hz (thumb).

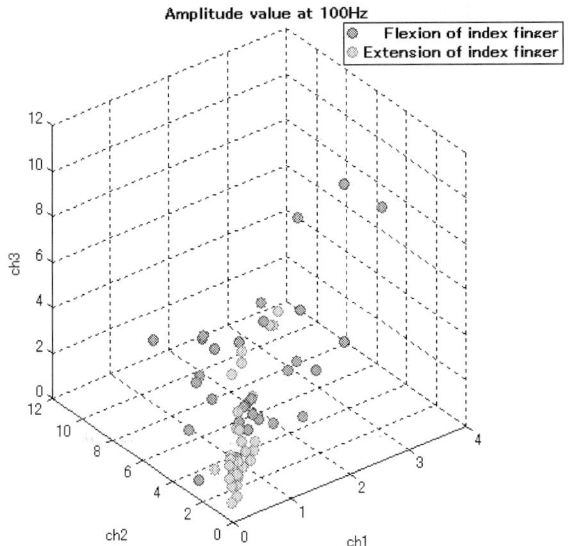

Fig. 9. Distribution of amplitude values at 100 Hz (index finger).

The amplitude values of flexion of the index finger are distributed over the whole space. Hence, it is anticipated that it is hard to distinguish the flexion of the index finger from extension of the index finger. However, in the case of the thumb, the distribution seems to be distinguishable to flexion or extension. Therefore, these values are used for recognition of the finger motions.

4. Recognition of finger motions

In this section, the identification methods of finger motions using neural network(s) are explained.

4.1 Recognition by one neural network

The recognition of the finger motions is performed via SEMG signals using neural network(s). First of all, an identifier which distinguishes four finger motions, namely 1) the flexion of the thumb in MP joint, 2) the extension of the thumb in MP joint, 3) the flexion of the index finger in MP joint, and 4) the extension of the index finger in MP joint, by only one neural network is constructed.

The input signals to the neural network are set of the amplitude values at 100Hz obtained through the signal processing explained in Section 3.2 for the SEMG signal measured in each electrode. The numerical values 1 to 4, 1 for flexion of the thumb, 2 for extension of the thumb, 3 for flexion of the index finger and 4 for extension of the index finger, are assigned as the teacher signals for each motion.

As for the structure of the neural network, the feedforward neural network was adopted. The number of the input layer and of the output layer is one, respectively, and the number of the hidden layer, each consisting of three neurons, is two. 20 set of pre-measured input signals for each finger motion were used for learning of the neural network. The error back propagation algorithm was used as the learning method of the neural network. The learning of the feedforward neural network was executed using the Neural Network Toolbox in MATLAB software. As a condition of the learning, the end of the learning was set with the repetition of 10000 times calculation. Hereafter, this identification method, namely recognition of the four finger motions by one neural network, is described as identifier (x).

After the learning of the neural network, simulation was carried out using the 30 set of pre-measured input signals for each finger motion, which differs from the input signals used for the learning, and its recognition rate was examined. The results are shown in Table 1.

Motion	Flexion of thumb	Extension of thumb	Flexion of index finger	Extension of index finger	Average
Recognition rate [%]	43.3	0.0	36.6	3.3	20.8

Table 1. Simulation results with identifier (x).

From Table 1, the successful recognition rate was only 20.8% on average for use of the identifier (x).

4.2 Improvement of identification method

The improvement of the recognition rate can be expected by modifying the identification method. By reducing the number of choice which one neural network should distinguish, and combining two or more neural networks in series or in parallel, much higher successful recognition rate will be obtained for each finger motion. In each neural network, the choice is given as alternative, namely the numerical values 0 and 1 are given as the teacher signals. The following three patterns of identifier shown in Fig.10 are considered.

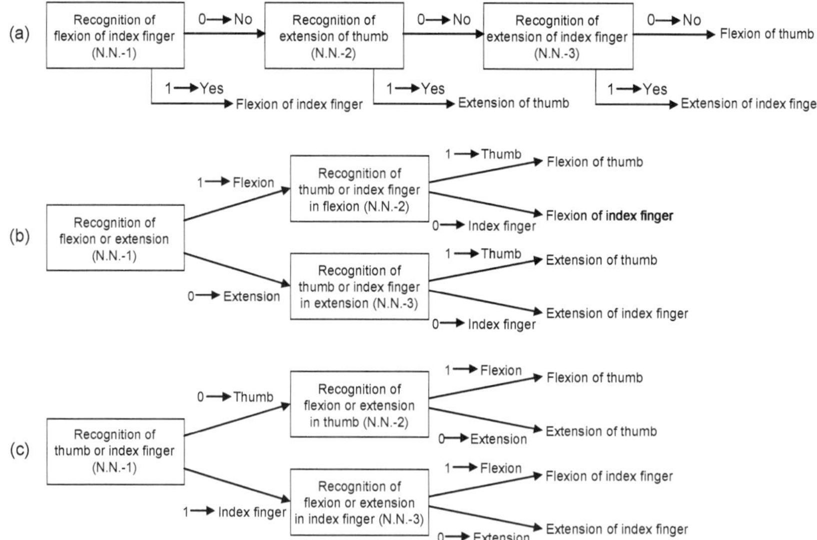

Fig. 10. Improved identification methods.

In identifier (a), the finger motion is distinguished by recognizing each one motion by one neural network in order with high successful recognition rate for each finger motion. The order of recognition by neural network was determined through simulation results, as flexion of the index finger, extension of the thumb and extension of the index finger. First, N.N.-1 identifies whether the finger motion is flexion of the index finger. N.N.-1 is trained to output 1 for flexion of the index finger and 0 for other finger motions by the learning. Therefore, N.N.-1 outputs 1 in the case where the finger motion is identified as flexion of the index finger. If the output of N.N.-1 is 0, likewise N.N.-2 identifies whether the finger motion is extension of the thumb. Again, if the output of N.N.-2 is 0, N.N.-3 identifies whether the finger motion is extension of the index finger. Finally, in the case where the finger motion was not identified as any of these three motions, it is finally recognized as flexion of the thumb. This identification method has drawback that incorrectly-identified finger motions are inevitably distinguished as flexion of the thumb.

In identifier (b), N.N.-1 is trained to output 1 for flexion of the thumb and the index finger and 0 for extension of the thumb and the index finger, and N.N.-2 and N.N.-3 are trained to output 1 for motion of the thumb and 0 for motion of the index finger. Firstly the flexion or the extension is distinguished, then motion of the thumb or motion of the index finger is distinguished. Thus, finally motion of the finger is distinguished to one of the finger motions.

In identifier (c), N.N.-1 is trained to output 1 for motion of the thumb and 0 for motion of the index finger, and N.N.-2 and N.N.-3 are trained to output 1 for the flexion and 0 for the extension. Firstly, motion of the thumb or motion of the index finger is distinguished, then the flexion or the extension is distinguished. Thus, finally motion of the finger is distinguished to one of the finger motions.

In identifier (b) and identifier (c), the first distinction is important, since if the first distinction is incorrect, subsequent distinction becomes meaningless. Therefore, high successful recognition rate of the first distinction is required.

The structure of each neural network and the learning method are same as the identifier (x). Namely, in each identifier, the error back propagation algorithm was used as the learning method for respective feedforward neural network.

4.3 Simulation with improved identification method

Simulation works for distinction of the finger motions were carried out for each identifier, and each recognition rate was examined. In each identifier, simulation was carried out using the 30 set of pre-measured input signals for each finger motion, which differs from the input signals used for the learning. The result using the identifier (a) is shown in Table 2.

Motion	Flexion of thumb	Extension of thumb	Flexion of index finger	Extension of index finger	Average
Recognition rate [%]	66.7	30.0	90.0	43.3	57.5

Table 2. Simulation results with identifier (a).

Table 2 shows that the recognition rate was improved compared with the identifier (x). However, since most of the incorrectly-identified motions are distinguished as flexion of the thumb, improvement is required.

Before simulating entire recognition rate using identifier (b), recognition rate of each neural network in identifier (b) was examined. In the simulation, the 30 set of input signals for each finger motion were used in N.N.-1. Each 30 set of input signals for flexion of the thumb and flexion of the index finger were used in N.N.-2, and each 30 set of input signals for extension of the thumb and extension of the index finger were used in N.N.-3. The results are shown in Table 3.

N.N.-1	Flexion	Extension	Average
Recognition rate [%]	83.3	76.7	80.0
N.N.-2	Flexion of thumb	Flexion of index finger	Average
Recognition rate [%]	93.3	86.7	90.0
N.N.-3	Extension of thumb	Extension of index finger	Average
Recognition rate [%]	53.3	73.3	63.3

Table 3. Partial recognition rate for each neural network in identifier (b).

From Table 3, the recognition rate of 80% on average was obtained at the first distinction. The result of entire recognition rate using identifier (b) is shown in Table 4.

Motion	Flexion of thumb	Extension of thumb	Flexion of index finger	Extension of index finger	Average
Recognition rate [%]	66.7	33.3	83.3	63.3	61.65

Table 4. Simulation results with identifier (b).

Table 4 shows that the entire recognition rate was improved as well as identifier (a).
Likewise, before simulating entire recognition rate using identifier (c), recognition rate of each neural network in identifier (c) was examined. In the simulation, the 30 set of input signals for each finger motion were used in N.N.-1. Each 30 set of input signals for flexion of the thumb and extension of the thumb were used in N.N.-2, and each 30 set of input signals for flexion of the index finger and extension of the index finger were used in N.N.-3. The results are shown in Table 5.

N.N.-1	Thumb	Index finger	Average
Recognition rate [%]	80.0	55.0	67.5
N.N.-2	Flexion of thumb	Extension of thumb	Average
Recognition rate [%]	80.0	100	90.0
N.N.-3	Flexion of index finger	Extension of index finger	Average
Recognition rate [%]	83.3	90.0	86.7

Table 5. Partial recognition rate for each neural network in identifier (c).

From Table 5, the recognition rate of the first distinction was only 67.5% on average, which is inferior to the case of the identifier (b). The result of entire recognition rate using identifier (c) is shown in Table 6.

Motion	Flexion of thumb	Extension of thumb	Flexion of index finger	Extension of index finger	Average
Recognition rate [%]	70.0	70.0	76.7	13.3	57.5

Table 6. Simulation results with identifier (c).

From Table 6, the entire recognition rate of 57.5% on average was obtained, which is almost same level as the identifier (a).

From the above results, it turned out that the recognition rate was improved in all identifiers (a), (b) and (c), compared with identifier (x). In addition, from the recognition rate of N.N.-1 in Table 3 and of N.N.-2 and N.N.-3 in Table 5, it can be said that distinction between the flexion and the extension is comparatively easy as compared with distinction between the thumb and the index finger.

5. Experiments

In this section, experimental results for online recognition and for online finger operation of the robot hand described in Section 2 are shown.

5.1 Experiment for online recognition

Experiment for online recognition of the finger motions was carried out using the identifier (b) which showed the most successful recognition rate on average, and recognition rate was examined. Each finger motion was performed in 1 second at intervals of about 10 seconds. The result of the recognition rate for 60 times of movements for each finger motion is shown in Table 7.

Motion	Flexion of thumb	Extension of thumb	Flexion of index finger	Extension of index finger	Average
Recognition Rate [%]	68.3	33.3	81.8	60.0	60.8

Table 7. Experimental results for online recognition with identifier (b).

Compared with the simulation results shown in Table 4, quite similar results were obtained.

5.2 Experiment for online control of robot hand

Experiment for online finger operation of the robot hand was executed. In the experiment, the SEMG of each electrode is measured online. Then, the start time of finger motion is detected as follows. Since the SEMG of ch1 has least noise and good response among the SEMG of ch1, ch2 and ch3, the SEMG of ch1 is rectified, and when the magnitude of the rectified SEMG exceeds a specified threshold, it is regarded as finger motion having begun. Synchronizing with the start of the finger motion, online recognition of the finger motion using the identifier (b) is carried out. In the experiment, the finger motion is performed in 1 second at intervals of about 5 seconds, respectively, and is performed in order with flexion of the thumb, extension of the thumb, flexion of the index finger, and extension of the index finger. The fingers of the robot hand are controlled based on the recognition result of the identifier.

The measured SEMG in each electrode when a series of finger motion was performed is shown in Fig.11.

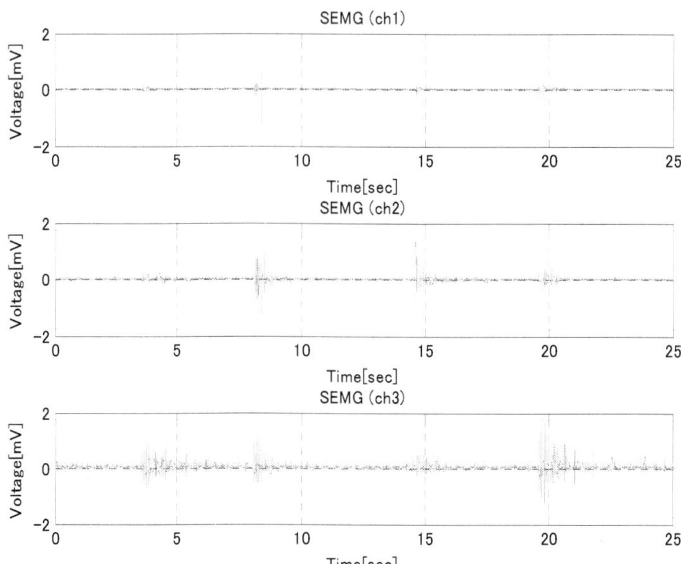

Fig. 11. SEMG in each electrode.

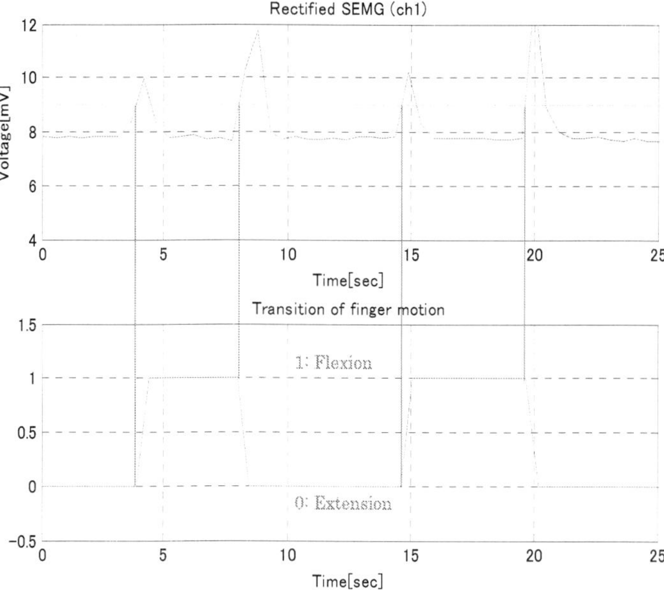

Fig. 12. Judgment of start of finger motion

The SEMG of ch1 in Fig.11 was rectified, and the time when the magnitude of the rectified SEMG exceeded the threshold determined as 9mV, was regarded as the start time of the finger motion. The rectified SEMG of ch1 is shown at the top of Fig.12, and the transition of the finger motion based on the judgment of the start of the finger motion is shown at the bottom of Fig.12, in which "output 1" shows the flexion and "output 0" shows the extension. Synchronizing with the start time of the finger motion shown in Fig.12, the input signal to the neural networks in the identifier is updated. The recognition result of each neural network in the identifier is shown in Fig.13.

N.N.-1 distinguishes the flexion or the extension. The output larger than 0.5, which is regarded as 1, is judged to be the flexion, and the output less than 0.5, which is regarded as 0, is judged to be the extension. N.N.-2 and N.N.-3 distinguish motion of the thumb or motion of the index finger. The output larger than 0.5, which is regarded as 1, is judged to be motion of the thumb, and the output less than 0.5, which is regarded as 0, is judged to be motion of the index finger.

In Fig.13, the blue dashed line shows the output from the neural network, and the red solid line shows the recognition result, in which "Output 1" shows the flexion and "Output 0" shows the extension for N.N.-1, and "Output 1" shows the motion of the thumb and "Output 0" shows the motion of the index finger for N.N.-2 and N.N.-3.

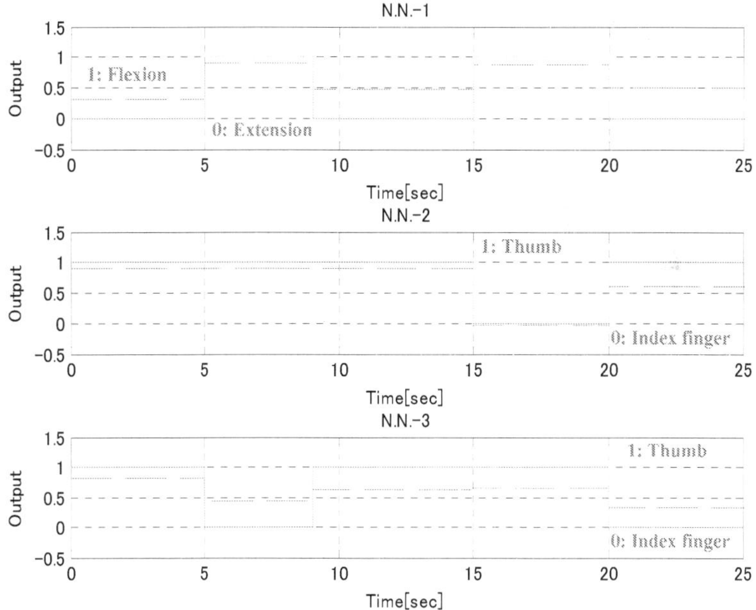

Fig. 13. Recognition result for each neural network.

When Fig.12 and Fig.13 are compared, a slight time delay is seen until the recognition result is obtained by the neural network after the start of the finger motion was judged, since it takes a slight time to calculate the input signal to the neural network due to the FFT processing and so on.

The entire recognition result of finger motion obtained from combination of the recognition results of the three neural networks in Fig.13 is shown in Fig.14, in which "Output 4" shows the flexion of the thumb, "Output 3" shows the extension of the thumb, "Output 2" shows the flexion of the index finger, and "Output 1" shows the extension of the index finger. Initial motion for 0 second to 5 seconds which has not operated any motion is judged as extension of the thumb due to the learning of the neural network.

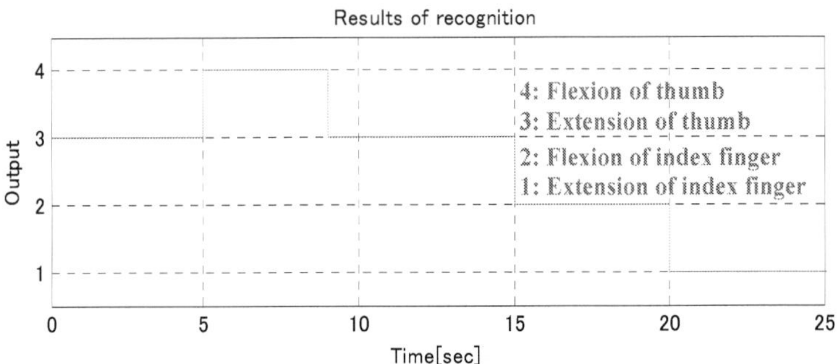

Fig. 14. Recognition result of finger motion.

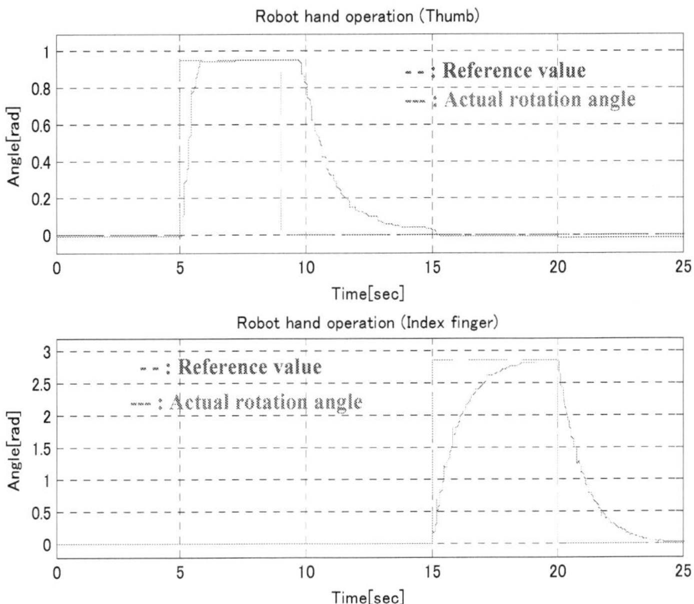

Fig. 15. Finger operation of robot hand.

From Fig.14, the recognition of the finger motion was performed correctly.

The operation of the robot hand is determined as follows. In both thumb and index finger, extension is in the state where the finger is lengthened, and flexion is in the state where the finger is bent at 90 degrees. In control of the thumb, the reference value for the flexion was set as 0.95 rad, which is the rotation angle of the pulley required to make the thumb bent at 90 degrees. On the other hand, the reference value for the extension was set as 0 rad in order to return the thumb to the original state. Likewise, in control of the index finger, the reference value for the flexion was set as 2.86 rad, which is the rotation angle of the pulley required to make the index finger bent at 90 degrees. On the other hand, the reference value for the extension was set as 0 rad in order to return the index finger to the original state. PI controllers were adopted in the construction of the servo systems, and the controller gains of the PI controllers were adjusted by trial and error through repetition of the experiments. The result of the operation of the robot hand is shown in Fig.15, in which the blue dashed line shows the reference value of the rotation angle of the pulley, and the red solid line shows the rotation angle of the actual pulley driven by motor.

The result shown in Fig.15 is the case where the recognition results of the finger motion for both the thumb and the index finger were successful, and both the thumb and the index finger of the robot hand were able to be operated as intended.

6. Conclusion

In this chapter, a robot hand for myoelectric prosthetic hand, which has thumb and index finger and can realize "fit grasp motion", was proposed and built. In order to control each finger of the developed myoelectric prosthetic hand independently, recognition of four finger motions, namely flexion and extension of the thumb and of the index finger in MP joint, was performed based on the SEMG using the neural network(s). First, recognition of these four finger motions by one neural network was executed. Then, the successful recognition rate was only 20.8% on average. In order to improve the recognition rate, three types of the improved identification methods were proposed. Simulation results for the recognition rate using each improved identification method showed successful recognition rate of more than 57.5% on average. Experiment for online recognition of the finger motions was carried out using the identification method which showed the most successful recognition rate, and similar recognition result as in the simulation was obtained. Experiment for online finger operation of the robot hand was also executed. In the experiment, the fingers of the robot hand were controlled online based on the recognition result by the identifier via SEMG, and both the thumb and the index finger of the robot hand were able to be operated as intended.

7. Acknowledgement

The author thanks Dr. T. Nakakuki for his valuable advices on this research, and thanks Mr. A. Harada for his assistance in experimental works.

8. References

Farry, K.A.; Walker, I.D. & Baraniuk, R.G. (1996). Myoelectric Teleoperation of a Complex Robotic Hand, *IEEE Transactions on Robotics and Automation*, Vol.12,No.5,pp.775-787

Graupe, D.; Magnussen, J. & Beex, A.A.M. (1978). A Microprocessor System for Multifunctional Control of Upper Limb Prostheses via Myoelectric Signal Identification, *IEEE Transactions on Automatic Control*, Vol.23, No.4, pp.538-544

Hudgins, B.; Parker, P.A. & Scott, R.N. (1993). A new strategy for multifunction myoelectric control, *IEEE Transactions on Biomedical Engineering*, Vol.40, No.1, pp.82-94

Ito, K.; Tsuji, T.; Kato, A.; & Ito, M. (1992) An EMG Controlled Prosthetic Forearm in Three Degrees of Freedom Using Ultrasonic Motors, *Proceedings of the Annual Conference the IEEE Engineering in Medicine and Biology Society*, Vol.14, pp.1487-1488

Kelly, M.F.; Parker, P.A. & Scott, R.N. (1990). The Application of Neural Networks to Myoelectric Signal Analysis: A preliminary study, *IEEE Transactions on Biomedical Engineering*, Vol.37, No.3, pp.211-230

Lee, S. & Saridis, G.N. (1984). The control of a prosthetic arm by EMG pattern recognition, *IEEE Transactions on Automatic Control*, Vol.29, No.4, pp.290-302

Weir, R. (2003). Design of artificial arms and hands for prosthetic applications, *In Standard Handbook of Biomedical Engineering & Design*, Kutz, M. Ed. New York: McGraw-Hill, pp.32.1–32.61

9

Self-Landmarking for Robotics Applications

Yanfei Liu and Carlos Pomalaza-Ráez
Indiana University – Purdue University Fort Wayne
USA

1. Introduction

This chapter discusses the use of self-landmarking with autonomous mobile robots. Of particular interest are outdoor applications where a group of robots can only rely on themselves for purposes of self-localization and camera calibration, e.g. planetary exploration missions. Recently we have proposed a method of active self-landmarking which takes full advantage of the technology that is expected to be available in current and future autonomous robots, e.g. cameras, wireless transceivers, and inertial navigation systems (Liu, & Pomalaza-Ráez, 2010a).

Mobile robots' navigation in an unknown workspace can be divided into the following tasks; obstacle avoidance, path planning, map building and self-localization. Self-localization is a problem which refers to the estimation of a robot's current position. It is important to investigate technologies that can work in a variety of indoor and outdoor scenarios and that do not necessarily rely on a network of satellites or a fixed infrastructure of wireless access points. In this chapter we present and discuss the use of active self-landmarking for the case of a network of mobile robots. These robots have radio transceivers for communicating with each other and with a control node. They also have cameras and, at the minimum, a conventional inertial navigation system based on accelerometers, gyroscopes, etc. We present a methodology by which robots can use the landmarking information in conjunction with the navigation information, and in some cases, the strength of the signals of the wireless links to achieve high accuracy camera calibration tasks. Once a camera is properly calibrated, conventional image registration and image based techniques can be used to address the self-localization problem.

The fast calibration model described in this chapter shares some characteristics with the model described in (Zhang, 2004) where closed-form solutions are presented for a method that uses 1D objects. In (Zhang, 2004) numerous (hundreds) observations of a 1D object are used to compute the camera calibration parameters. The 1D object is a set of 3 collinear well defined points. The distances between the points are known. The observations are taken while one of the end points remains fixed as the 1D object moves. Whereas this method is proven to work well in a well structured scenario it has several disadvantages it is to be used in an unstructured outdoors scenario. Depending on the nature of the outdoor scenario, e.g. planetary exploration, having a moving long 1D object might not be cost effective or even feasible. The method described in this chapter uses a network of mobile robots that can communicate with each other and can be implemented in a variety of outdoor environments.

2. Landmarks

Humans and animals use several mechanisms to navigate space. The nature of these mechanisms depends on the particular navigational problem. In general, global and local landmarks are needed for successful navigation (Vlasak, 2006; Steck & Mallot, 2000). As their biological counterparts, robots use landmarks that can be recognized by their sensory systems. Landmarks can be natural or artificial and they are carefully chosen to be easy to identify. Natural landmarks are those objects or features that are already in the environment and their nature is independent of the presence or not of a robotic application, e.g. a building, a rock formation. Artificial landmarks are specially designed objects that are placed in the environment with the objective of enabling robot navigation.

2.1 Natural landmarks

Natural landmarks are selected from some salient regions in the scene. The processing of natural landmarks is usually a difficult computational task. The main problem when using natural landmarks is to efficiently detect and match the features present in the sensed data. The most common type of sensor being used is a camera-based system. Within indoor environments, landmark extraction has been focused on well defined objects or features, e.g. doors, windows (Hayet et al., 2006). Whereas these methods have provided good results within indoor scenarios their application to unstructured outdoor environments is complicated by the presence of time varying illumination conditions as well as dynamic objects present in the images. The difficulty of this problem is further increased when there is little or no *a priori* knowledge of the environment e.g., planetary exploration missions.

2.2 Artificial landmarks

Artificial landmarks are manmade, fixed at certain locations, and of certain pattern, such as circular (Lin & Tummala, 1997; Zitova & Flusser, 1999), patterns with barcodes (Briggs et al., 2000), or colour pattern with symmetric and repetitive arrangement of colour patches (Yoon & Kweon, 2001). Compared with natural landmarks, artificial landmarks usually are simpler; provide a more reliable performance; and work very well for indoor navigation. Unfortunately artificial landmarks are not an option for many outdoor navigation applications due to the complexity and expansiveness of the fields that robots traverse. Since the size and shape of the artificial landmarks are known in advance their detection and matching is simpler than when using natural landmarks. Assuming that the position of the landmarks is known to a robot, once a landmark is recognized, the robot can use that information to calculate its own position.

3. Camera calibration

Camera calibration is the process of finding: (a) the internal parameters of a camera such as the position of the image centre in the image, the focal length, scaling factors for the row pixels and column pixels; and (b) the external parameters such as the position and orientation of the camera. These parameters are used to model the camera in a reference system called world coordinate system.

The setup of the world coordinate system depends on the actual system. In computer vision applications involving industrial robotic systems (Liu et al., 2000), a world coordinate system for the robot is often used since the robot is mounted on a fixed location. For autonomous mobile robotic network, there are two ways to incorporate a vision system. One is to have a distributed camera network located in fixed locations (Hoover & Olsen, 2000; Yokoya et al., 2008). The other one is to have the camera system mounted on the robots (Atiya & Hager, 2000). Either of these two methods has its own advantages and disadvantages. The fixed camera network can provide accurate and consistent visual information since the cameras don't move at all. However, it has constraints on the size of the area being analysed. Also even for a small area at least four cameras are needed to form a map for the whole area. The camera-on-board configurations do not have limitations on how large the area needs to be and therefore are suited for outdoor navigation.

The calibration task for a distributed camera network in a large area is challenging because they must be calibrated in a unified coordinate system. In (Yokoya et al., 2008), a group of mobile robots with one robot equipped with visual marker were developed to conduct the calibration. The robot with the marker was used as the calibration target. So as long as the cameras are mounted in fixed locations a fixed world coordinate system can be used to model the camera. However, for mobile autonomous robot systems with cameras on board, a still world coordinate system is difficult to find especially for outdoor navigation tasks due to the constantly changing robots' workspace. Instead the camera coordinate system, i.e. a coordinate system on the robot, is chosen as the world coordinate system. In such case the external parameters are known. Hence the calibration process in this chapter only focuses on the internal parameters.

The standard calibration process has two steps. First, a list of 3D world coordinates and their corresponding 2D image coordinates is established. Second, a set of equations using these correspondences is solved to model the camera. A target with certain pattern, such as grid, is often constructed and used to establish the correspondences (Tsai, 1987). There is a large body of work on camera calibration techniques developed by the photogrammetry community as well as by computer vision researchers. Most of the techniques assume that the calibration process takes place on a very structured environment, i.e. laboratory setup, and rely on well defined 2D (Tsai, 1987) or 3D calibration objects (Liu et al., 2000). The use of 1D objects (Zhang, 2004; Wu et al., 2005) as well as self calibration techniques (Faugeras, 2000) usually come at the price of an increase in the computation complexity. The method introduced in this chapter has low numerical complexity and thus its computation is relatively fast even when implemented in simple camera on-board processors.

3.1 Camera calibration model

The camera calibration model discussed in this section includes the mathematical equations to solve for the parameters and the method to establish a list of correspondences using a group of mobile robots. We use the camera pinhole model that was first introduced by the Chinese philosopher Mo-Di (470 BCE to 390 BCE), founder of Mohism (Needham, 1986).

In a traditional camera, a lens is used to bend light waves into a narrow beam that produces an image on the film. With a pinhole camera, the hole acts like a lens by only allowing a narrow beam of light to enter. The pinhole camera produces the same type of upside-down, reversed image as a modern camera, but with significantly fewer parts.

3.1.1 Notation

For the pinhole camera model (Fig. 1) a 2D point is denoted as $\boldsymbol{a}_i = [a_{ix} \quad a_{iy}]^T$. A 3D point is denoted as $\boldsymbol{A}_i = [A_{ix} \quad A_{iy} \quad A_{iz}]^T$. In Fig. 1 $\boldsymbol{p} = [p_y \quad p_y]^T$ is the point where the principal axis intersects the image plane. Note that the origin of the image coordinate system is in the corner. f is the focal length.

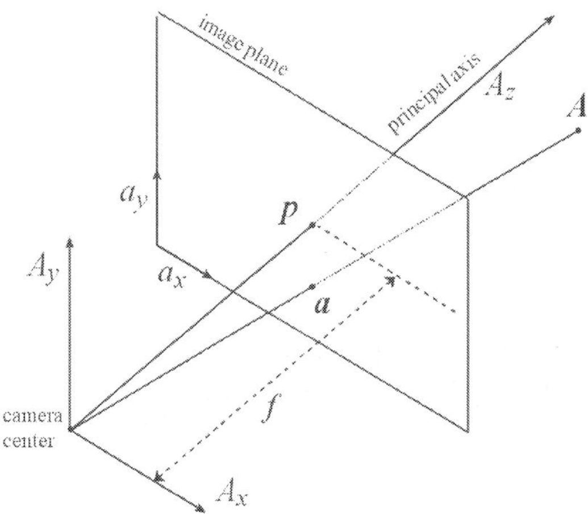

Fig. 1. Normalized camera coordinate system.

The augmented vector $\tilde{a}_i$ is defined as $\tilde{a}_i = [a_{ix} \quad a_{iy} \quad 1]^T$. In the same manner $\tilde{A}_i$ is defined as $\tilde{A}_i = [A_{ix} \quad A_{iy} \quad A_{iz} \quad 1]^T$. The relationship between the 3D point A_i and its projection a_i is given by,

$$z_{A_i}\tilde{a}_i = K[R \quad t]\tilde{A}_i \tag{1}$$

where K stands for the camera intrinsic matrix,

$$K = \begin{bmatrix} \alpha & \gamma & u_0 \\ 0 & \beta & v_0 \\ 0 & 0 & 1 \end{bmatrix} \tag{2}$$

and

$$K^{-1} = \begin{bmatrix} \dfrac{1}{\alpha} & -\dfrac{\gamma}{\alpha\beta} & \dfrac{\gamma v_0 - u_0\beta}{\alpha\beta} \\ 0 & \dfrac{1}{\beta} & -\dfrac{v_0}{\beta} \\ 0 & 0 & 1 \end{bmatrix} \tag{3}$$

$[u_0 \quad v_0]$ are the coordinates of the principal point in pixels, α and β are the scale factors for the image u and v axes, and γ stands for the skew of the two image axes. $[R \quad t]$ stands for the extrinsic parameters and it represents the rotation and translation that relates the world coordinate system to the camera coordinate system. In our case, the camera coordinate system is assumed to be the world coordinate system, $R = I$ and $t = 0$. If $\gamma = 0$ as it is the case for CCD and CMOS cameras then,

$$K = \begin{bmatrix} \alpha & 0 & u_0 \\ 0 & \beta & v_0 \\ 0 & 0 & 1 \end{bmatrix} \tag{4}$$

and

$$K^{-1} = \begin{bmatrix} \dfrac{1}{\alpha} & 0 & -\dfrac{u_0}{\alpha} \\ 0 & \dfrac{1}{\beta} & -\dfrac{v_0}{\beta} \\ 0 & 0 & 1 \end{bmatrix} \tag{5}$$

The K matrix can also be written as,

$$K = \begin{bmatrix} m_x f & 0 & m_x p_x \\ 0 & m_y f & m_y p_y \\ 0 & 0 & 1 \end{bmatrix} \tag{6}$$

Where m_x, m_y are the number of pixels per meter in the horizontal and vertical directions. It should be mentioned that CMOS based cameras can be implemented with fewer components, use less power, and/or provide faster readout than CCDs. CMOS sensors are also less expensive to manufacture than CCD sensors.

3.1.2 Mathematical model

The model described in this section is illustrated in Fig. 2. The reference camera is at position R_i while the landmark is located at position A_i.

The projection of the landmark in the image plane of the reference camera changes when the camera moves from position 0 to position 1 as illustrated in Fig. 2.

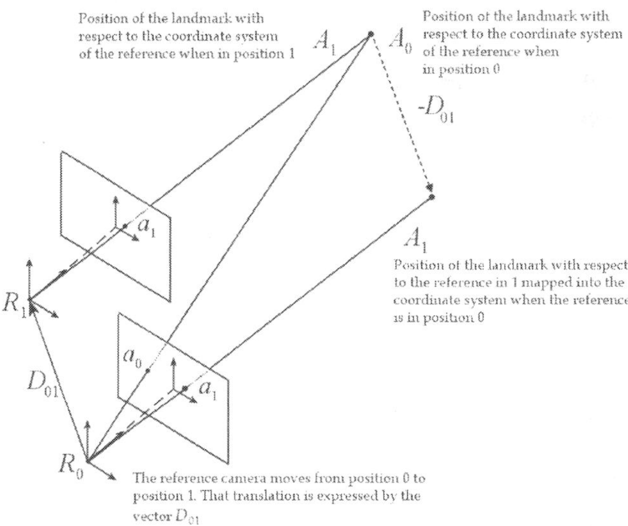

Fig. 2. Changes in the image coordinates when the reference camera or the landmark moves.

This motion is represented by the vector D_{01}. If instead the landmark moves according to $-D_{01}$, as shown in Fig. 2, and the reference camera does not move, then both the location of the landmark, A_1, and its projection on the image, a_1, would be the same as in the case when the reference camera moves.

For any location of the landmark, A_i, and its projection on the image, a_i.

If $\tilde{a}_i = \begin{bmatrix} a_{ix} & a_{iy} & 1 \end{bmatrix}^T$ with $R = I$ and $t = 0$ from eq. (1), then

$$A_i = z_{A_i} K^{-1} \tilde{a}_i \tag{7}$$

also define

$$D_{ij} = A_j - A_i = \begin{bmatrix} d_{ijx} & d_{ijy} & d_{ijz} \end{bmatrix}^T \tag{8}$$

The magnitudes of A_i, A_j, and D_{ij} (L_{A_i}, L_{A_j}, and $L_{D_{ij}}$, respectively) can be estimated using the strength of the received signal. Also for D_{ij} it is possible to estimate $L_{D_{ij}}$ using the data from the robot navigational systems. Both estimation methods, signal strength on a wireless link and navigational system data, have certain amount of error that should be taken into account in the overall estimation process.

$$L_{A_j}^2 = A_j^T A_j = \left(A_i^T + D_{ij}^T\right)\left(A_i + D_{ij}\right)$$

$$= A_i^T A_i + A_i^T D_{ij} + D_{ij}^T A_i + D_{ij}^T D_{ij} \tag{9}$$

$$L_{A_j}^2 = L_{A_i}^2 + L_{D_{ij}}^2 + 2 D_{ij}^T A_i \tag{10}$$

$$D_{ij}^T A_i = \frac{L_{A_j}^2 - L_{A_i}^2 - L_{D_{ij}}^2}{2}$$

$$= D_{ij}^T z_{A_i} \begin{bmatrix} \frac{1}{\alpha} & 0 & -\frac{u_0}{\alpha} \\ 0 & \frac{1}{\beta} & -\frac{v_0}{\beta} \\ 0 & 0 & 1 \end{bmatrix} \begin{bmatrix} a_{ix} \\ a_{iy} \\ 1 \end{bmatrix} \tag{11}$$

Define δ_{ij} as,

$$\delta_{ij} = \frac{L_{A_j}^2 - L_{A_i}^2 - L_{D_{ij}}^2}{2} =$$

$$= z_{A_i} \begin{bmatrix} d_{ijx} & d_{ijy} & d_{ijz} \end{bmatrix} \begin{bmatrix} M_1 & 0 & M_2 \\ 0 & M_3 & M_4 \\ 0 & 0 & 1 \end{bmatrix} \begin{bmatrix} a_{ix} \\ a_{iy} \\ 1 \end{bmatrix} \tag{12}$$

where,

$$M_1 = \frac{1}{\alpha} \quad M_2 = -\frac{u_0}{\alpha} \quad M_3 = \frac{1}{\beta} \quad M_4 = -\frac{v_0}{\beta} \tag{13}$$

for $0 \le i < j \le N$

Where N is the number of locations where the landmark moves to

$$\delta_{ij} = z_{A_i} \begin{bmatrix} a_{ix} d_{ijx} M_1 + d_{ijx} M_2 + a_{iy} d_{ijy} M_3 + d_{ijy} M_4 + d_{ijz} \end{bmatrix} \tag{14}$$

At the end of this section it is shown that z_{A_i} can be separately estimated from the values of the M_i parameters. Assuming then that z_{A_i} has been estimated,

$$\frac{\delta_{ij}}{z_{A_i}} - d_{ijz} = \begin{bmatrix} a_{ix}d_{ijx} & d_{ijx} & a_{iy}d_{ijy} & d_{ijy} \end{bmatrix} \begin{bmatrix} M_1 \\ M_2 \\ M_3 \\ M_4 \end{bmatrix} \tag{15}$$

Let's define λ_{ij} as,

$$\lambda_{ij} = \frac{\delta_{ij}}{z_{A_i}} - d_{ijz} \tag{16}$$

for $0 \leq i < j \leq N$
Then,

$$\lambda_{ij} = c_{ij}^T x \tag{17}$$

where $c_{ij} = \begin{bmatrix} a_{ix}d_{ijx} & d_{ijx} & a_{iy}d_{ijy} & d_{ijy} \end{bmatrix}^T$ and $x = [M_1 \; M_2 \; M_3 \; M_4 \;]^T$
If the landmark moves to N locations, $A_1, A_2, \cdots, A_N$, the corresponding equations can be written as,

$$\Lambda = Cx \tag{18}$$

where $\Lambda = \begin{bmatrix} \lambda_{12} & \lambda_{13} & \cdots & \lambda_{1N} & \lambda_{23} & \cdots & \lambda_{(N-1)N} \end{bmatrix}^T$
and $C = \begin{bmatrix} c_{12} & c_{13} & \cdots & c_{1N} & c_{23} & \cdots & c_{(N-1)N} \end{bmatrix}^T$
The N locations are cross-listed to generate a number of $N(N-1)/2$ pair of points (as shown in Fig. 3) in the equations.

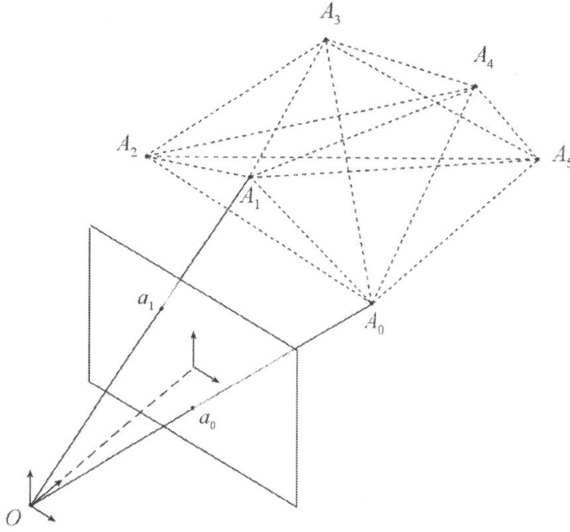

Fig. 3. Cross-listed locations.

The least-squares solution for x is,

$$x = [\mathbf{C}^T\mathbf{C}]^{-1}\mathbf{C}^T\Lambda \tag{19}$$

Once x is estimated the camera intrinsic parameters can be easily computed. Next we will describe two ways to compute z_{A_i}. $A_{iz} = z_{A_i}$, is the projection of the vector A_i on the z-axis. Also $z_{A_i} = z_{A_0} + S_i$ where,

$$S_i = \sum_{j=1}^{i}\delta_{(j-1)jz} \quad \text{for} \quad 1 \le i \le N \tag{20}$$

Thus one way to compute z_{A_i} is to first estimate z_{A_0} and then to use the robot navigation system to obtain the values of $\delta_{(j-1)jz}$ (the displacement along the z-axis as the robot moves) to compute S_i in equation (20). The value of z_{A_0} itself can be using the navigation system as the robot takes the first measurement position.

A second way to compute z_{A_i} relying only on the distance measurement is as follows. From equation (12),

$$\delta_{ij} = z_{A_i}\begin{bmatrix}d_{ijx} & d_{ijy} & d_{ijz}\end{bmatrix}\begin{bmatrix}M_1 a_{ix} + M_2 \\ M_3 a_{iy} + M_4 \\ 1\end{bmatrix} \tag{21}$$

$$\delta_{ij} = \begin{bmatrix}d_{ijx} & d_{ijy} & d_{ijz}\end{bmatrix}\begin{bmatrix}z_{A_i}(M_1 a_{ix} + M_2) \\ z_{A_i}(M_3 a_{iy} + M_4) \\ z_{A_i}\end{bmatrix} = c_{ij}^T x \tag{22}$$

As the landmark moves to N locations $A_1, A_2, \cdots, A_N$, the corresponding equations can be written as

$$\Delta = \mathbf{C}x \tag{23}$$

where

$$\Delta = \begin{bmatrix}\delta_{i1} & \delta_{i2} & \delta_{ij} & \cdots & \delta_{iN}\end{bmatrix}^T \quad \text{for } j \ne i \tag{24}$$

and

$$\mathbf{C} = \begin{bmatrix}c_{i1} & c_{i2} & \cdots & c_{iN}\end{bmatrix}^T \quad \text{for } j \ne i \tag{25}$$

The least-squares solution for x is,

$$x = [\mathbf{C}^T\mathbf{C}]^{-1}\mathbf{C}^T\Delta$$

Once x is estimated, z_{A_i} is also estimated.

4. Self-landmarking

Our system utilizes the paradigm where a group of mobile robots equipped with sensors measure their positions relative to one another. This paradigm can be used to directly address the self-localization problem as it is done in (Kurazume et al., 1996). In this chapter we use self-landmarking as the underlying framework to develop and implement the fast camera calibration procedure described in the previous section. In our case a network of mobile robots travel together as illustrated in Fig. 4. It is assumed that the robots have cameras and are equipped with radio transceivers that allow for communications among them and with a control node.

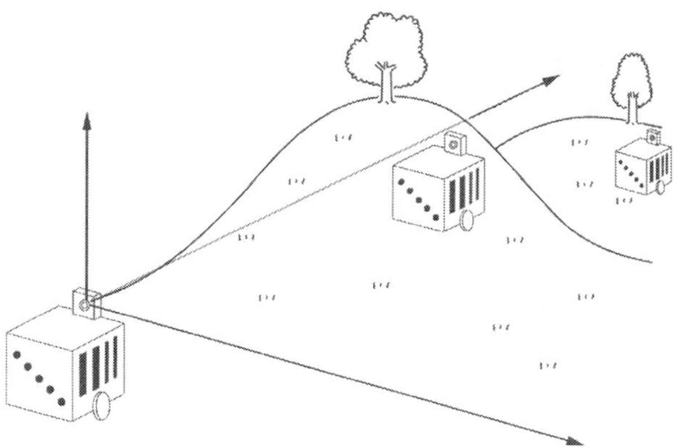

Fig. 4. Self-landmarking mobile robots.

Having decided on using the robots themselves as landmarks to each other the next step is to choose the type of artificial landmark that can be mounted on a robot's body. One possible choice is to use passive landmarks with invariant features such as circular shapes (Zitova & Flusser, 1999) or with simple patterns (Briggs et al., 2000) that are quickly recognizable under a variety of viewing conditions. Whereas these methods have provided good results within indoor scenarios their application to unstructured outdoor environments is complicated by the presence of time varying illumination conditions as well as dynamic objects present in the images. To overcome these drawbacks we have proposed the use of active landmarks.

The current state of LED technology allows for low-power and relative high luminance from these devices. Depending on the constraints imposed by the robot's shape and dimensions one or more LEDs can be located on its outer surface. Since the robots have communication capabilities they can schedule when the LEDs can be turned on and off to match the periods when the cameras are capturing images for image differencing. Power can thus be saved by having the LEDs ON intervals as short as possible. Short ON intervals can also greatly simplify the detection and estimation of the landmarks (the LEDs) in the images since it minimizes the effects of time varying illumination conditions and the motion of other objects in the scene.

Further savings in power can be achieved by using smart cameras. These cameras feature camera-on-a-chip integration (Rinner et al., 2008). For distributed sensing applications this feature allows the cameras to perform a fair amount of on-chip image processing before the information is sent to a central node, e.g. through a wireless channel. For applications where the communications' bandwidth is limited the image processing and data fusion operations carried out on the cameras need to be fast and efficient (Rinner & Wolf, 2008). For our work the detection and location estimation of the landmarks can be reduced to the analysis of a binary image that is obtained by thresholding the difference of the images just before the landmark is turned on and then when it is on. A blob finding algorithm (Liu & Pomalaza-

Ráez, 2010b) can be applied to the task of detecting the landmark. This algorithm is very efficient, i.e. low-complexity, and can be performed on the camera processor. The on-camera chip only needs to report the location (pixel coordinates) of the blob.

5. Wireless localization

Measurements of the strength of the received radio signals can be used to estimate the distance between a transmitter and a receiver. The received signal strength indicator (RSSI) is a measurement that is readily available even in the simple transceivers used in a variety of wireless sensor networks (WSNs). Another common measurement is the link quality indicator (LQI). Both RSSI and LQI can be used for localization by correlating them with distance values. However most of the methods using those estimates have relative large errors in particular within indoor environments (Luthy et al., 2007; Whitehouse et al., 2005).

By combining signal time-of-flight and phase measurements and making use of the full ISM (Industrial, Scientific, Medical) spectrum band it is possible to have estimation errors of less than 20 cm with standard deviations less than 3 cm when using IEEE 802.15.4 devices (Schwarzer et al., 2008). This latter method requires the addition of a low-cost hardware/software that is not part of the 802.15.4 standard. Likewise for IEEE 802.11 devices it is possible, with an extra hardware, to achieve distance estimation measurements with errors less than one meter (Bahillo et al., 2009). The consensus of most researchers is that it is very difficult to guarantee distance estimation errors of less than 10 cm when using 802.11 (Wi-Fi) or 802.15.4 (ZigBee) devices in both indoor and outdoor environments that have many feature rich objects.

Communication systems using Ultra-wide Bandwidth (UWB) signals have shown excellent accuracy in terms of distance measurements (Shimuzi & Sanada, 2003). Using time of arrival (ToA) methods several researchers have reported estimation errors of less than 5 cm in a variety of outdoor and indoor environments (Falsi et al., 2006). UWB signals have been proposed for the detection of vegetation, which can be very useful for outdoor navigation (Liang et al., 2008), and of people behind walls (Zetik et al., 2006) which can be useful in rescue missions. It is then expected that in many applications mobile robots will be equipped with UWB transceivers. It should be noted that the accuracy of common GPS devices is usually more than 10 meters. The Wide Area Augmentation System (WAAS) developed by the Federal Aviation Administration uses a network of GPS reference receivers to increase the accuracy to around 1 meter.

6. Experimental validation

The paradigm described in this chapter is suitable for a wireless mobile robotic network. In principle, the distance from a landmark to a reference camera can be estimated using the wireless communication transceiver that each robot is equipped with. Depending on the type of transceiver, the error in this estimation can be in the order of meters, e.g. for the IEEE 805.15.4 protocol, or in centimeters, e.g. when using UWB technology. Errors in the order of meters are not acceptable for any camera calibration method, including the one presented here. We want to test the calibration method independent of a particular wireless transceiver technology. Thus in order to have measurements with errors in the centimeters range we used common construction tools, such as tape measures, rulers, plumb-blob, to carefully measure the coordinates of each landmark location in the camera coordinate system and the distances between the reference camera and the landmark. Using a laser range finder we estimated that the errors incurred using those construction tools are in the order of ± 2 or 3 cm, which are in the same range the estimation errors one has when using UWB technology (Dardari et al., 2009).

Unless a global localization method is used, such as using GPS devices, the actual coordinates at each location of a roaming robot is usually not known. In our mathematical model (Section 3.1.2), the variables assumed to be known are the vectors between the landmarks' locations in the reference camera coordinate system and the image coordinates of each landmark location. In our experiments, the measurements of the landmarks' coordinates are not used directly in the calculations. These measurements are used to calculate the vectors between the landmarks. When this calibration method is used in a mobile robotic network, these vectors can be obtained from the robots' navigation system. It should be noted that with current GPS technology localization errors in the order of centimeters is not possible unless additional hardware is included.

The CMUcam3 used for the experiments was mounted in a regular office environment. A wood frame was built to support the camera in a way that the Z axis (principle axis) is in the horizontal direction. Fig. 5 shows the front and top view of the CMUcam3 and the mounting structure.

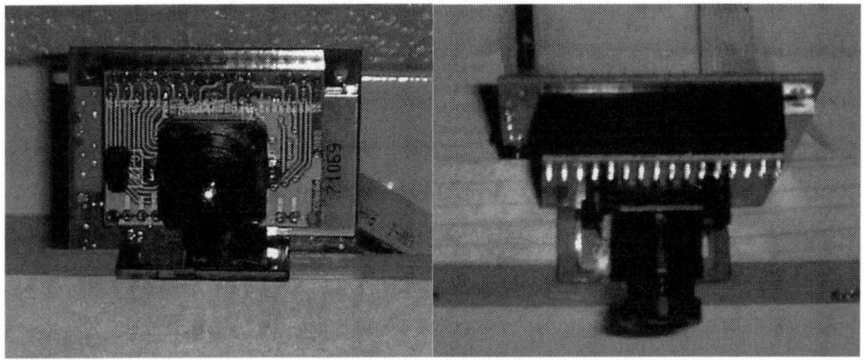

Front view   Top view

Fig. 5. The CMUcam3.

The active landmark was built using the metal structure parts from the VEX robotics design system (Cass, 2006) and LEGO bricks with holes. With the wireless communication capabilities, the robots can turn on and off the LEDs whenever needed to form visible landmarks. Fig. 6 shows the pictures of the robot frame where the LEDs are in ON and OFF state.

The metal frame with the active landmark was placed in different locations in the room. For our experiment twelve locations were chosen so that the landmarks were spread out in the image plane. A newly developed efficient blob finding algorithm (Liu & Pomalaza-Ráez, 2010b) was used to automatically find the landmark anywhere in a scene and then calculate the centroid of the landmark. Fig. 7 shows the picture of one of the landmark locations and the output from the blob finding algorithm.

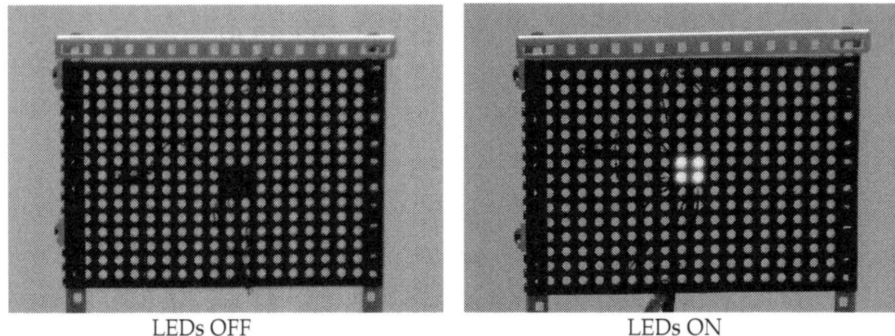

LEDs OFF   LEDs ON

Fig. 6. Active landmarks.

The measurements of the landmarks in the twelve locations are shown in Table 1. A_x, A_y, and A_z are the coordinates in the camera coordinate system. L is the magnitude of the A vector, a_x and a_y are the image coordinates. To make full use of the measurements $n(n-1)/2$ equations can be generated from the n locations as shown in Fig. 3.

Landmark OFF   Landmark ON   Centroid of the landmark

Fig. 7. Pre- and post-processed images by the CMUcam3.

	A_x (cm)	A_y (cm)	A_z (cm)	L (cm)	a_x (pixels)	a_y (pixels)
1	-3.4	-41.0	184.5	188.0	166	45
2	-61.0	-41.0	189.0	199.0	47	49
3	-21.0	-41.0	230.0	232.0	145	63
4	-66.0	-41.0	229.0	239.0	61	65
5	-10.0	2.0	176.6	176.6	153	149
6	-74.0	2.0	177.8	189.6	11	147
7	-12.0	2.0	224.0	224.0	151	148
8	-67.0	2.0	228.4	235.2	56	147
9	33.0	-46.0	264.4	271.5	228	70
10	29.0	-3.0	287.0	287.7	218	143
11	50.0	31.5	272.5	281.3	251	197
12	-66.0	36.5	223.5	232.7	60	215

Table 1. Measurements.

Thus the twelve points listed in Table 1 can be used to generate a maximum of (12x11)/2=66 equations. In order to compare the results of the calibration model using different numbers of measurements and their corresponding equations, our calculation used 5 to 12 locations that generate 10 to 66 equations. The calculation results are shown in Table 2.

In the datasheet of the CMUcam3, the range of values for the focal length f is (2.8~4.9mm). With the value of m_x and m_y, we can calculate the range for α to be (311.1~544.4) and for β to be (341.5~597.6). It is difficult to know what the exact value of f is, thus the exact values of α and β cannot be known either. However, the ratio of α/β is known and is equal to m_x/m_y (0.91). The relative errors of the estimation of the intrinsic parameters are shown in Fig. 8. The estimation results show that the estimates of the parameters converge to the correct values as more measurements are used.

No. of data points	Image sets used	α	β	u_0	v_0	α/β
10	$1 \rightarrow 5$	391.9	440.0	174.6	143.0	0.89
15	$1 \rightarrow 6$	392.6	368.1	175.4	128.3	1.07
21	$1 \rightarrow 7$	394.0	386.1	175.5	132.0	1.02
28	$1 \rightarrow 8$	393.3	371.2	175.3	130.0	1.06
36	$1 \rightarrow 9$	395.0	452.6	175.2	145.3	0.87
45	$1 \rightarrow 10$	396.0	451.0	175.0	145.4	0.88
55	$1 \rightarrow 11$	397.5	442.0	175.3	144.4	0.90
66	$1 \rightarrow 12$	399.7	442.5	176.0	144.0	0.90
Intrinsic parameters				176	143	0.91

Table 2. Calibration results.

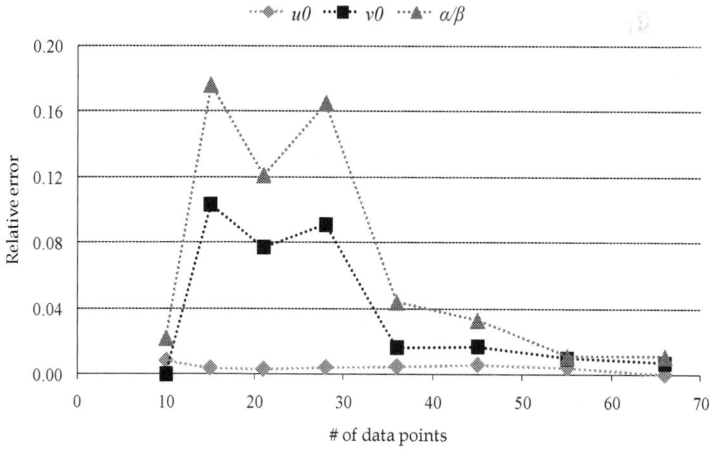

Fig. 8. Relative errors when computing the camera intrinsic parameters.

More than one landmark suited robot can be used in this model to collect a larger number of samples without increasing the amount of time needed to have enough measurements. The mathematical model itself is unchanged when using multiple robots. One advantage when using two or more robots is that it is possible to estimate the distance between the various locations by just using the measurements of the strength of the wireless communication signals between the robots. This type of estimation is possible if for each pair of location points one can position a robot at each point. A minimum of two mobile robots is then needed to obtain a set of measurements.

7. Future developments

The active self-landmarking described in this chapter requires energy efficient LED devices. Currently there is a lot of interest in organic LEDs (OLED). They can be fabricated on flexible substrates which can better fit a variety of robot shapes. Once OLEDs are at the stage to be used in outdoors they will be good candidates for active-landmarking applications. UWB transceivers have shown to provide distances estimation accuracy with errors less than 5 cm which makes them ideal for many localization applications. There are only few commercial suppliers of UWB devices for particular applications. Research in UWB antennas and signal processing is still an active area. It is expected that in the coming years UWB transceivers suitable for robotics applications will be readily available.

To further our research in autonomous mobile robots we are currently building two platforms, equipped with cameras and wireless communication capabilities. Unlike other robot platforms which usually have a computer on board, each of these robots has a single-board RIO (reconfiguration I/O) based microcontroller. The wireless router integrated with the robot is Linksys WRT160N, which is 802.11b/g/n compatible. The choice of cameras is still not finalized. Our goal is to have real-time mobile robot platforms. In order to fully control the image grabbing and transmitting process, we have decided to build the vision system on our own by integrating a FIFO memory with the camera.

8. Conclusion

In this chapter, a new method for fast camera calibration is presented and tested using a smart-camera, the CMUcam3 camera. This method can be easily implemented in a camera-equipped wireless mobile robotic network, where the robots use each other as landmarks. The distances between the robots can be estimated using the wireless signals supported by standard communication protocols. Active landmarks made of LEDs are proposed. The LEDs can be turned on and off through wireless communications commands. One of the limitations of this method is that it relies on the ranging accuracy of the wireless signals measurements. Signals using the protocol 802.15.4 will give errors in the order of meters (not acceptable for calibration tasks). UWB technology, which has errors in the order of centimetres, is more appropriate for this type of application.

9. Acknowledgment

The authors would like to thank the Indiana Space Consortium for their support of the work presented in this chapter.

10. References

Atiya, S. & Hager, G. D.(2000) . Real-time vision-based robot localization, *IEEE Transactions on Robotics and Automation*, pp. 785-800, Vol. 9, No. 6, ISSN : 1042-296X, Dec 1993 .

Bahillo, A.; Prieto, J.; Mazuelas, S.; Lorenzo, R.; Blas, J. & Fernández, P. (2009). IEEE 802.11 Distance Estimation Based on RTS/CTS Two-Frame Exchange Mechanism. *Proceedings of the 2009 IEEE 69th Vehicular Technology Conference*, pp. 1-5, ISBN: 978-1-4244-2517-4, Barcelona, Spain, April 26-29, 2009

Briggs, A.; Scharstein, D.; Braziunas, D.; Dima, C. & Wall, P. (2000). Mobile Robot Navigation Using Self-Similar Landmarks, *Proceeding of the 2000 IEEE International Conference on Robotics and Automation* (ICRA 2000), pp. 1428-1434, ISBN: 0-7803-5889-9 San Francisco, California, USA, April 24-28, 2000.

Cass, S (2006). Getting Vexed, *IEEE Spectrum*, Vol. 43, No. 5, pp. 68-69, ISSN: 0018-9235, May 2006.

Dardari, D.; Conti, A.; Ferner, U.; Giorgetti, A. & Win (2009). Ranging With Ultrawide Bandwidth Signals in Multipath Environments, *Proceedings of the IEEE*, Vol. 97, No. 2, pp. 404-426, February 2009

Falsi, C.; Dardari, D.; Mucchi, L. & Win, M. (2006). Time of Arrival Estimation for UWB Localizers in Realistic Environments, *EURASIP Journal on Applied Signal Processing*, Vol. 2006, 01 January, pp. 1-13, ISSN:1110-8657.

Faugeras, O.; Quan, L. & Strum, P. (2000). Self-calibration of a 1D projective camera and its application to the self-calibration of a 2D projective camera, *IEEE Transactions on Pattern Analysis and Machine Intelligence*, Vol. 22, No. 10, pp. 1179-1185, ISSN: 0162-8828, October 2000

Hoover, A. & Olsen, B. (2000). Sensor network perception for mobile robotics, *Proceedings of the 2000 IEEE International Conference on Robotics and Automation (ICRA 2000)*, Vol. 1, pp. 342-347, ISBN: 0-7803-5886-4, San Francisco, California, USA, April 24-28, 2000

Kurazume, R.; Hirose, S.; Nagata, S. & Sahida, N. (1996). Study on Cooperative Positioning System: Basic Principle and Measurement Experiment, *Proceedings of the 1996 IEEE International Conference on Robotics and Automation*, Vol. 2, pp. 1421-1426, ISBN: 0-7803-2988-0, Minneapolis, Minnesota, USA, April 22-28, 1996

Liang, Q.; Samn, S. & Cheng, X. (2008). UWB Radar Sensor Networks for Sense-through-Foliage Target Detection, *Proceedings of the 2008 IEEE International Conference on Communications*, pp. 2228-2232, ISBN: 978-1-4244-2075-9, Beijing, China, May 19-23, 2008

Lin, C. & Tummala, R. (1997). Mobile Robot Navigation Using Artificial Landmarks, *Journal of Robotics Systems*, Vol. 14, pp. 93-106, February 1997

Liu, Y.; Hoover, A. & Walker, I. (2000). Sensor Network Based Workcell for Industrial Robots, *Proceedings of the IEEE/RSJ International Conference on Intelligent Robots and Systems*, pp. 1434-1439, ISBN: 0-7803-6348-5, Hawaii, USA, October 2001.

Liu, Y. & Pomalaza-Ráez, C. (2010a). Application of Active Self-Landmarking to Camera Calibration, *Proceedings of the International Conference on Machine Vision (ICMV 2010)*, ISBN: 978-1-4244-8888-9, Hong Kong, China, December 28-30, 2010

Liu, Y. & Pomalaza-Ráez, C. (2010b). On-Chip Body Posture Detection for Medical Care Applications Using Low-Cost CMOS Cameras, *Journal of Integrated Computer-Aided Engineering*, Vol. 17, No. 1. pp. 3-13, ISBN: 1069-2509, 2010

Luthy, K.; Grant, E. & Henderson, T. (2007). Leveraging RSSI for Robotic Repair of Disconnected Wireless Sensor Networks, *Proceedings of the 2007 IEEE International*

Conference in Robotics and Automation (ICRA), pp. 3659-3664, ISBN: 1-4244-0601-3, Roma, Italy, April, 2007

Needham, J. (1986). *Science and Civilization in China: Volume 4, Physics and Physical Technology, Part 1, Physics*, Taipei, Caves Books Ltd

Rinner, B.; Winkler, T.; Schriebl, W.; Quaritsch, M. & Wolf, W. (2008). The Evolution from Single to Pervasive Smart Cameras, *Proceedings of the ACM/IEEE International Conference on Distributed Smart Cameras (ICDSC-08)*, pp. 1-10, ISBN: 978-1-4244-2664-5, Stanford University, California, USA, September 7-11, 2008

Rinner, B. & Wolf, W. (2008). An Introduction to Distributed Smart Cameras, *Proceedings of the IEEE*, Vol. 96, pp. 1565-1575, ISSN: 0018-9219, October 2008

Shimizu, Y. & Sanada, Y. (2003). Accuracy of Relative Distance Measurement with Ultra Wideband System, *Proceedings of 2003 IEEE Conference on Ultra Wideband Systems and Technologies*, pp. 374-378, ISBN: 0-7803-8187-4, Reston, Virginia, USA, November 16-19, 2003

Steck, S. & Mallot, H. (2000). The Role of Global and Local Landmarks in Virtual Environment Navigation, *Presence: Teleoperators and Virtual Environments*, Vol. 9, pp. 69-83, ISSN: 1054-7460, February 2000

Schwarzer, S.; Vossiek, M.; Pichler, M. & Stelzer, A. (2008). Precise Distance Measurement with IEEE 802.15.4 (ZigBee) Devices, *Proceedings of the 2008 IEEE Radio and Wireless Symposium*, pp. 779-782, ISBN: 1-4244-1463-6, Orlando, Florida, USA, January 22-24, 2008

Tsai, R. (1987). A Versatile Camera Calibration Technique for 3D Machine Vision, *IEEE Journal of Robotics & Automation*, Vol. 3, No. 4, pp.323-344, ISSN: 0882-4967, August 1987.

Vlasak, A. (2006). The Relative Importance of Global and Local Landmarks in Navigation by Columbian Ground Squirrels (*Spermophilus Columbianus*), *Journal of Comparative Psychology*, Vol. 120, pp. 131-138

Whitehouse, K.; Karlof, C.; Woo, F. Jiang, F. & Culler, D. (2005). The Effects of Ranging Noise on Multihop Localization: An Empirical Study, *Proceedings of the 4th International Symposium on Information Processing in Sensor Networks*, pp. 73-80, ISBN:0-7803-9202-7, Los Angeles, California, USA, April 25-27, 2005

Wu, F.; Hu, Z. & Zhu, H. (2005). Camera calibration with moving one-dimensional objects, *Pattern Recognition*, Vol. 38, No. 5, pp. 755-765, ISSN: 0031-3203, May 2005.

Yokoya, T.; Hasegawa, T. & Kurazume, R. (2008). Calibration of distributed vision network in unified coordinate system by mobile robots, *Proceedings of the 2008 International Conference on Robotics and Automation*, pp. 1412-1417, ISBN 978-1-4244-1647-9, Pasadena, California, USA, May 19-23, 2008

Yoon, K. & Kweon, I. (2001). Landmark design and real-time landmark tracking for mobile robot localization, *Proceedings of The International Society for Optical Engineering (SPIE)*, Boston, USA, Vol. 4573-21, pp. 219-226, October 29-November 2, 2001

Zetik, R.; Crabbe, S.; Krajnak, J.; Peyerl, P.; Sachs, J. & R. Thomä, R. Detection and localization of persons behind obstacles using M-sequence through-the-wall radar, *Proceeding of the SPIE Defense and Security Symposium*, Vol. 6201, ISBN: 9780819462572, 2006, Orlando, Florida, USA, April 17-21, 2006.

Zhang, Z. (2004). Camera Calibration with One-Dimensional Objects, *IEEE Trans. Pattern Analysis and Machine Intelligence*, 26(7):892-899, 2004.

Zitova, B. & Flusser, J. (1999). Landmark recognition using invariant features, *Pattern Recognition Letters*, Vol. 20, pp. 541-547

10

Robotic Waveguide by Free Space Optics

Koichi Yoshida[1], Kuniaki Tanaka[1] and Takeshi Tsujimura[2]
[1]NTT Corporation,
[2]Saga University
Japan

1. Introduction

Road construction and work on the water supply often require the relocation of aerial/underground telecommunication cables. Each optical fiber leading from an optical line terminal (OLT) in a telephone office to a customer's optical network unit (ONU) must be cut and reconnected. Customers expect real-time transmission for high-quality communications to continue uninterrupted, especially for video transmission services.

Some electrical transmission apparatus can maintain communication without interruption, even when optical cables are temporarily cut. The system is complicated and any transmission delay during O/E conversion is fatal to real-time communication. Although it is desirable to directly switch the transmission medium itself, it had been thought that some data bits would inevitably be lost during the replacement of optical fibers. An optical fiber cable transfer splicing system [1] has been developed to minimize the disconnection time. It takes 30 ms to switch a transmission line, and more than 2 seconds * to restore communications with, for example, GE-PON [2].

We have developed an interruption-free replacement method for in-service telecommunication lines, which can be applied to the current PON system equipped with conventional OLTs and ONUs [3, 4]. Two essential techniques were presented; a measurement method and a system for adjusting the transmission line length. The latter continuously lengthens/shortens the line over very long distances without losing transmitted data based on free space optics (FSO). The former distinguishes the difference between the duplicated line lengths by analyzing signal interference. The mechanism that automatically coordinates both these two functions and referred to as a robotic waveguide in this paper compensates for the traveling time difference of a transmitted pulse. Interferometry is the technique of diagnosing the properties of two or more lasers or waves by studying the pattern of interference created by their superposition. It is an important investigative technique in the fields of astronomy, fiber optics, optical metrology and so on. Studies on optical interferometry are reported to improve tiny optical devices [5-7]. We have applied the technique to measure length of several kilometers of optical fibers with a 10 mm resolution.

This paper describes the design of our robotic waveguide system. An optical line length measurement method is studied to distinguish the difference of two lines by evaluating interfered optical pulses. An optical line switching procedure is designed, and a line length adjustment system is prototyped. Finally, we applied the proposed system to a 15 km GE-

PON optical fiber network while adding a 10 m extension to show the efficiency of this approach when replacing in-service optical cables.

2. Optical line duplication for switching over

Figure 1 shows an individual optical line in a GE-PON transmission system with a single star configuration. Optical pulse signals at two wavelengths are bidirectionally transmitted through a regular line between customers' ONU and an OLT in a telephone office via a wavelength independent optical coupler, WIC1, and a 2x2 optical splitter, 2x2 SP, respectively.

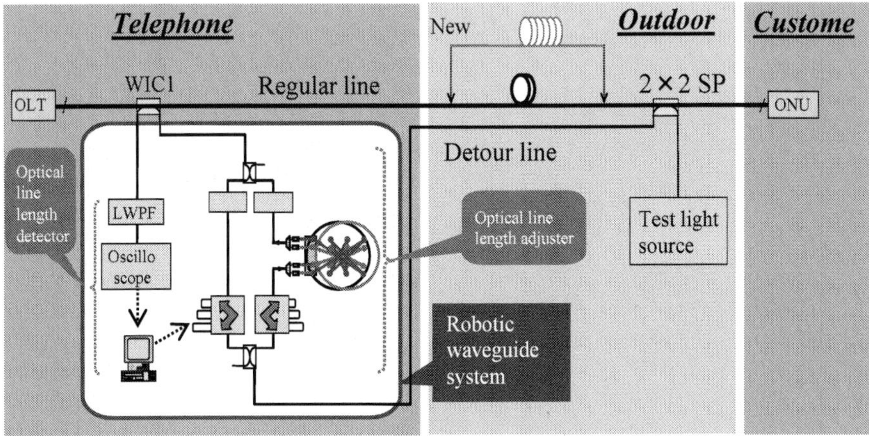

Fig. 1. Robotic waveguide system for optical fiber transmission line.

We have designed a robotic waveguide system, and a switching procedure for three wavelengths, namely 1310 and 1490 nm for GE-PON transmission, and 1650 nm for measurement. A robotic waveguide system is installed in a telephone office. It is composed of an optical line length detector and an optical line length adjuster. A test light at a wavelength different from those of the transmission signals is sent from one of the optical splitter's ports to the duplicated lines. An oscilloscope is connected to the optical coupler to detect the test light through a long-wavelength pass filter (LWPF). The optical line length adjuster is an FSO application. Some optical switches (SWs) and optical fiber selectors (FSs) control the flow of the optical signals managed by a controller. The optical pulses are compensated by 1650 and 1310/1490 nm amplifiers [8]. The proposed method temporarily provides a duplicate transmission line as shown in Fig. 1 to replace optical fiber cables. A detour line is prepared in advance through which to divert signals while the existing line is replaced with a new one. This system transfers signals between the two lines. Signals are duplicated at the moment of changeover to maintain continuous communications. The signals travel separately through the two lines to a receiver. A difference in the line lengths leads to a difference in the signals' arrival times. A communication fault occurs if, as a result of their proximity, the waveforms of the two arriving signals are too blurred for the signals to be identified as discrete. Thus it is important to adjust the lengths of both lines precisely.

Experiments determined that the tolerance of the difference in line length is 80 mm with regard to the GE-PON transmission system.

The proposed system controls the adjustment procedure so that the difference in length between the detour and regular lines is adjusted within 80 mm.

3. Optical line length difference detection

We use laser pulses at a wavelength of 1650 mm to detect the optical path length difference. They are introduced from an optical splitter, duplicated, and transmitted toward the OLT through the active and detour lines. They are distributed by an optical coupler just in front of the OLT, and observed with an oscilloscope. The conventional measurement method evaluates the arrival time interval between the duplicated signals, and converts it to the difference between the lengths of the regular line and the detour line at a resolution of 1 m. The difference in line length, ΔL is described as

$$\Delta L = c \cdot \Delta t / n, \tag{1}$$

where c is the speed of light, Δt is the difference between the signal arrival times for the regular and detour lines, and n is the refractive index of optical fiber.

Figure 2 shows the received pulses observed with an oscilloscope. When the detour is 99 m shorter than the active line, pulses traveling through the detour line reach the oscilloscope about 500 ns earlier than through the regular line. The former pulse approaches the latter as shown in Fig. 2(b), while the system lengthens the detour line using the optical path length adjuster. This method fails if the difference between the line lengths is less than 1 m, because the two pulses combine as shown in Fig. 2(c).

We also developed an advanced technique for measuring a difference of less than 1 m between optical line lengths. Interferometry enables us to obtain more detailed measurements when the optical pulses combine. A chirped light source generates interference in the waveform of a unified pulse.

Each pulse, $E(L_j, t)$ is expressed as

$$E(L_j, t) = A_j \exp[-i(k \cdot n \cdot L_j - \omega_j \cdot t + \phi_0)], \tag{2}$$

where j represents the regular line, 1, or the detour line, 2. And, A_j, k, n, L_j, ω_j, t, ϕ_0 denote amplitude, wavenumber in a vacuum, refractive index of optical fiber, line length, frequency, time, and initial phase, respectively. The intensity of a waveform with interference, I, is calculated by taking the square sum as

$$I = |E(L_1, t) + E(L_2, t)|^2 = A_1{}^2 + A_2{}^2 + 2A_1A_2 \cos(k \cdot n \cdot \Delta L - \Delta\omega \cdot t), \tag{3}$$

where ΔL and $\Delta\omega$ represent the differences between line lengths and frequencies, respectively.

The waveform with interference depends on the delay between the pulses' arrival times. Time-domain waveforms are shown in Fig. 3. When the gap was 0.5 m, the waveform contained high-frequency waves as shown in Fig. 3(a). The less the gap became, the lower-frequency the interfered waveform was composed of. When the lengths of two lines coincided, a quite low-frequency waveform was observed as Fig. 3(d).

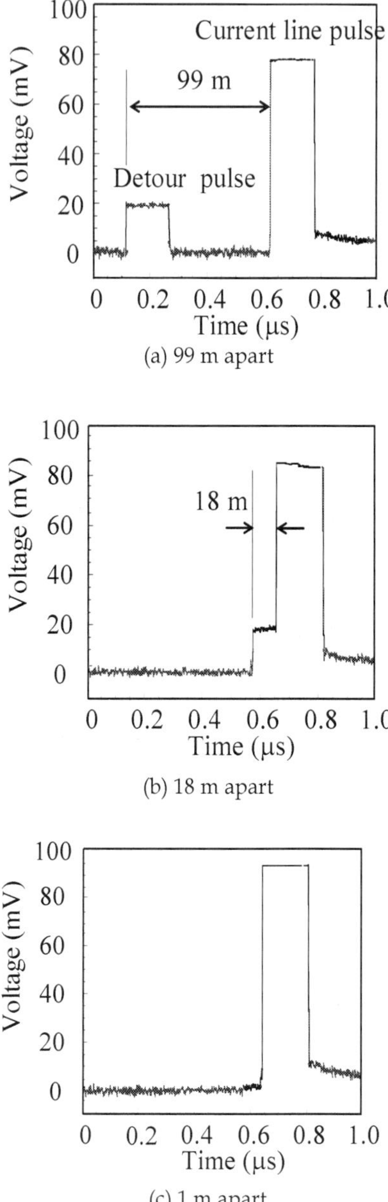

Fig. 2. Time-domain optical line length measurement when difference in line length is more than 1 m.

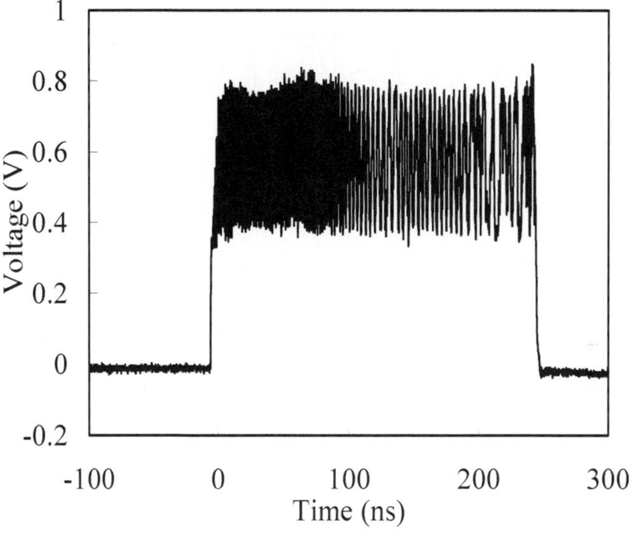

Fig. 3. (a) 0.5 m apart

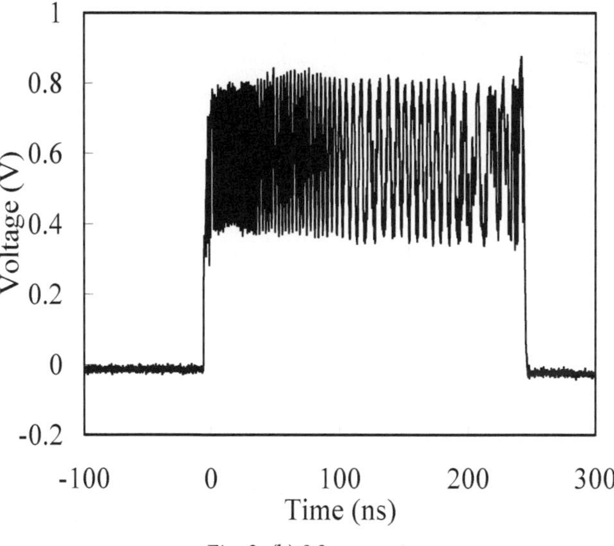

Fig. 3. (b) 0.3 m apart

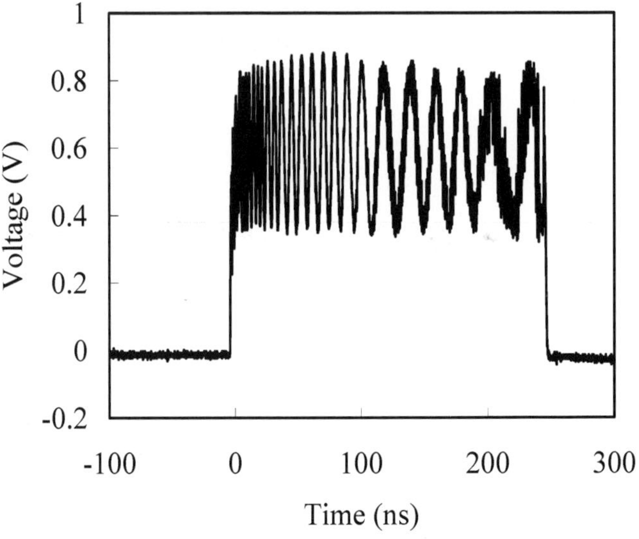

Fig. 3. (c) 0.1 m apart

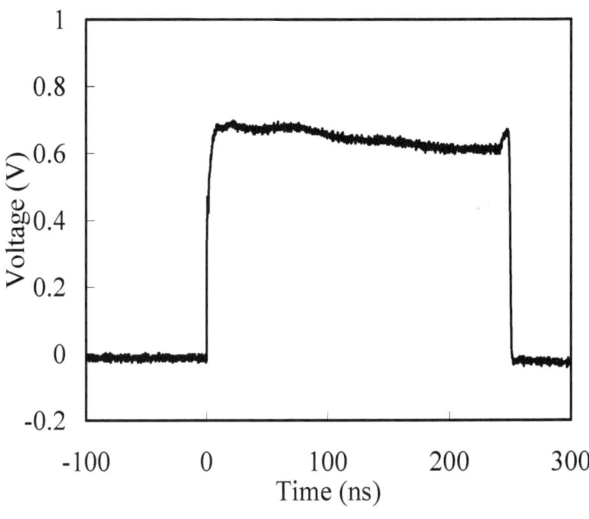

Fig. 3. (d) 0 m apart

Fig. 3. Time-domain optical line length measurement when difference in line length is less than 1 m.

A Fourier-transform spectrum reveals the characteristics. When the gap was 0.5 m, the waveform with interference was composed of the power spectrum shown in Fig. 4(a). The peak power indicated that the major frequency component was around 600 MHz. Figure 4(b) and (c) indicate that the peak powers for gaps of 0.3 and 0.1 m were 360 and 120 MHz, respectively. It became difficult to determine the peak for smaller gaps, because the frequency peak became so low that it was hidden by the near direct-current part of the frequency component. When the lengths of duplicated lines coincided, the power spectrum was obtained as Fig. 4(d).

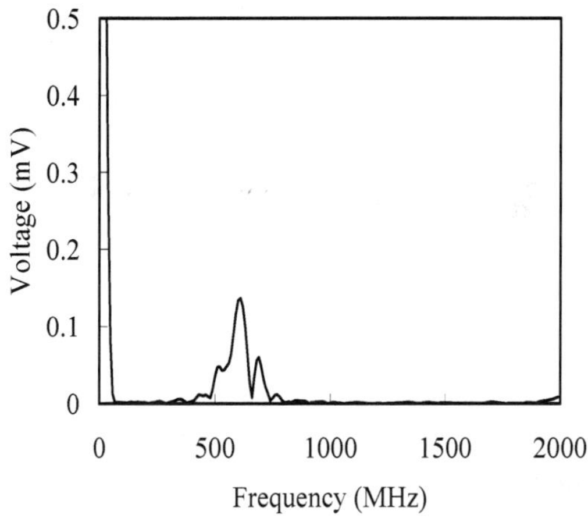

Fig. 4. (a) 0.5 m apart

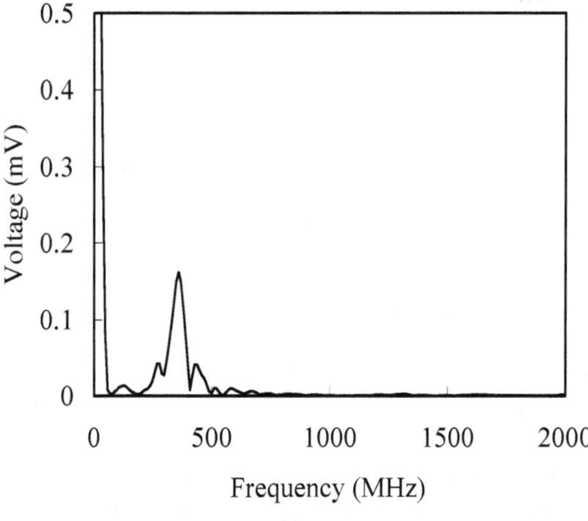

Fig. 4. (b) 0.3 m apart

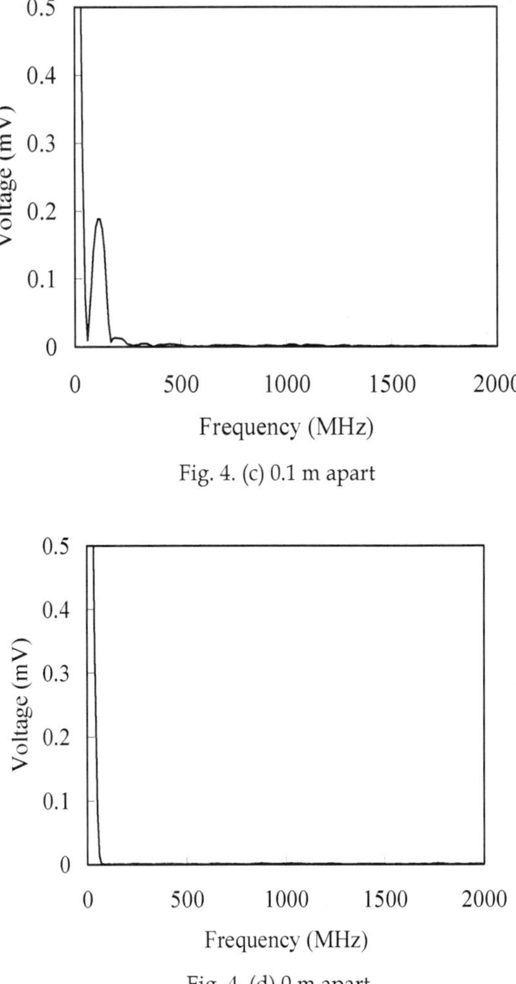

Fig. 4. (c) 0.1 m apart

Fig. 4. (d) 0 m apart

Fig. 4. Frequency-domain optical line length measurement.

An evaluation of the frequency characteristics in the interfered waveforms showed that the peak frequencies are proportional to the difference between the line lengths from -1 to 1 m as shown in Fig. 5. This result helps us to determine the optimal position for adjustment. The optimal position where the line lengths coincide can be estimated by extrapolating the data.

We have established a technique for distinguishing the difference between line lengths to an accuracy of better than 10 mm by analyzing interfering waveforms created by chirped laser pulses.

We have realized a complete length measurement for optical transmission lines from 100 m to 10 mm.

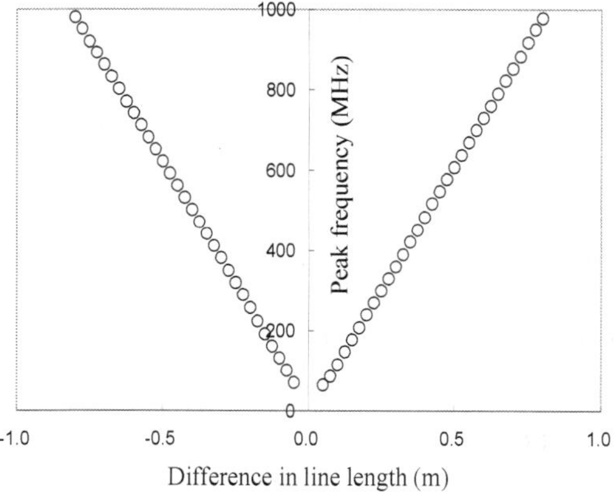

Fig. 5. Estimation of line length coincidence.

4. Robotic waveguide system

We designed a prototype of the robotic waveguide system to apply to a GE-PON optical fiber line replacement according to the procedure described below.

An optical line length adjuster, shown in Photo 1, is installed along the detour line. The adjuster is equipped with two retroreflectors, which directly face each other as shown in Fig. 6. A retroreflector consists of three plane mirrors, each of which is placed at right angles to the other two. And it accurately reflects an incident beam in the opposite direction regardless of its original direction, but with an offset distance. The vertex of the three mirrors in the retroreflector is in the middle of a common perpendicular of the axes of the incoming and outgoing beams as shown in Fig. 6. The number of reflections is determined based on the retroreflector arrangement. A laser beam travels 10 times between the retroreflectors in our prototype, and are introduced into the other optical fiber. Optical pulses are transmitted through an optical fiber, divided into three wavelengths by wavelength division multiplexing (WDM) couplers, and discharged separately into the air from collimators. The focuses of a pair of collimators corresponding for a wavelength is best tuned for the wavelength to achieve the minimum coupling loss. The collimators for multiple wavelenghts are arranged to share the two retroreflectors as shown in Fig. 7.

The detour line between the retroreflectors consists of an FSO system [9]. The detour line length can be easily adjusted by controlling the retroreflector interval with a resolution of 0.14 mm. Optical pulses travel n-times faster in the air than in an optical fiber, where n is the refractive index of the optical fiber. Thus the optical line length adjuster lengthens/shortens the corresponding optical fiber length, L, by $k\Delta x/n$, where k, Δx, n are the number of journeys between the retroreflectors, the retroreflector interval variation, and the refractive index of optical fiber, respectively. The FSO lengthens the optical line length up to L_0.

Photo 1. Free-space optics line length adjuster.

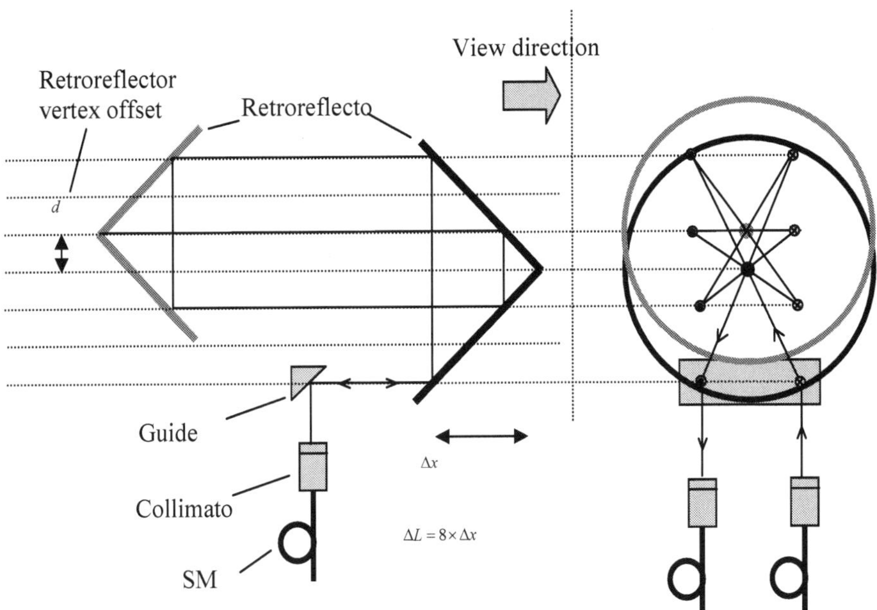

Fig. 6. Free-space optics line length adjuster.

$$L_0 = k \, \Delta x_{max} / n, \tag{5}$$

where Δx_{max} is the maximum range of the retroreflector interval variation. The maximum range of our prototype, Δx_{max}, is around 0.3 m, the refractive index, n, of the optical fiber is 1.46, the number of journeys, k is 10, and the optical line span, L_0, tuned by the adjuster is 2 m.

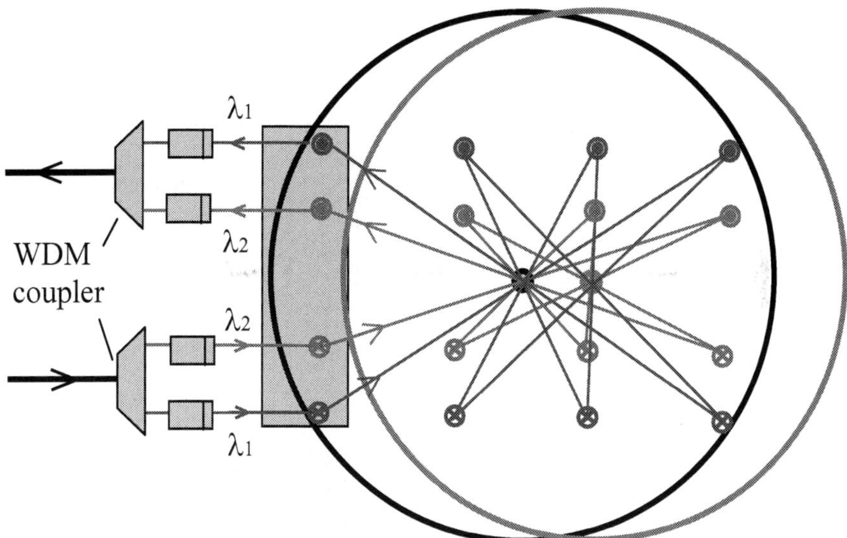

Fig. 7. Collimator arrangement for use of multiple wavelengths.

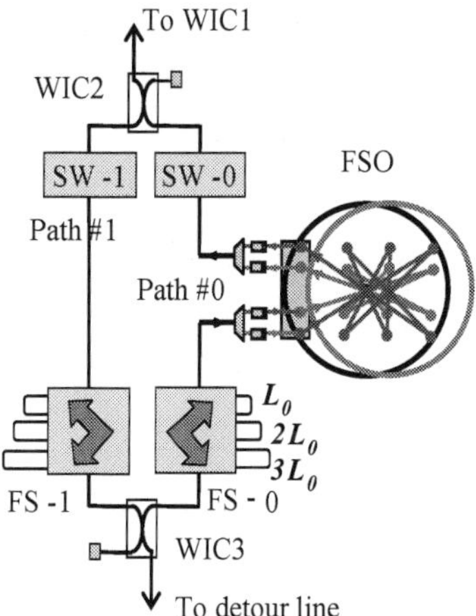

Fig. 8. FSO system with optical path length accumulation mechanism.

The limit of the adjustable range is a practical problem when this system is applied to several kilometers of access network. Therefore, we employ optical line length accumulators. The optical line length adjuster contains two optical paths, #0 and #1 as shown in Fig. 1 or Fig. 8. An optical switch and an optical fiber selector are installed in each path. Optical switches control the optical pulse flow. Each optical fiber selector is equipped with various lengths of optical fiber, for example L_0, $2L_0$ and $3L_0$. The path length can be discretely changed by choosing any one of them.

The optical line length adjuster can extend the detour line as much as required using the following operation as shown in Fig. 9. First, the FSO system lengthens path #0 by L_0 by gradually increasing the retroreflector interval. After the optical fiber selector has selected an optical fiber of length L_0, the active line is switched from path #0 to path #1. The FSO system then returns to the origin, and the optical fiber selector   selects an optical fiber of length L_0 instead to keep the length of path #0 at L_0. The FSO system increases the retroreflector interval again to repeat the same operation. In this way the adjuster accumulates spans extended by the FSO system. The scanning time of our prototype is 10 seconds, because the retroreflector operates at 30 mm/s.

The optical line length adjuster enables us to lengthen/shorten the detour line while continuing to transmit optical signals.

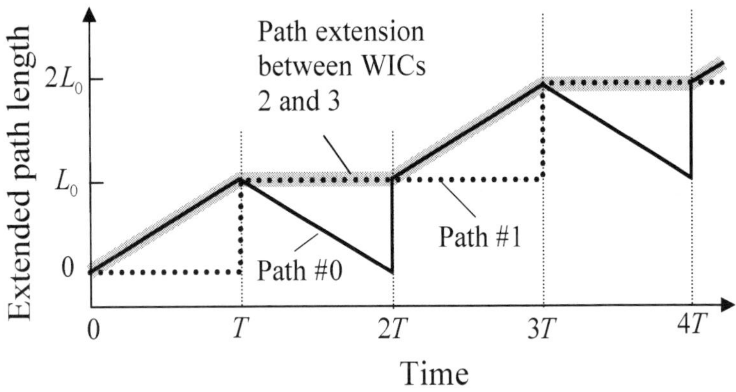

Fig. 9. Time chart of operation for optical path length accumulation.

5. Experiments on optical line replacement

The optical line replacement procedure, shown in Fig. 10 where a 2x8 optical splitter is used instead of a 2x2 splitter, is as follows:

1. A detour line is established between a WIC and a 2x8 optical splitter.
2. The detour line length is measured with a 1650 nm test light using an optical line length measuring technique, and is adjusted to the same length as the regular line using an optical line length adjusting technique. These techniques are described in the preceding sections.

3. Once the lengths of the two lines coincide, the transmission signals are also launched into the detour line.
4. The regular line is cut and replaced with a new line, while the signals are being transmitted through the detour line. A long-wavelength pass filter (LWPF) is temporarily installed in the new line.
5. The test light measures the lengths of the new line and the detour line. The detour line is adjusted to the new line while communications are maintained. The LWPF prevents only the optical transmission pulses from traveling through the new line.
6. The LWPF is then removed and the transmission is duplicated. The detour line is finally cut off.

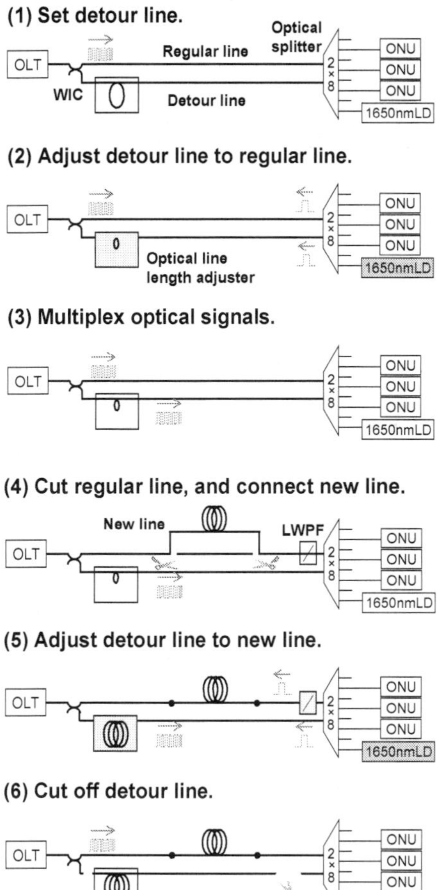

Fig. 10. Optical line replacement procedure.

We investigated the tolerance of the multiplexed signal synchronicity in advance. The transmission quality is observed by changing the difference between the duplicated line lengths. The results show that the transmission linkage is maintained if the difference is within 80 mm as with GE-PON. A multiplexed signal cannot be perceived as a single bit when the duplicated line lengths have a larger gap for 1 Gbit/s transmission. Because these characteristics depend on the periodic length of a transmission bit, the requirement is assumed to be severe when the method is applied to higher-speed communication services.

Next, we constructed a prototype of the robotic waveguide system shown in Fig. 1, and applied it to a 15 km GE-PON optical transmission line replacement. A 10 m optical fiber extension was added to the transmission line, while optical signals were switched between the duplicated lines during transmission.

Figure 11 shows the frame loss that occurred during optical line replacement, which we measured with a SmartBit network performance analyzer. No frame loss was observed at any switching stage if the difference between the duplicated line lengths was less than 80 mm. If the difference exceeded 80 mm, signal multiplexing caused frame loss in stages (a) and (d). We confirmed that the optical signals can be completely switched between the regular, detour, and new lines on condition that the line length is adjusted with sufficient accuracy.

The experimental results proved that our proposed system successfully relocated an in-service broadband network without any service interruption.

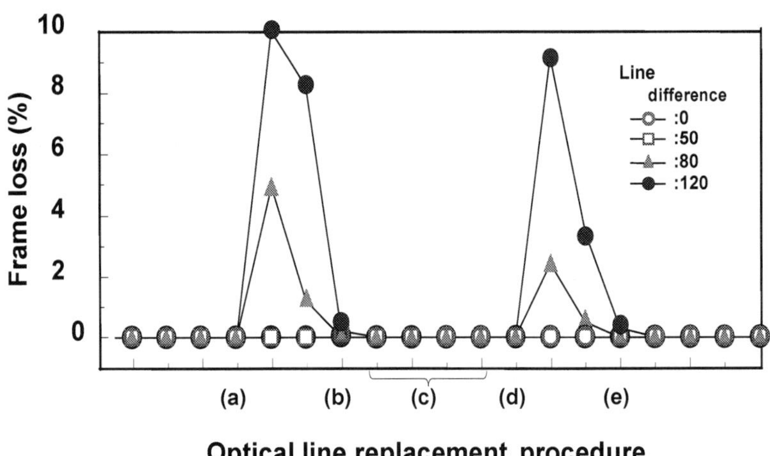

Fig. 11. Frame loss while replacing transmission line according to the procedure; (a) Multiplex signals of current line and detour line, (b) Cut current line, (c) Extend detour line, (d) Multiplex signals of detour line and new line, (e) Cut off detour line.

6. Conclusion

We proposed a new switching method for in-service optical transmission lines that transfers live optical signals. The method exchanges optical fibers instead of using electric apparatus to control transmission speed. The robotic waveguide system is designed to apply to duplicated optical lines. An optical line length adjuster, designed based on an FSO system, continuously lengthened the optical line up to 100 m with a resolution of 0.1 mm. An optical line length measurement technique successfully evaluated the difference in length between the duplicated lines from 100 m to 10 mm. An interferometry measurement distinguished the difference between line lengths to an accuracy of better than 10 mm by analyzing interfering waveforms created by chirped laser pulses. We applied this system to a 15 km GE-PON network and succeeded in replacing the communication lines without inducing any frame loss.

7. References

Tanaka, K.; Zaima, M.; Tachikura, M. & Nakamura, M. (2002). Downsized and enhanced optical fiber cable transfer splicing system, *Proceedings of 51st IWCS*, 2002

Azuma, Y.; Tanaka, K. ; Yoshida, K. ; Katayama, K.; Tsujimura, T.; & Shimizu, M. (2008). Basic investigation of new transfer method for optical access line, *IEICE Technical Report Japan*, OFT2008-52, pp. 27-31, 2008

Tanaka, K.; Tsujimura, T.; Yoshida, K. ; Katayama, K.; & Azuma, Y. (2009). Frame-loss-free Line Switching Method for In-service Optical Access Network using Interferometry Line Length Measurement, *OFC2009*, paper PDPD6, 2009

Tanaka, K.; Tsujimura, T.; Yoshida, K. ; Katayama, K.; Azuma, Y. M.; & Shimizu, M. (2010). Frame-Loss-Free Optical Line Switching System for In-Service Optical Network, *J. Lightw. Technol.*, Vol.28, No.4, pp. 539-546, 2010.

Tsujimura, T.; Tanaka, K.; Yoshida, K. ; Katayama, K.; & Azuma, Y. (2009). Infallible Layer-one Protection Switching Technique for Optical Fiber Network, *Proceedings of 14th European Conference on Networks and Optical Communications/ 4th Conference on Optical Cabling and Infrastructure*, 2009

Saunders, R.; King, J. & Hardcastle, I. (1994). Wideband chirp measurement technique for high bit rate sources, *Electronics Letters*, Vol.30, No.16, pp. 1336-1338, 1994.

Cao, S. & Cartledge, J. (2002). Measurement-Based Method for Characterizing the Intensity and Phase Modulation Properties of SOA-MZI Wavelength Converters, *IEEE Photonics Technology Letters*, Vol.14, No.11, pp. 1578-1580, 2002

Torregrosa, A. et al., Return-to-Zero Pulse Generators Using Overdriven Amplitude Modulators at One Fourth of the Data Rate, *IEEE Photonics Technology Letters*, Vol.19, No.22, pp. 1837-1839, 2007

Fukada, Y.; Suzuki, K.; Nakamura, H.; Yoshimoto, N. & Tsubokawa, M. (2008). First demonstration of fast automatic-gain-control (AGC) PDFA for amplifying burst-mode PON upstream signal, ECOC2008, paper WE.2.F.4, 2008.

Yoshida, K.; Tsujimura, T. & Kurashima, T. (2008). Seamless transmission between single-mode optical fibers using free space optics system, *SICE Annual Conference*, pp. 2219-2222, 2008

Yoshida, K. & Tsujimura, T. (2010). Seamless Transmission Between Single-mode Optical Fibers Using Free Space Optics System, *SICE J. Contr. Measure. Syst. Integr. (JCMSI)*, Vol.3, No.2, pp.94-100, 2010.

Surface Reconstruction of Defective Point Clouds Based on Dual Off-Set Gradient Functions

Kun Mo[1] and Zhoupin Yin[2]

[1]*Dongfang Electric Corporation, Research & Develop Centre, Intelligent Equipment & Control Technology Institute*
[2]*State Key Laboratory of Digital Manufacturing Equipment and Technology Huazhong University of Science and Technology China*

1. Introduction

Surface reconstruction is an interesting and challenging task in extensively applied fields including rapid prototype manufacturing, computer vision, virtual reality and computer aided design (CAD). A typical reconstruction procedure begins with scanning, in which the point data are sampled from physical objects by digitizing measurement systems (such as laser-range scanners and hand-held digitizers). And then, the point data are generated as a smooth, water-tight and proper resulting surface by a suitable reconstruction method. In industry the most difficulty comes from the defective samples that are subject to the noise, holes and overlapping regions. The defective samples are often unavoidable due to the sampling inaccuracy, scan mis-registration and accessibility constraints of scanning device. They often make most existing reconstruction methods not practical for engineering application because the oriented or neighbour information of points, which the most methods are highly based on, are hard to evaluate. For instance, many methods rely on consistent normals, or pose the demand on triangular meshes generated from point data. However, the holes and overlapping samples confuse the point's neighbour relationship, some jagged, self-intersect regions could exist in the corresponding triangular mesh or the estimation of consistent normals becomes an ill-posed problem. Only a few methods need not such specific information, but they have to resort to some complex or time-consuming steps, like re-sampling, distance-computing, mesh-smooth or deformable models. Even if these methods can generate a water-tight resulting surface, the reasonableness of fitting overlapping samples and holes is not guaranteed. In fact, such issues, especially "bad-scanning" data, often lead long scanning time, massive manual work and poor model quality.

Given these challenges, this paper propose a novel surface reconstruction method that takes as input defective point clouds without any specific information and output a smooth and water-tight surface. The main idea is that (1) this technique is based on implicit function, because implicit reconstruction is convenient to guarantee a water-tight result; (2) the approach is indirect, two off-set surfaces are generated to best fit the point clouds instead of direct approximation. As shown in Fig.1 (1D situation for simple expression), the point clouds are represented as origin of coordinates (Fig. 1 (a)). The space is divided into inside part (positive axis in Fig. 1 (a)) and outside part (negative axis in Fig. 1 (a)). If the point clouds are defective, it is very hard to reconstruct final implicit function with w-width(real line in Fig.1 (b)) directly, including reducing noise, filling holes and merging overlapping samples. Therefore, this method constructs dual off-set functions to approximate the inner and outer level set of the final implicit surface (real lines in Fig.1 (c)). The dual water-tight surfaces form a minimal crust surrounding the point data. Based on the dual relative functions, a novel energy function is defined. By minimizing the energy function, the resulting surface (dash dot line in Fig.1 (c)) is finally obtained and visualized.

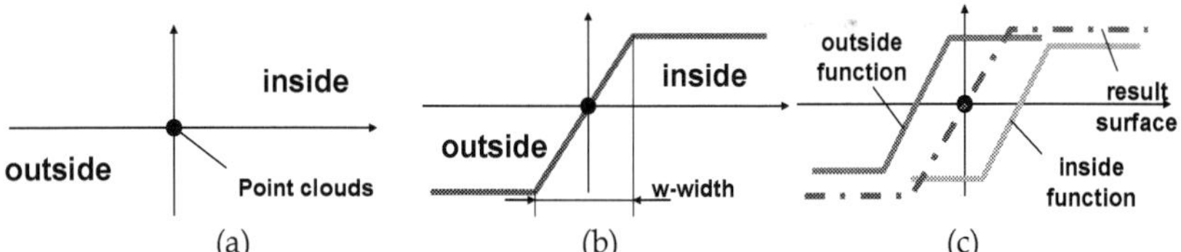

Fig. 1. The main idea of this method: (a) Point clouds. (b) Implicit resulting surface. (c) Off-set functions and resulting surface.

The dual relative functions are built on volumetric grids by extending some sophisticated 2D grey image processing algorithms into 3D space, including morphology operation and weighted vector median filter algorithm. The method needs not any specific information and also has the advantages that (1) the implementation needs not time-consuming steps, like computing distances between each point which is performed normally by most existing methods; (2) the dual gradient functions provide global constrains to the resulting surface, the holes could be filled smoothly and the overlapping samples could be fitted much reasonably. (3) the method can successfully construct "bad-scanning" point data which could not be handled by many methods. The reminder of the paper is organized as follows, after the previous works are reviewed and compared in section 2, the process and details of this method is described in section 3. To demonstrate the effectiveness, extensive numerical implementations are discussed in section 4. Finally, the conclusion and future work is summarized in section 5.

2. Related work

The previous algorithms of surface reconstruction can be generally classified into two categories: explicit methods and implicit methods. Most explicit methods employ Delaunay-triangulation or Voronoi diagrams, like alpha shapes (Edelsbrunner & Mücke 1994), crust method (Nina et al. 1998), triangular-sculpting (Jean-Daniel 1984), mesh growing (Li et al. 2009) and their developed version(Veltkamp 1995; Baining et al. 1997; Attali 1998; Amenta et al. 2000; Amenta et al. 2001; Yang et al. 2009). But the noise and overlapping samples could make the resulting surface jagged. Smoothing (Ravikrishna et al. 2004), refitting (e.g. (Chandrajit et al. 1995; Shen et al. 2005; Shen et al. 2009)) or blending (LA Piegl 1997) of subsequent processing are required.

In contrast, the implicit methods are much efficacious to infer topology of points, blend surface primitives, tolerate noise, and fill holes automatically. A popular algorithm is based on blending locally fitted implicit primitives, such as Radial basis functions (RBF) method (Carr et al. 2001; Greg & James 2002), multi-level partition of unity (MPU) method (Yutaka et al. 2003), products of univariate B-splines (Song & Jüttler 2009) and tri-cubic B-spline basis functions (Esteve et al. 2008) on voxelization. However they either need the consistent normals as aided information or the point clouds with fewer defective samples. The local primitives also include polynomial of point set surface (Alexa et al. 2001; Gael & Markus 2007) with moving least squares (MLS) approximation. MLS methods have to employ normal computation and projection operators, which could lead to low efficiency and need certain extra procedure to improve (Anders & Marc 2003; Marc & Anders 2004). Some methods need pre- or post-processing, like oriented estimation (Vanco & Brunnett 2004; David & Guido 2005), smooth operation (e.g. (Yukie et al. 2009)) or holes filling. For instance, the method proposed in (Davis et al. 2002) uses diffusion to fill holes on reconstructed surfaces. The approach essentially solves the homogeneous partial differential equation (PDE) given boundary conditions to create an implicit surface spanning boundaries. Poisson method (Michael et al. 2006) is also a PDE-based reconstruction algorithm with oriented point clouds. Several approaches use combinatorial structures, such as signed distance function and Voronoi diagram as (Boissonnat & Cazals 2002) (Hybrid method). But the normal information of point clouds is still required.

Only a few methods could demand little restriction on point data. Hoppe's method (Hoppe et al. 1992) is a typical method of this category. It creates the object surface by locally fitting planes to generate a signed distance function and triangulate its zero-level set. The signed distance function can also be cumulated into volumetric grids as proposed in (Brian & Marc 1996). But the two methods are troubled by the noisy or sparse data, which make the connection relationship of these regions hard to confirm. The method proposed in (Alliez et al. 2007) employs Voronoi diagram to estimate the un-consistent normals and solves a generalized eigenvalue problem to construct resulting surface. However, it has to suffer from low computation efficiency. Level set method (Zhao et al. 2000), a typical deformable models, reconstructs the surface by solving corresponding level set equation defined on point data. It is a time-consuming method since it requires a process of re-initialization and needs updating all the nodes of compute grids in very time step. The reconstruction method also employ voting algorithm (Xie et al. 2004) to cluster points into local groups, which are then blended to produce a signed distance field using the modified Shepard's method. But it needs to compute the medial axis transformation and perform an active contour growing process, like deformable models. The methods in (Esteve et al. 2005) proposes DMS operation on volumetric grids to fill holes by detecting the incursions to the interior of the surface and approximates them with a bounded maximum distance. It is an improvement of (Song & Jüttler 2009), but a post process has to be introduced to the low density data zones.

The typical implicit methods mentioned above are summarized in Table 1 with four respects that whether the methods need specific information and have the effectiveness of reducing noise, filling holes and merging overlapping samples. Although all the implicit methods can guarantee water-tight results that the holes could be filled, some methods (like Level set method) could not fill holes smoothly. Many methods have certain low efficient steps, like solving density matrix equation (e.g. RBF), projection procedure (e.g. MLS) and compute the distance function among all the point clouds for neighbor information. All the methods can reduce noise, but the effectiveness is not the same. For instance, MPU and Hoppe's method could be influenced by noise more than others, if the noise is too much the resulting surface is still not smooth. All the methods do not address how to merge overlapping samples, especially the "bad-scanning" points, which is a common problem in practice.

Method name	Specific information	Reduce noise	Fill holes	Merge overlapping samples
RBF (Carr et al. 2001; Greg & James 2002)	Yes	Yes	Yes	— —
MPU (Yutaka et al. 2003)	Yes	Yes	Yes	— —
MLS (Marc et al. 2001; Gael & Markus 2007)	No	Yes	Yes	— —
Poisson method (Michael et al. 2006)	Yes	Yes	Yes	— · —
Hybrid method (Boissonnat & Cazals 2002)	Yes	Yes	Yes	— —
Hoppe's method (Hoppe et al. 1992)	No	Yes	Yes	— —
Voronoi method (Alliez et al. 2007)	No	Yes	Yes	— —
Level set method (Zhao et al. 2000)	No	Yes	Yes	— —
Voting algorithm (Xie et al. 2004)	No	Yes	Yes	— —
DMS method (Esteve et al. 2005)	No	Yes	Yes	— —
This method	No	Yes	Yes	Yes

Table 1. Comparison between this method and typical implicit methods.

Rather than constructing final surface directly, it is much easier to confirm the off-set surfaces of point clouds. Some methods have focused on the respect, like duplex fitting method (Liu & Wang 2009) or dual-RBF method (Lin et al. 2009). But they need the consistent normals for accurately fitting. Recently, some robust and efficient methods in other research areas (like image processing (Peng & Loftus 1998; Peng & Loftus 2001) and statistics (Roca-Pardinas et al. 2008)) have been introduced in reverse engineering. Inspired by the two ideas, this paper describes a novel reconstruction method using the dual off-set surface by extending morphology operation and weighted vector median filter algrithms. The comparison between this method and typical implicit methods is added in Table 1. This method provides a convenient and efficient manner for reconstruction and addresses the issues of overlapping points and "bad-scanning" samples.

3. Method description

The main process of the proposed method is illustrated in Fig.2. Let $\mathbf{P} = \{p_1, p_2, ... p_m\}$ represent defective point clouds (black points in Fig.2) sampled from a real object surface $\mathbf{S}$ (black line in Fig.2). The goal of this method is to reconstruct an implicit function $\phi(\mathbf{x})$ whose zero level set approximates S reasonably. The general intermediate step is to approximate $\phi_{\pm\delta}$, which denote the offset surfaces to $\mathbf{S}$ with slight distance $\pm\delta$. Finally, $\phi(\mathbf{x})$ is generated by blending $\phi_{\pm\delta}$ according to a minimal model. The mathematic description can be defined as follow: given the position coordinates of discrete point sets P around 3D local region $\mathbf{V}$, find an implicit function ϕ that satisfy,

$$\begin{cases} \phi(\mathbf{x}) > 0 & \text{for } \mathbf{x} \in \mathbf{V} \\ \phi(\mathbf{x}) < 0 & \text{for } \mathbf{x} \notin \mathbf{V} \\ \phi(\mathbf{x}) = 0 & \text{for } \mathbf{x} \in \Gamma = \partial\mathbf{V} \end{cases} \qquad (1)$$

by approximating the offset function $f_{in} \approx \phi_{+\delta}$ and $f_{out} \approx \phi_{-\delta}$. $\Gamma = \partial\mathbf{V}$ in Eq. (1) represents the interface of the region $\mathbf{V}$.

In the first step, f_{out} and f_{in} are constructed on volumetric grids respectively (represented as two red lines in Fig.2 (b)). The dual functions are water-tight and guarantee a minimal narrowband surrounding the point clouds. This step is accomplished by the morphology operation, a powerful theory in image processing (Maragos et al. 1996). In the second step, by blurring the boundary of f_{out} and f_{in} , two monotonic functions are obtained and transformed as gradient fields (arrows in Fig. 2 (c)). This paper extends weighted vector median filter algorithm to reduce the noise influence. In the third step, the surface reconstruction problem is formulated as solving a minimal energy model by blending the dual gradient fields (arrows in Fig. 2 (d)). By deriving and solving the corresponding Euler-Lagrange equation, the implicit resulting function is obtained. Finally, the reconstructed surface is extracted by marching cube method (William & Harvey 1987) and visualized (dashed line in Fig. 2 (e)).

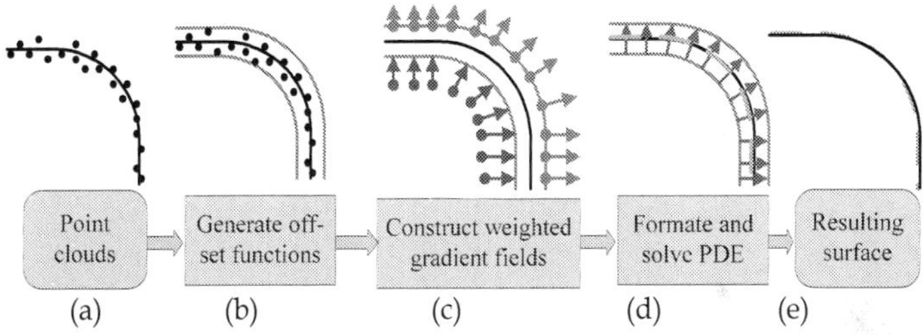

Fig. 2. The main process of the proposed method.

3.1 Generate off-set functions

The point data are first need divide into volumetric grids (voxelization). A best voxelization is that the size of voxel cooperates with the data density, where one grid only contains one point. As shown in Fig. 3, the black points represent defective points and the real line represents the reasonable resulting surface. If the point clouds are uniform, the voxel-building step is immediate. In industry, the uniform samples are common data because the original scanning data are numerous, a regular sampling for reduction are often needed before reconstruction. If the point data have a very irregular sampling density, the ratio, number of voxels which contain two or more points divided by the total number, should be calculated. If the valve is too high, the size of the volumetric grids is re-calculated by decreasing the length of grids. The most difficulty is how to guarantee off-surface, f_{in} and f_{out} , water-tight. It needs determining the inside/outside of the grids near defective samples, especially holes and overlapping regions (circles in Fig.3 (a)). This method extends the basic operations of mathematical morphology, dilation and erosion, to confirm the sign of f_{in} and f_{out} near the holes and overlapping samples (dash dot line in Fig.3 (b)).

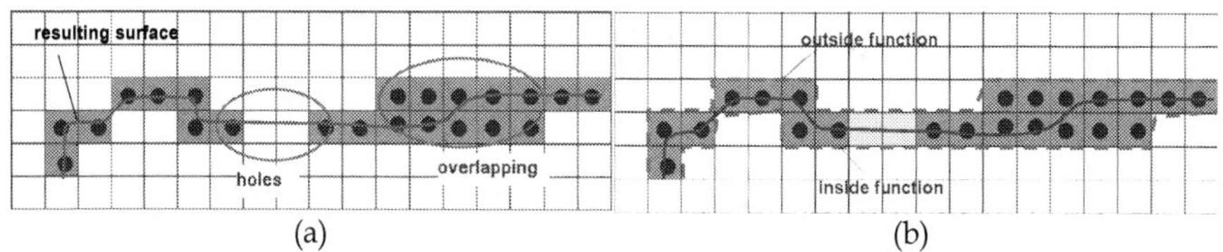

(a)   (b)

Fig. 3. The main idea of constructing dual off-set functions. (a) Defective point clouds. (b) off-set surface and resulting surface.

To express simply, 2D uniform point clouds of arbitrary shape are designed (Fig.4 (a)). This shape contains lines, fillets and free curves according to the practical products. Some random noise, overlapping regions and holes are added. Fig.4 (b) shows the voxelization results. The node (white rectangles) of the grids within points (black points) is labeled as value 1 and other nodes are labeled as value -1. The dilation operation of morphology is then used to construct a rough crust (as shown in Fig. 4 (d)). Let $F(x,y,z)$ denote the point image of voxelization and $\mathbf{B}$ denote structuring element. Dilation of the image $F(x,y,z)$ by an element $\mathbf{B}$ is denoted by

$$F \oplus \mathbf{B}(x,y,z) = \max\left\{F(x-u,y-v,z-w) + \mathbf{B}(u,v,w)\right\} \tag{1}$$

The best choice of $\mathbf{B}$ should preserve the shape of point clouds and perform dilation with less times. According to the research (Maragos et al. 1996), a disk shape (as shown in Fig.4 (c)) in 3D space is a good choice. If $\mathbf{B}$ is a rectangle or sphere shape, some shape details may be blurred. The structuring element size should be little larger than the noise distribution. If error of noise is defined as ε_r, the length of $\mathbf{B}$ should be set as

$$l(\mathbf{B}) > \text{int}(\frac{\varepsilon_r}{h}) \tag{2}$$

where $l(\mathbf{B})$ represents the length of $\mathbf{B}$, h denotes the grid size according to the density of points, $\text{int}(\cdot)$ is rounded down function. For instance, if the random noise with $\varepsilon_r \sim N(0,0.1^2)$ is added in points and the grid size is defined as $0.05mm$, the structuring element size should be more than 2. If the size of $\mathbf{B}$ is too large, although the dilation times is less, the shape of points could be blurred. A suitable selection is given by

$$\text{int}(\frac{\varepsilon_r}{h}) < l(\mathbf{B}) < \text{int}(3 \cdot \frac{\varepsilon_r}{h}) \tag{3}$$

In this paper, it is generally set as median size $3 \times 3 \times 3$ (Fig. 4 (c)), if the point clouds are much dense with little noise, the size can be smaller.

By the close crust, the inside and outside of the point data can be roughly separated. The inside part is then filled (see Fig. 4 (e)) by a simple flood-fill algorithm. It starts at a node (E.g. the middle gird node) known to be inside, those nodes accessible from initial node are labeled "inside", and the remaining nodes are labeled "outside". Each node of resulting image is therefore classified as lying inside the object (value 1) and outside/on the object (value -1). When sparse samples or large holes exist, the dilation should be executed for several times until a water-tight crust is constructed. The flood-fill step can check if the crust

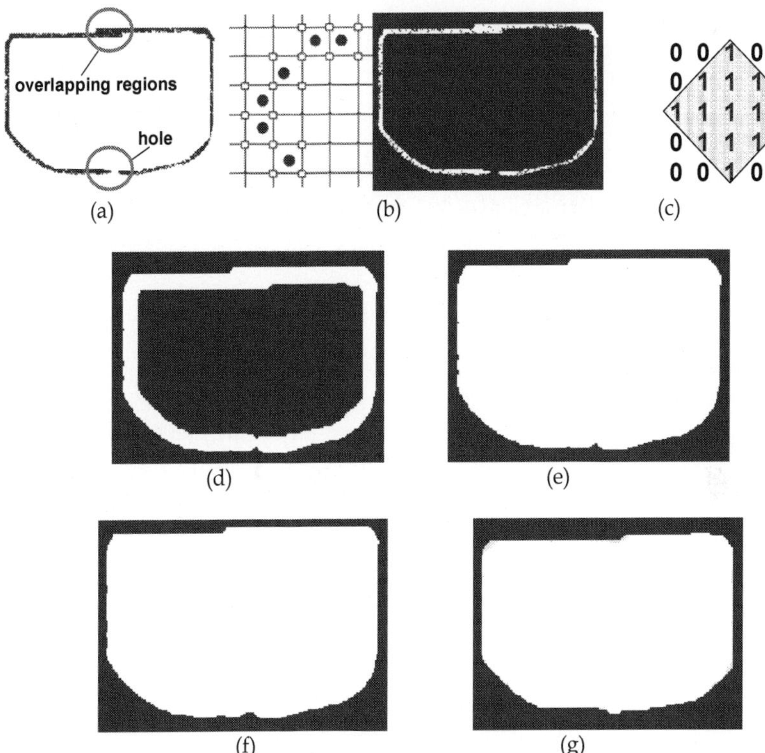

Fig. 4. Generation of off-set surfaces. (a) Defective point clouds. (b) Voxel image. (c) Disk shape with $3 \times 3 \times 3$ size. (d) Close crust. (e) Image after filling inside. (f) Outside function. (g) Inside function.

is water-tight. If the crust is not close, the flood-fill operation could cover the whole space grids. Let $\mathbf{F}_f$ donates the function after the flood-fill step, $\mathbf{F}_{all}$ denotes the whole volumetric image, if $\mathbf{F}_f = \mathbf{F}_{all}$, the crust is not close. Thus, image $\mathbf{F}$ needs be dilated and check again. Due to the voxelization and the choice of $\mathbf{B}$, the dilation of examples in the paper performed less than 3 times (except Fig.18). The resulting image is shown in Fig. 4 (e) by iteratively performing the operation. And then the erosion operation is used to restore the image, expressed by

$$f_{out} = \mathbf{F}_f \ominus \mathbf{B}(x,y,z) = \min\left\{\mathbf{F}_f(x+u,y+v,z+w) - \mathbf{B}(u,v,w)\right\} \qquad (4)$$

structuring element $\mathbf{B}$ and the erosion times must be the same as the dilation step otherwise it cannot recover the image. The restored image is treated as the outside function $f_{out} \approx \phi_{-\delta}$ (Fig. 4 (f)).

The inside function f_{in} is generated by dilation from inside part as follow, where the intermediate result $\mathbf{F}_f$ is adopted,

$$f_{in} = (\mathbf{F}_f - \mathbf{F} \oplus \mathbf{B}) \oplus \mathbf{B} \tag{5}$$

where $\mathbf{B}$ is the same as in the process of constructing f_{out}. The result of $f_{in} \approx \phi_{+\delta}$ is shown in Fig. 4 (g).

As shown in Fig.3 (b) the relative functions f_{out} and f_{in} construct a narrowband. This process needs not complex computation just set operation. Theoretically, this step could be also used in the situation of non-manifold surface, which divides the space into more than two pieces. But it needs to confirm the start points for flood-fill operation artificially, because without any pre-information, the topology of point clouds is not unique and may be confused. In industrial application, this kind of products is a rare case.

Actually if many small details of points need to preserve, f_{out} and f_{in} could be updated on non-uniform grids by a subdividing process with two general steps (Fig.5). In Fig.5 (a), the dotted lines represent the dual functions on rough uniform grids, the real line represents the old resulting surface. First, the grid containing more than two points is subdivided as next level of 8 neighbor grids (see Fig.5 (a)) and the points within are inserted in new subdivided grids. The iterative subdividing process stops until no grid contains two or more points. Then, a local flood-fill process is performed in the new subdivided grids. For instance, for generating new inside function f_{in}, the filling operation starts at a node of last level known to be inside (real rounds at nodes in Fig.5 (b)). It performed on each subdivided levels hierarchically and stop until all the inside/outside of new subdivided grids are confirmed. Since f_{in} and f_{out} are digitally based, they could be easily updated. With non-uniform grids, the new resulting surface (red dotted line in Fig.5 (b)) could preserver more details. However, the subdividing process is not often necessary because like other reconstruction methods based on fitting local primitives, the details of corresponding resulting surface are influenced deeply by the noise. So it could be used at the situation that the point clouds are dense enough without much noise.

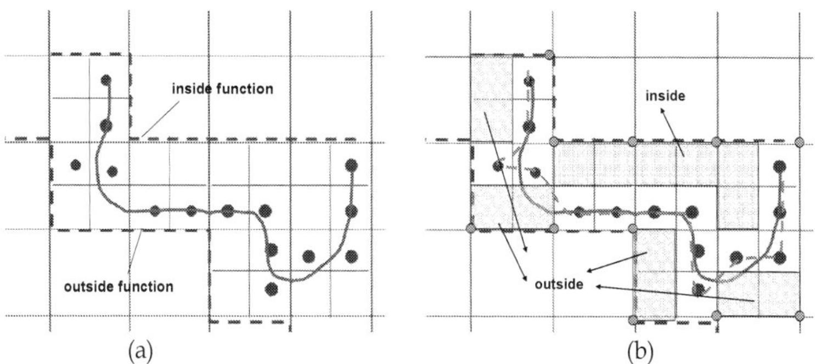

Fig. 5. Process of constructing subdivided grids. (a) Subdivided grids and old dual functions. (b) Updated dual functions.

3.2 Construct weighted gradient fields

f_{in} and f_{out} generated in last step are rough approximation to the off-set surface of $\phi(\mathbf{x})$ due to the noise influence. This step is to reduce the affection and reconstruct a smooth and reasonable off-set surface. The weighted vector median filter in 2D grey image processing is extended into 3D space. Since f_{in} and f_{out} are Heaviside-like functions, they need to transform as gradient fields by two sub-steps. First, they are transformed as monotonic function by blurring their boundaries within neighbor space. Second, the corresponding gradient functions are computed. In this paper, a $3\times3\times3$ kernel of Gaussian function $G(x,y,z)$ with standard deviation σ is employed to produce non-integer node values,

$$G(x,y,z) = (1/\sqrt{2\pi}\cdot\sigma)\cdot\exp(-(x^2+y^2+z^2)/2\sigma^2) \tag{1}$$

f_{in} and f_{out} are thus transformed as,

$$\begin{cases} g_{in} = f_{in}*G \\ g_{out} = f_{out}*G \end{cases} \tag{2}$$

The blurring results by Eq.(2) are shown in Fig. 6. It shows that noise influence still exist in some place. g_{in} and g_{out} are then transformed as gradient functions $\mathbf{v}_{in}$ and $\mathbf{v}_{out}$.

$$\begin{cases} \mathbf{v}_{in} = -\nabla\cdot g_{in}/|\nabla\cdot g_{in}| \\ \mathbf{v}_{out} = -\nabla\cdot g_{out}/|\nabla\cdot g_{out}| \end{cases} \tag{3}$$

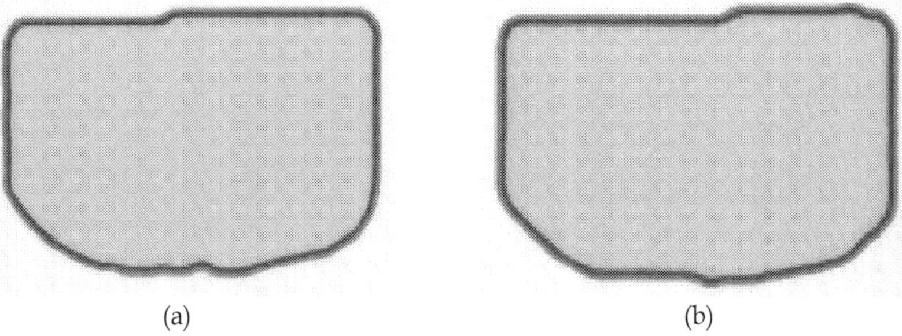

(a)   (b)

Fig. 6. Construction of gradient functions. (a) Outside gradient function. (b) Inside gradient function.

Although the computation of Eq.(3) amplifies the noise influence, it is much suitable of weighted vector median filter to reduce the noise in function $\mathbf{v}_{in}$ and $\mathbf{v}_{out}$, which is more effective and robust to denoise in image process (Barner et al. 2001). The algorithm needs to define the metric and the relationship between the elements in neighbor grid region. Let $\mathbf{v}_i, i=1,2...n$ represents the vector in a neighbor region Ω, which contains n vectors. The metric $M(\mathbf{v}_i,\mathbf{v}_j)$ of two vectors $\mathbf{v}_i$ and $\mathbf{v}_j$ can be defined as the L_p norm of the difference between them,

$$M(\mathbf{v}_i,\mathbf{v}_j) = L_p(\mathbf{v}_i,\mathbf{v}_j) = \left\|\mathbf{v}_i - \mathbf{v}_j\right\|_p \tag{4}$$

The relationship between each vector, represented as $R(\mathbf{v}_i, \mathbf{v}_j)$ should changes according to the metric. According to (Nie & Barner 2006), it can be expressed by following constraints,

$$\begin{cases} M(\mathbf{v}_i, \mathbf{v}_j) \to 0, R(\mathbf{v}_i, \mathbf{v}_j) = 1 \\ M(\mathbf{v}_i, \mathbf{v}_j) \to \infty, R(\mathbf{v}_i, \mathbf{v}_j) = 0 \\ M(\mathbf{v}_i, \mathbf{v}_j) \leq M(\mathbf{v}_k, \mathbf{v}_l), R(\mathbf{v}_i, \mathbf{v}_j) \geq R(\mathbf{v}_k, \mathbf{v}_l) \end{cases} \tag{5}$$

Usually, $R(\mathbf{v}_i, \mathbf{v}_j)$ can adopted the Gaussian function coupled with metric $M(\mathbf{v}_i, \mathbf{v}_j)$,

$$R(\mathbf{v}_i, \mathbf{v}_j) = e^{-M(\mathbf{v}_i, \mathbf{v}_j)^2 / 2\sigma^2} \tag{6}$$

$R(\mathbf{v}_i, \mathbf{v}_j)$ is adopted to modify the median vector in local region Ω for local geometric constraints. Therefore, the output of the modified vector $\mathbf{v}_f$ is defined as

$$\mathbf{v}_f = W(\mathbf{v}) = \frac{\sum_{i=1}^{n} \mathbf{v}_i R(\mathbf{v}_i, \mathbf{v}_m)}{\sum_{i=1}^{n} R(\mathbf{v}_i, \mathbf{v}_m)} \tag{7}$$

where $\mathbf{v}_m$ is the median vector defined by,

$$\mathbf{v}_m = \arg\min \sum_{i=1}^{n} M(\mathbf{v}, \mathbf{v}_i) \tag{8}$$

The neighbor region Ω is set as $5 \times 5 \times 5$ neighbor space, a little larger than the structuring elements $\mathbf{B}$ in last step. Because if it is less than $\mathbf{B}$, the noise could not be reduced, if it is too large, the computation is not efficiency, since it needs to compute the each vector $\mathbf{v}_i$ to all other vectors in neighbor region of n elements to find the median vector $\mathbf{v}_m$. Norm L_p is adopted as L_2 in the rest of the paper. According to the convolution, function $\mathbf{v}_{in}$ and $\mathbf{v}_{out}$ are redefined as

$$\begin{cases} \mathbf{v}_{in}^* = W * \mathbf{v}_{in} \\ \mathbf{v}_{out}^* = W * \mathbf{v}_{out} \end{cases} \tag{1}$$

$\mathbf{v}_{in}^*$ and $\mathbf{v}_{out}^*$ are visualized by extracting the zero-level lines from their integer function $\int \mathbf{v}_{in}^* d\Omega$ and $\int \mathbf{v}_{out}^* d\Omega$ (Fig.7).

The details of the results demonstrate that the noise influence is effectively rejected according to the comparison of one noise region (on the right in Fig.7 (a) and Fig.7 (b)). The overlapping regions don't lead any jagged errors or self-intersection. Actually, only one of the dual functions, either $\mathbf{v}_{in}^*$ or $\mathbf{v}_{out}^*$, can be used as for the surface reconstruction by gradient computation with large kernel size. But the holes can not be filled flatly, some concave parts exist (at the bottom in Fig.7 (a) and Fig.7 (b)). Since outside function f_{out} and inside function f_{in} are obtained by erosion and dilation respectively, which are opposite with each other, the reasonable resulting could be obtained by blending the dual functions.

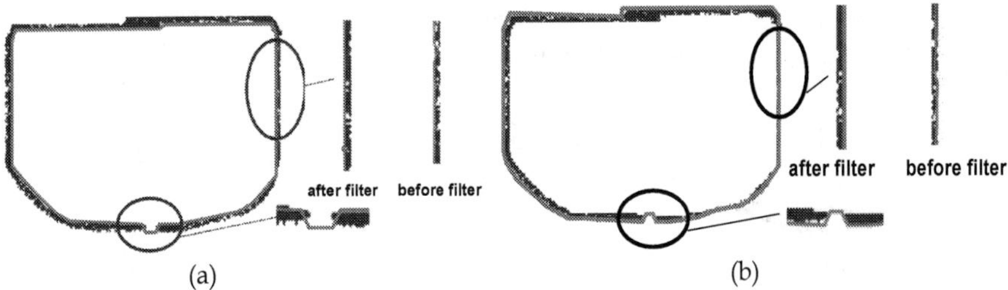

Fig. 7. The final gradient fields and details. (a) Inside gradient function. (b) Outside gradient function.

3.3 Formulate and solve PDE

Based on the dual gradient functions, a minimal energy model is proposed. The gradient of resulting surface $\phi(\mathbf{x})$ should best approximate a combined field generated in last step. The differences between them are defined as the square of L_2 norm. The object function is expressed as

$$E = \int_{\Omega} \left\| \nabla\phi - (\lambda_1 \frac{\mathbf{v}_{in}^*}{\left|\mathbf{v}_{in}^*\right|} + \lambda_2 \frac{\mathbf{v}_{out}^*}{\left|\mathbf{v}_{out}^*\right|}) \right\|_2^2 d\Omega \tag{2}$$

where λ_1 and λ_2 are positive constants for adjusting the influence of g_{in}^* and g_{out}^* to ϕ. Thus, the corresponding Euler-Lagrange equation is derived as,

$$\Delta\phi - \nabla(\lambda_1 \frac{\mathbf{v}_{in}^*}{\left|\mathbf{v}_{in}^*\right|} + \lambda_2 \frac{\mathbf{v}_{out}^*}{\left|\mathbf{v}_{out}^*\right|}) = 0 \tag{3}$$

The boundary condition is $\phi_{\partial V} = 0$. Actually, this method treats the two gradient functions $\mathbf{v}_{in}$ and $\mathbf{v}_{out}$ equally, thus, the positive parameter is set as $\lambda_1 = \lambda_2 = 1$. The PDE is a typical Poisson equation, there are many methods to solve the classic equation. This paper adopts Fast Fourier Transform (FFT) method, since it only needs to transform the equation as Fourier series and solve a linear equation.

To visualize the resulting surface, the level set valve of ϕ should be confirmed. Theoretically, zero-level set is the object surface, however, due to the weighted combination, the valve of the object level set changes little from zero. This paper adopts the approximation valve of the positions of the input samples. It is confirmed by evaluating ϕ at the sample positions and using the average of the values for iso-surface extraction,

$$\partial\phi = \{p \in R^3 \,|\, \phi(p) = \gamma\} \; with \; \gamma = \frac{1}{m}\sum_{i=1}^{m}\phi(p_i) \tag{4}$$

where γ is iso-value, P is the point sets. The marching cubes method (William & Harvey 1987) is employed to extract the iso-surface. The details are shown in Fig. 8.

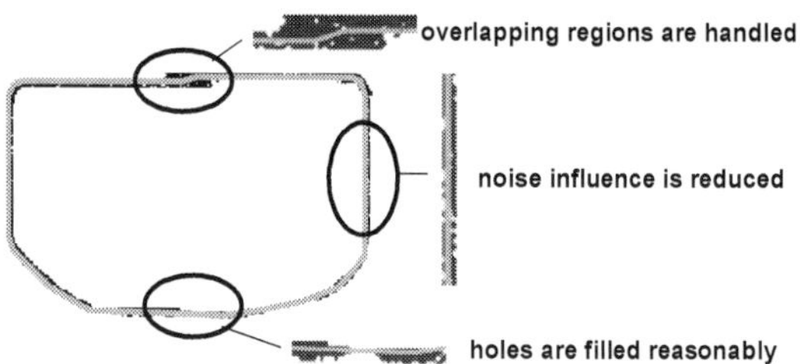

Fig. 8. Result surface and details.

The dual functions $\mathbf{v}_{in}$ and $\mathbf{v}_{out}$ give reasonable constraints to guarantee the global shape of the resulting surface. The parameters $\lambda_1 = \lambda_2 = 1$ make the overlapping regions and holes naturally adopt the flat result. Compared with the result in Fig.7, besides the noise influence, the holes are filled flatly and the surface patch in overlapping samples is much reasonably.

4. Numerical examples and analysis

In this paper, the implementation employs PC CPU 2G Hz and 1G main RAM with the soft platform Matlab coupled with C++ API. To demonstrate the effectiveness and robustness, this paper takes the point clouds series sampled from a fan disk model (Fig. 9 (a)) as the examples. The point clouds (Fig. 9 (b)) are sampled with uniform density $0.01mm$, thus the length of the voxel is set as $h = 0.01mm$.

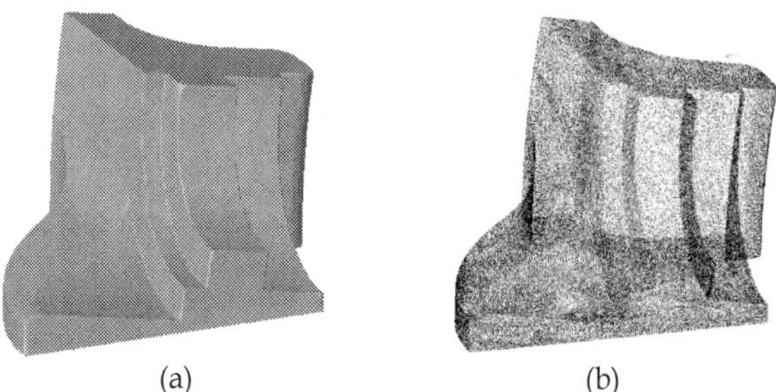

(a)   (b)

Fig. 9. Fan disk point clouds without noise. (a) Geometric model. (b) Original point clouds.

This paper gives all the situations of the defective samples as shown in the left row of Fig.10, including sparse point clouds (Fig.10 (a)), point clouds with random noise ($\varepsilon_r \sim N(0,0.1^2)$, Fig.10 (c)), point clouds with holes (Fig. 10 (e)), with overlapping samples (Fig.10 (g)) and the hybrid point clouds (Fig.10 (i)) which contains all the defective situations. The point clouds with holes are generated by reducing some random places on the original samples (Fig.9 (b)). The overlapping samples are generated by mis-registration (the error is set as $0.05mm$) which often make the resulting surface have some scallops. The right row is the corresponded resulting surfaces and the details (from Fig. 10 (b) to Fig. 10 (j)). The propose method is much convenient to implement, the hybrid defective samples (Fig.10 (i)) needs not any extra steps, the holes of resulting surface (Fig. 10 (j)) are filled smoothly and no self-intersections exist in overlapping samples.

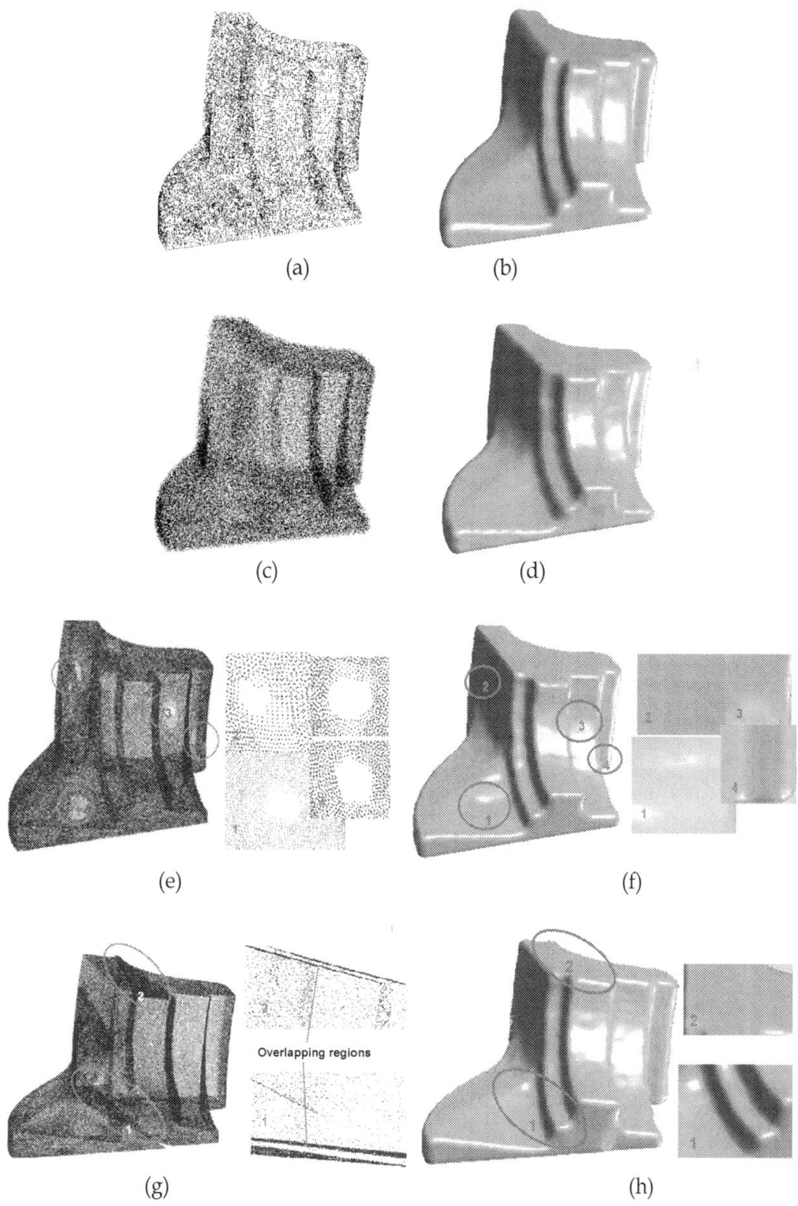

(a)

(b)

(c)

(d)

(e)

(f)

(g)

(h)

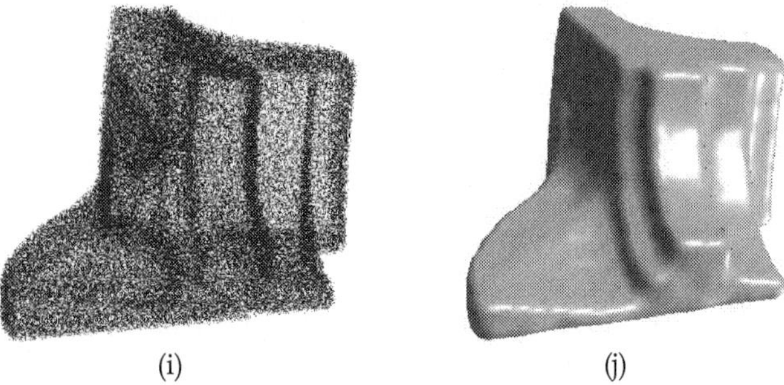

Fig. 10. Examples of fan disk. (a) Sparse point clouds. (b) Resulting surface of sparse point clouds. (c) Noisy point clouds. (d) Resulting surface of noisy point clouds. (e) Point clouds with holes and details. (f) Resulting surface of (e) and details. (g) Point clouds containing overlapping regions and details. (h) Resulting surface of (g) and details. (i) Hybrid defective samples. (j) Resulting surface of Hybrid defective samples.

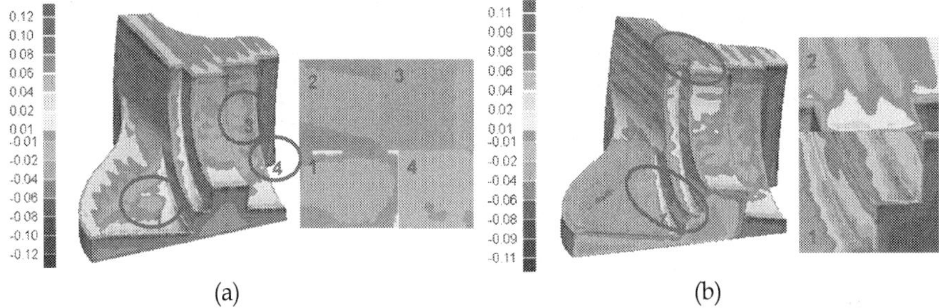

Fig. 11. Error destruction with color map. (a) Result of point clouds with holes. (b) Result of point clouds with overlapping samples.

The numerical details of all the examples about fan disk are shown in Table.2. The time complexity of the method generally includes three main components, dilation-erosion ($O(N)$), weighted vector median filter ($O(m)$) and FFT ($O(N\log(N))$). As weighted vector median filter adopts a fixed window of neighbor space, its time complexity is only relevant to point number m, which is far less than grid number N. Therefore, the computing time mainly depends on the resolution of space grids, which is related with the point density. Since the point clouds are all generated from same original model, the computing time is not difference too much except for the sparse points. The whole time of all the examples is within 80 seconds. This paper adopts the average errors (between resulting surfaces and point clouds) as the main accuracy standard for evaluation. The average errors of all examples are lower than $0.05mm$, which is accuracy enough to satisfy the practical application. The results also demonstrate that the noise and overlapping regions can cause more errors than other defective samples since they often influence some sharp corners of resulting surface. Besides the average errors to the point clouds, this paper gives the comparison between the resulting surface and original model (Fig.9 (a)) with error distribution of colour map (Fig.11) Fig. 11 (a) is the result of point clouds with holes. the hole are filled smoothly and reasonably, thus the errors of the holes are nearly the same as their neighbor regions. Fig. 11 (b) is the result of point clouds with overlapping samples. Of course, the errors are larger than other regions due to mis-registration, but he overlapping samples are merged reasonably and they don't cause any impulse changes in resulting surface, thus their errors change smoothly.

Point clouds of fan disk	Number of points	Grid resolutions	Compute time (s)	Average errors (mm)
Original points	100448	$163 \times 163 \times 163$	75.6355	0.014
Sparse points	10908	$126 \times 88 \times 76$	48.6522	0.031
Noisy points	100448	$163 \times 163 \times 163$	74.9853	0.045
Points with holes	99155	$163 \times 163 \times 163$	75.6654	0.016
Points with overlapping	125567	$163 \times 163 \times 163$	76.1203	0.035
Hybrid defective samples	124221	$163 \times 163 \times 163$	75.2368	0.046

Table 2. Details of the example about fan disk model.

In fact, the length of voxel could not follow the density of point clouds strictly, but if it is not set suitably, the resulting surface becomes over-fit or over-smooth cases. Two examples with "bad" grid size are shown in Fig. 12, which are both the resulting surface of noisy point clouds (Fig. 10 (c)). Fig. 12 (a) is the over-fit resulting example with grid $h = 0.004mm$. Fig. 10 (b) is the over-smooth case with larger grid size $h = 0.3mm$. This paper suggests a suitable grid size as $0.8p \leq h \leq 1.2p$, where p is the average density of point clouds.

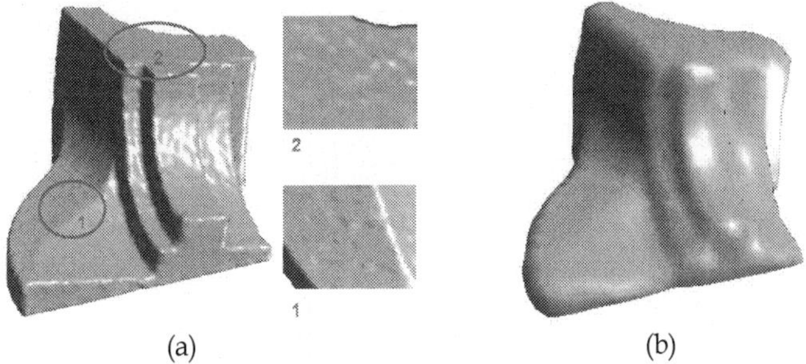

(a)   (b)

Fig. 12. Reconstruction with different grid size. (a) Resulting surface and details with smaller grid size. (b) Final resulting surface with larger grid size.

Beside the theoretical model of fan disk, this paper also adopts some practical examples since in real case the overlapping regions and holes are complicated. The following practical point clouds are scanned by the hand-held digitizer (type number: Cimcore Infinite Sc2.4). Fig.13 (a) shows the point clouds of a mechanical part. It is the example containing much overlapping samples (details labeled in circles). The resulting surface (Fig.13 (b)) demonstrates the overlapping regions can be reasonably fitted and smoothed. The next example is the point clouds of piston rod. In the middle bottom of Fig.14 (a) is the points within a section plane, where overlapping samples exist. The detail of a hole is shown in the lower right (Fig.14 (b)). In practice, the sparse points often exist in un-uniform point data. Fig.15 (a), point clouds of engine outtake ports from an automobile in real case, shows the situation. Because the density is not uniform, this paper could adopt the average density to decide the grid size. The result of smooth and water-tight surface is shown in Fig.15 (b).

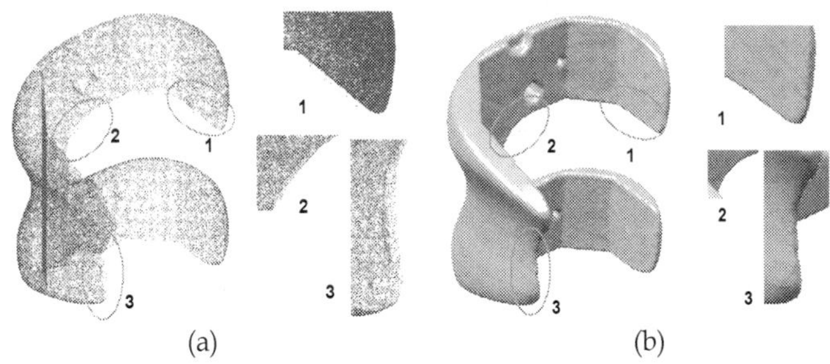

(a)   (b)

Fig. 13. Reconstruction of mechanical part. (a) Point clouds and details. (b) Final resulting surface and details.

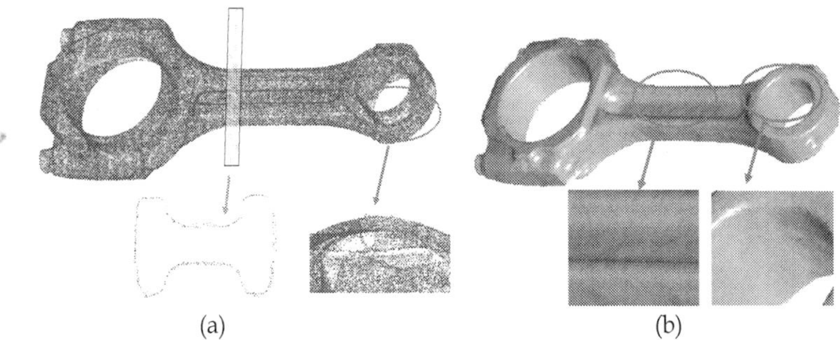

(a)   (b)

Fig. 14. Reconstruction of piston rod. (a) Point clouds and details. (b) Final resulting surface and details.

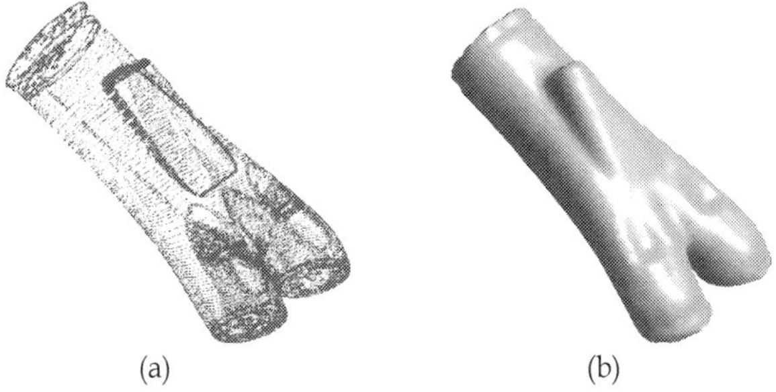

(a)   (b)

Fig. 15. Reconstruction of engine outtake ports. (a) Point clouds and details. (b) Final resulting surface and details.

Fig.16 shows the example of an ancient cup which contains much free-form details for preserve. Since the point clouds have no holes, just little noise but many overlapping samples due to multi-scanning, the non-uniform grids are employed. Detail 3 in Fig.16 is the little noisy part, Details 1 and details 2 are the overlapping regions because the cup bottom is hard to scan within only once. The point clouds are density enough, the structuring element **B** could be chosen with small size. In the example, the structuring element size is set as $1 \times 1 \times 1$, just only for merging overlapping samples. The resulting surface and details are shown in Fig. 16 (b). The resulting surface is smooth, the noise influence and overlapping regions are reduced. Although only fewer small details are blurred due to the dilation-erosion operation, but most shape features are preserved.

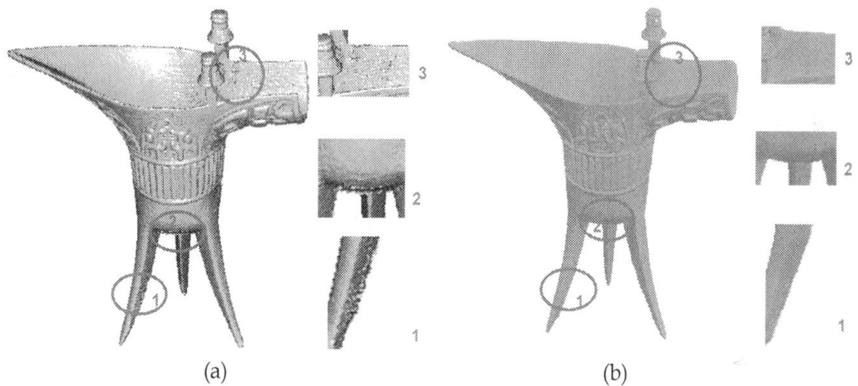

(a)   (b)

Fig. 16. Reconstruction of ancient cup. (a) Point clouds and details. (b) Final resulting surface and details.

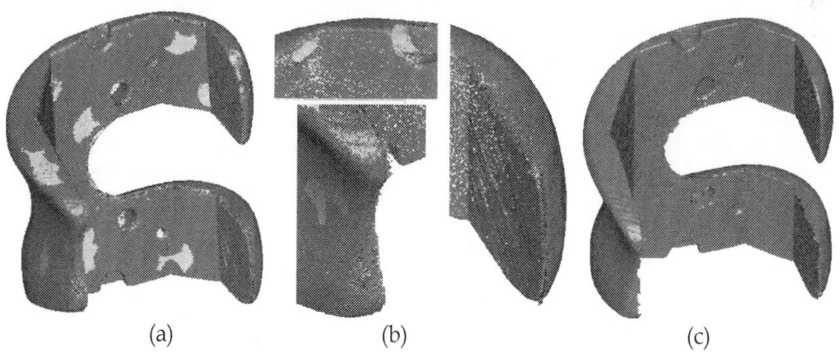

(a)   (b)   (c)

Fig. 17. Triangulation of mechanical part. (a) Triangular meshes. (b) Details of meshes. (c) Meshes after artificial holes-filling.

Because this method needs not any triangulation, the efficiency is highly improved more than the methods which need triangulation. The mechanical part (Fig. 13 (a)) is taken as an example for comparison. This paper uses Geomagic (version 8.0) which is a widely used commercial software utility in reverse engineering domain. It performs reconstruction based on building triangular meshes, as most conventional methods do. The result is shown in Fig. 17 (a). As the influence by the noise and non-uniform regions in the point clouds, it is difficult to construct a water-tight triangular mesh (Fig. 17 (a)). Some holes and rough place exist (Fig. 17 (b)). Thus, the artificial work of filling holes has to be performed (Fig. 17 (c)). The time of triangulation and reconstruction by this method is given in Table 3, where the time of Goemagic contains no artificial filling time. The comparison demonstrates that if the number of point clouds increase, or much noise exist, the triangulation time increases more than the time of surface reconstruction by this method.

Number of points	Time of triangulation (s)	Reconstruction time of this method (s)
60000	30.3324	———
120000	66.5633	———
250000	81.9623	———
500000	171.9932	———
550000	180.6897	76.5367

Table 3. Time comparison between triangulation and reconstruction of this method with different sample point clouds.

The proposed method is especially useful to deal with the point clouds of "bad-scanning" like the example shown in Fig.18. It is a resin mould of engine intake ports from an automobile in real case (Fig. 18 (a)) and the original CAD model is shown in Fig.18 (b). Because the resin mould is too soft to fix, the point clouds by multi-scanning have larger errors of mis-registration (e.g. labeled by circle 1 in Fig. 18 (c)) than all examples above. The holes (e.g. labeled by circle 2 in Fig. 18 (c)) are much large and the noise appears everywhere. The points are much disadvantageous to obtain the specific information, especially triangular mesh with high quality. This paper gives the results by two typical methods based on Delaunay triangulation. Fig.18 (d) is the triangular mesh reconstructed by Raindrop Geomagic. The defective samples obviously influence the results: the holes need to be filled by other artificial work and the overlapping regions lead to some jagged triangles. Fig. 18 (e) is the resulting surface by power crust method (Amenta et al. 2001). Although the resulting surface is watertight, the surface is rugged and overlapping regions are self-intersect. The two results need some complex post-process. Fig. 18 (f) shows the resulting surface by level set method(Zhao et al. 2000), which need no specific information. The resulting surface is much better, but the surface is not smooth enough, especially the regions of holes. The final surface by this method is shown in Fig. 18 (f). This method guarantees a smooth and water-tight surface, holes are filled flatly and no self-intersections in overlapping samples. The "bad-scanning" samples need more dilation-erosion operations, the resulting surface is acceptable and convenient for certain post-process. But if the input data have too large holes or serious overlapping samples, the details of the resulting surface may be blurred due to too many dilation-erosion times.

The numerical details of the practical examples are shown in Table. 4. Since the point data of engine intake ports contain so many defective samples, the resulting surface has the largest average errors than other examples. Even if the ancient cup example has nearly 1 million points, the compute time is only 90 seconds.

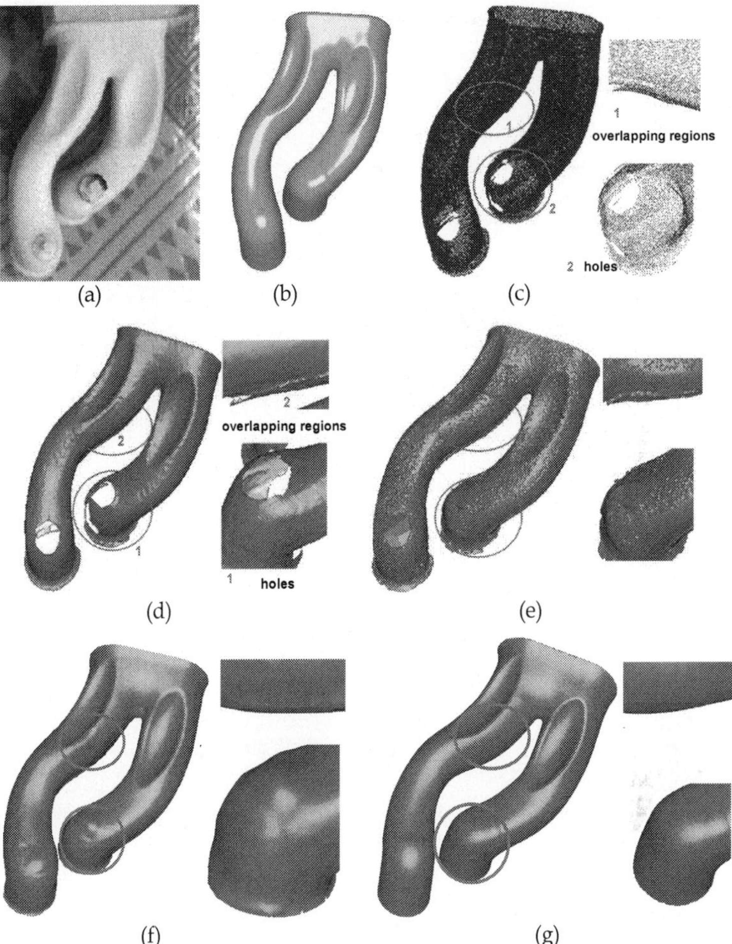

Fig. 18. Example of engine intake ports. (a) Resin mould of engine intake ports. (b) Original CAD model. (c) Point clouds and details. (d) Triangular mesh by Geomagic software. (e) Triangular mesh by power crust method. (f) Resulting surface and details by level set method. (g) Resulting surface and details by proposed method.

Name of point clouds	Number of points	Grid resolution	Average errors (mm)
Mechanical part	537925	136×188×154	0.022
Piston rod	412847	242×106×150	0.021
Engine outtake ports	98175	66×63×41	0.035
Ancient cup	999944	144×307×344	0.019
Engine intake ports	154890	94×126×84	0.087

Table 4. Details of practical examples.

5. Conclusions and future work

This paper presents a novel implicit surface reconstruction method based on dual off-set gradient functions. Its core idea is to construct the dual off-set functions and generate the resulting surface indirectly. Through the extensive examples, it is indicated that the proposed method is robust to reconstruct discrete point sets, especially practical point data or "bad-scanning" data. The method need not any specify information, thus it can skip complex pre-process and make the reconstruction process much efficient. The morphology operation is based on set calculation so it is much faster to make off-set surfaces water-tight. The weighted vector median filter algorithm is extended into 3D space for reducing the noise influence and making the final surface much smooth. The dual relative functions construct a minimal crust surrounding the points from dual side, which can guarantee the holds and overlapping samples are fitted reasonably.

In future research, the problem of preserving details from "bad-scanning" points would be well-studied. Some advanced hierarchical data structures would be discussed for more efficiency implementation. The choice of structuring elements in morphology is a world classic problem in the field of image processing. Some improved structuring elements would be discussed in future work, such as combinational shape of structuring elements. The method would also be improved to handle some complex non-manifold point clouds. Some other image processing methods would also be extended in surface reconstruction, like some adaptive filter algorithms. To generate a suitable surface from defective point data, it is much important to employ industry prior design knowledge. With such prior but general knowledge, the resulting surface could be much reasonable than what are reconstructed only based on geometric information. The research would also focus on this respect.

6. Acknowledgment

The research work is supported by Dongfang Electric Corporation Research & Develop Centre, Intelligent Equipment & Control Technology Institute and the Natural Science Fund of China (NSFC) (Projects No. 50835004 and 50625516).

7. References

Alexa, M., et al. (2001). Point set surfaces, *Proceedings of the conference on Visualization '01*, pp. 21-28, ISBN 0-7803-7200-X, San Diego, California

Alliez, P., et al. (2007). Voronoi-based variational reconstruction of unoriented point sets, *Proceedings of the fifth Eurographics symposium on Geometry processing*, pp. 39 - 48, Barcelona, Spain

Amenta, N., et al. (2000). A simple algorithm for homeomorphic surface reconstruction, *Proceedings of the sixteenth annual symposium on Computational geometry*, pp. Clear Water Bay, Kowloon, Hong Kong

Amenta, N., et al. (2001). The power crust, unions of balls, and the medial axis transform. *Computational Geometry*, Vol. 19, No. 2-3, pp. 127-153

Anders, A. & A. Marc. (2003). Approximating and intersecting surfaces from points, *Proceedings of the 2003 Eurographics/ACM SIGGRAPH symposium on Geometry processing*, pp. 230-239, ISBN 1-58113-687-0, Aachen, Germany

Attali, D. (1998). r-regular shape reconstruction from unorganized points. *Computational Geometry*, Vol. 10, No. 4, pp. 239-247

Baining, G., et al. (1997). Surface Reconstruction Using Alpha Shapes. *Computer Graphics Forum*, Vol. 16, No. 4, pp. 177-190

Barner, K. E., et al. (2001). Fuzzy ordering theory and its use in filter generalization. *EURASIP Journal on Applied Signal Processing*, Vol. 4, No. pp. 206-218

Boissonnat, J. D. & F. Cazals. (2002). Smooth surface reconstruction via natural neighbour interpolation of distance functions. *Computational Geometry: Theory and Applications*, Vol. 22, No. 1-3, pp. 185-203

Brian, C. & L. Marc. (1996). A volumetric method for building complex models from range images, *Proceedings of the 23rd annual conference on Computer graphics and interactive techniques*, pp. 303 - 312, ISBN 0-89791-746-4, New York, USA

Carr, J. C., et al. (2001). Reconstruction and representation of 3D objects with radial basis functions, *Proceedings of the 28th annual conference on Computer graphics and interactive techniques*, pp. 67 - 76, ISBN 1-58113-374-X, New York, USA

Chandrajit, L. B., et al. (1995). Automatic reconstruction of surfaces and scalar fields from 3D scans, *Proceedings of the 22nd annual conference on Computer graphics and interactive techniques*, pp. 109 - 118, ISBN 0-89791-701-4, New York, USA

David, B. & B. Guido. (2005). An extended concept of voxel neighborhoods for correct thinning in mesh segmentation, *Proceedings of the 21st spring conference on Computer graphics*, pp. 119-125, ISBN 1-59593-204-6, Budmerice, Slovakia

Davis, J., et al. (2002). Filling holes in complex surfaces using volumetric diffusion, *3D Data Processing Visualization and Transmission, 2002. Proceedings. First International Symposium on*, pp. 428-861, ISBN 0-7695-1521-5, Padova, Italy

Edelsbrunner, H. & E. P. Mücke. (1994). Three-dimensional alpha shapes. *ACM Transactions on Graphics (TOG)*, Vol. 13, No. 1, pp. 72

Esteve, J., et al. (2008). Piecewise algebraic surface computation and smoothing from a discrete model. *Computer Aided Geometric Design*, Vol. 25, No. 6, pp. 357-372

Esteve, J., et al. (2005). Approximation of a Variable Density Cloud of Points by Shrinking a Discrete Membrane. *Computer Graphics Forum*, Vol. 24, No. 4, pp. 791-807

Gael, G. & G. Markus. (2007). Algebraic point set surfaces, *ACM SIGGRAPH 2007 papers*, pp. 23, ISBN 0730-0301, San Diego, California

Greg, T. & F. O. b. James. (2002). Modelling with implicit surfaces that interpolate. *ACM Trans. Graph.*, Vol. 21, No. 4, pp. 855-873

Hoppe, H., et al. (1992). Surface reconstruction from unorganized points. *Computer Graphics(ACM)*, Vol. 26, No. 2, pp. 71-78

Jean-Daniel, B. (1984). Geometric structures for three-dimensional shape representation. *ACM Trans. Graph.*, Vol. 3, No. 4, pp. 266-286

LA Piegl, W. T. (1997). *The Nurbs Book*. Splinger-Verlag, ISBN 3-540-6i545-8 Berlin

Li, X., et al. (2009). On surface reconstruction: A priority driven approach. *Computer-Aided Design*, Vol. 41, No. 9, pp. 626-640

Lin, Y., et al. (2009). Dual-RBF based surface reconstruction. *The Visual Computer*, Vol. 25, No. 5, pp. 599-607

Liu, S. & C. C. L. Wang. (2009). Duplex fitting of zero-level and offset surfaces. *Computer-Aided Design*, Vol. 41, No. 4, pp. 268-281

Maragos, P., et al. (1996). *Mathematical Morphology and Its Applications to Image and Signal Processing*. Kluwer Academic, ISBN 0792397339 9780792397335, Boston, USA

Marc, A. & A. Anders. (2004). On normals and projection operators for surfaces defined by point sets, *Proc. Eurographics Symposium on Point-based Graphics*, pp. 149 - 155, ISBN

Marc, A., et al. (2001). Point set surfaces, *Proceedings of the conference on Visualization '01*, pp. San Diego, California

Michael, K., et al. (2006). Poisson surface reconstruction, *Proceedings of the fourth Eurographics symposium on Geometry processing*, pp. 61 - 70, Cagliari, Sardinia, Italy

Nie, Y. & K. E. Barner. (2006). The fuzzy transformation and its applications in image processing. *IEEE Transactions on Image processing*, Vol. 15, No. 4, pp. 910-927

Nina, A., et al. (1998). A new Voronoi-based surface reconstruction algorithm, *Proceedings of the 25th annual conference on Computer graphics and interactive techniques*, pp. 415 - 421, ISBN 0-89791-999-8, New York, USA

Peng, Q. & M. Loftus. (1998). A new approach to reverse engineering based on vision information. International Journal of Machine Tools and Manufacture, Vol. 38, No. 8, pp. 881-899

Peng, Q. & M. Loftus. (2001). Using image processing based on neural networks in reverse engineering. *International Journal of Machine Tools and Manufacture*, Vol. 41, No. 5, pp. 625-640

Ravikrishna, K., et al. (2004). Spectral surface reconstruction from noisy point clouds, *Proceedings of the 2004 Eurographics/ACM SIGGRAPH symposium on Geometry processing*, pp. 11 - 21, ISBN 3-905673-13-4, Nice, France

Roca-Pardinas, J., et al. (2008). From laser point clouds to surfaces: Statistical nonparametric methods for three-dimensional reconstruction. *Computer-Aided Design*, Vol. 40, No. 5, pp. 646-652

Shen, J., et al. (2005). Accurate correction of surface noises of polygonal meshes. *International Journal for Numerical Methods in Engineering*, Vol. 64, No. 12, pp. 1678-1698

Shen, J., et al. (2009). Spectral moving removal of non-isolated surface outlier clusters. *Computer-Aided Design*, Vol. 41, No. 4, pp. 256-267

Song, X. & B. Jüttler. (2009). Modeling and 3D object reconstruction by implicitly defined surfaces with sharp features. *Computers & Graphics*, Vol. 33, No. 3, pp. 321-330

Vanco, M. & G. Brunnett. (2004). Direct segmentation of algebraic models for reverse engineering. *Computing*, Vol. 72, No. 1, pp. 207-220

Veltkamp, R. C. (1995). Boundaries through Scattered Points of Unknown Density. *Graphical Models and Image Processing*, Vol. 57, No. 6, pp. 441-452

William, E. L. & E. C. Harvey. (1987). Marching cubes: A high resolution 3D surface construction algorithm, *Proceedings of the 14th annual conference on Computer graphics and interactive techniques*, pp. 163 - 169 ISBN 0-89791-227-6,

Xie, P., et al. (2004). Surface Reconstruction of Noisy and Defective Data Sets, *Proceedings of the conference on Visualization '04*, pp. 259 - 266, ISBN 0-7803-8788-0, Washington DC, USA

Yang, Z., et al. (2009). Adaptive triangular-mesh reconstruction by mean-curvature-based refinement from point clouds using a moving parabolic approximation. *Computer-Aided Design*, Vol. No. pp.

Yukie, N., et al. (2009). Smoothing of Partition of Unity Implicit Surfaces for Noise Robust Surface Reconstruction. *Computer Graphics Forum*, Vol. 28, No. 5, pp. 1339-1348

Yutaka, O., et al. (2003). Multi-level partition of unity implicits. *ACM Trans. Graph.*, Vol. 22, No. 3, pp. 463-470

Zhao, H.-K., et al. (2000). Implicit and Nonparametric Shape Reconstruction from Unorganized Data Using a Variational Level Set Method. *Computer Vision and Image Understanding*, Vol. 80, No. 3, pp. 295-314

Part 3

Other Applications and Theory

12

Advanced NO$_x$ Sensors for Mechatronic Applications

Angela Elia[1], Cinzia Di Franco[1], Adeel Afzal[2],
Nicola Cioffi[2] and Luisa Torsi[2]
[1]CNR-IFN U.O.S. Bari, Bari,
[2]Department of Chemistry, University of Bari, Bari,
Italy

1. Introduction

Vehicle emissions represent an increasing contributor to air pollution in urban and country areas which gives rise to a range of environmental problems related to air quality. This has led to increasingly stringent regulatory laws on exhaust emissions level and composition.

The US regulations, signed in Dec. 2000, set new standards for measurement of automotive exhaust emissions. The emission limits for regulated species, such as carbon oxides (CO and CO_2), hydrocarbons (THC), and nitrogen oxides, i.e. NO, NO_2, N_2O (referred as NO_x), are being reduced, and new components such as methanol and formaldehyde are being added to the list of monitored species. In Japan, diesel emission standards require that in-use on-road light commercial vehicles should meet NO_x emission of 0.25 g/km starting from the end of 2005 and achieve full implementation by 2011 (Nakamura et al., 2006). The European Commission has also introduced a series of regulations, the so called EURO Emission directives, from Euro I in 1991 to tighter limitations in 2009 (Euro V 80 g/km) and 2014 (Euro VI 0.08 g/Km), to meet the air quality standards stated by the international agencies (van Asselt & Biermann, 2007). The standard and validated on-line technologies for regulated emissions are effective at monitoring few components, but are limited in their use for measuring other gases. Single-wavelength non-dispersive infrared filters for CO and CO_2 monitoring cannot be used for other species due to interferences from water and other molecules. Chemiluminescence analyzers, which traditionally are used to measure NO_x compounds, cannot differentiate NO from NO_2 in the same test, nor identify the other NO_x gases. Flame ionization detectors cannot differentiate individual hydrocarbons. To measure raw exhaust, each technique requires a cold trap for water vapour, which can affect the concentrations of other gases. In addition, calibrations are necessary for each analysis, avoiding on-line and real-time emission monitoring.

To sample other gases in the exhaust, bag samples must be collected and taken to a laboratory for further analysis. Expensive dilution equipment must be used to prevent water condensation in the bag. Each bag sample gathers exhaust for several minutes, therefore the final result of the test is an integrated average of the gas concentrations and all time resolution is lost. Methanol and formaldehyde samples are collected with impingers, which dissolve the gases by passing them through a water-based solution. The extract is then analyzed using high-performance liquid chromatography (HPLC). Gas chromatography

(GC) can measure many of the other hydrocarbons in a bag sample with high resolution and accuracy. While HPLC and GC are effective and widely used, they are unable to track transient concentration information.

The ideal analyzer for emissions testing would combine the advantages of on-line, real-time analysis with the accuracy and speciation of laboratory-based methods. Changes in emissions chemistry that occur with changes in engine speed and torque need to be measured every second. The analyzer must be able to identify the point at which exhaust levels are reduced by the catalyst as it heats up.

New emissions regulations are forcing auto, diesel and catalyst producers to better understand combustion and emission reduction processes. The optimization of these processes requires sub-second analytical data and new methodologies to monitor gas species such as NH_3 and N_2O that are not currently monitored by traditional analyzers.

Among the different pollutants present in vehicle exhaust, nitric oxide (NO) detection is of particular interest. NO is emitted from the exhausts of both gasoline and diesel engine vehicles and is generated during high temperature combustion processes from the oxidation of nitrogen in the air or fuel. NO contributes to ground-level ozone (Alving et al., 1993), acid rains and a variety of adverse human health effects (Seinfeld & Pandis, 1998), which have led to increasingly stringent regulatory mandates on the emission of NO.

Thus, accurate and reliable real-time NO sensors are required as an important part of any control system. In particular, the measurement of NO directly in the engine or in the exhaust could link into engine management systems for its control and its in situ measurement could also be used to test the catalytic converter efficiency. Over the years, a huge number of optical and electronic technologies have been developed and applied to produce sensors with remarkable sensitivity toward nitric oxide and efficient response (and recovery) kinetics. In this chapter, particular emphasis will be given on the recent achievements and ground-breaking results obtained in the past few years, namely 2008-2011, in the field of both optical and electronic NO_x sensors.

NO_x optical sensor technology is among the fastest growing for mechatronic applications, as a result of its high sensitivity and selectivity, high speed, accuracy, and capability of multi-species detection. On the other hand, they need sophisticated and cumbersome equipments. In this chapter, novel spectroscopic sensing schemes suitable for the integration in high performance automated inspection systems will be reviewed.

Electronic metal oxide devices offer the advantages of low cost, low cross sensitivity to humidity and an output signal which is easy to be read and processed. Disadvantages, however, include limitations for operating at high temperatures, signal drift over time, limited selectivity or sensitivity as well as high power consumption. For these reasons, the use of novel materials such as innovative nanostructures, in place of metal-oxides, is being widely investigated in chemiresistors and transistor based device structures. Improvements have been seen for selectivity, but operation at high temperature is still an open issue.

2. Mid infrared absorption spectroscopy

Optical absorption methods represent a valid alternative to traditional extractive methods having the potential for fast, sensitive, accurate, selective and in situ measurements even in the presence of harsh conditions in terms of temperature, pressure, gas composition and presence of particulate.

Different spectroscopic schemes, based on the absorption of electromagnetic radiation by the gas sample, have been developed. The empirical relationship relating the absorption of light to the composition of the gas mixture, when the light travels through, is given by the Lambert-Beer law. In the absence of optical saturation and particulate-related scattering, the absorbance is proportional to the concentration of gas. So, the gas concentration can be obtained from the absorption spectrum with a predetermined correlation.

Mid-infrared spectroscopic schemes show particular promise owing to the potential for accessing strong infrared (IR) absorption transitions. In the mid-infrared region of the electromagnetic spectrum 2.5 - 25 μm (fingerprint region), most molecular species exhibit a unique spectral signature, i.e. a characteristic series of fundamental absorption lines due to transitions between rotational-vibrational states, characterized by very large cross-sections.

Engine exhaust emissions can be successfully analyzed by Fourier Transform Infrared (FTIR) and laser based spectroscopic techniques. Their advantages over more traditional measurement techniques include direct sampling of raw exhaust, measurement of many compounds with one analyzer and highly-stable calibrations.

In the following paragraphs we will report on innovative solutions for sensitive and selective NO monitoring in vehicle exhaust based on IR spectroscopic techniques.

2.1 Fourier transform infrared spectroscopy

FTIR spectroscopy is known as a reliable, self-validating and powerful tool for vehicle emission analysis due to its multicomponent detection capability, sensitivity and time resolution (Adachi, 2000; Durbin, 2002). It is especially well suited for monitoring non-regulated pollutants when testing alternative fuels and newer emission-control technologies. The FTIR technique is a well established methodology which has been validated by several regulatory and standardization agencies for extractive gas-sampling analysis (EPA, 1998; Reyes, 2006). Many highly reactive species in the exhaust can be simultaneously measured by FTIR even below part-per-million levels, replacing several discrete analyzers which may require complicated calibration procedures.

The analyzer developed by the Thermo Fischer Scientific Company, Antaris IGS, allows concentrations of up to 40 gases to be calculated simultaneously at one-second intervals (Thermo, 2007) and requires minimal maintenance and recalibration.

The MultiGas™ 2030 HS by MKS Instruments (Tingvall et al., 2007), has been designed to measure both traditional and non-traditional combustion emissions gases. The system incorporates a patented fast scanning FTIR capable of providing high resolution (0.5 cm^{-1}) data at 5 Hz frequency. The system was also configured to allow combustion exhaust to flow through the 200 mL gas cell at rates up to 100 L/min to prevent diffusion and measurement delay. The software and computer hardware provided with each system are optimized to allow the simultaneously quantification of 20 gases.

Reyes and co-workers developed a method to acquire valuable information on the chemical composition and evolution of emissions from typical driving cycles of a Toyota Prius hybrid vehicle in the Mexico City Metropolitan Area (Reyes, 2006). The analysis of the gases is performed by passing a constant flow of a sample gas from the tail-pipe into a 10 L multi-pass cell. The absorption spectra within the cell are obtained using an FTIR spectrometer at 0.5 cm^{-1} resolution along a 13.1 m optical path. Additionally, the total flow from the exhaust is continuously measured from a differential pressure sensor on a Pitot tube installed at the exit of the exhaust. This configuration aims to obtain a good speciation capability by co-adding spectra during 30 s and reporting the emission of NO and other non-regulated pollutants.

2.2 Quantum cascade laser-based optical techniques

The development of mid infrared detection laser-based techniques has received a significant boost from the invention and   optimization of efficient mid-infrared semiconductor laser sources which can effectively substitute optical methods based on the study of overtones and combination of lines, falling in the near-infrared spectral region, where the absorption cross sections drop by orders of magnitudes.

Among the absorption techniques, direct absorption spectroscopy, cavity enhancement approaches and photoacoustic (PA) spectroscopy have the potentiality to give sensitive and selective sensors for on-line and in-situ applications. The first two techniques take advantage of long optical path length absorption in multi-pass cells and high finesse optical cavities, respectively. Although they are characterized by high sensitivity, they need sophisticated and cumbersome equipments. More compact and transportable sensors can be obtained by using the photoacoustic spectroscopy, which presents many advantages, i.e. high sensitivity (up to parts per billion detection limits), compact set-up, fast time response and portability.

In the mid infrared region traditional source options include gas lasers (CO, CO_2), lead-salt diode lasers, coherent sources based on difference frequency generation (DFG) and optical parametric oscillators (OPOs). While these lasers have allowed effective spectrometers with trace-level sensitivity, they have several disadvantages: lack of continuous wavelength tunability and large size and weight of gas lasers, low output power and cooling requirement of lead salt diode lasers, inherently low infrared power and finite line width of nonlinear optical devices.

The development of innovative mid infrared laser sources, the quantum cascade lasers (QCLs), has given a new impulse to infrared laser-based trace gas sensors. QCLs are unipolar semiconductor lasers based on intersubband transitions in a multiple quantum-well heterostructure. They are designed by means of band-structure engineering and grown by molecular beam epitaxy techniques (Faist et al., 1994).

The benefit of this approach is a widely variable transition energy primarily regulated by the thicknesses of the quantum well and barrier layers of the active region rather than the band gap as in diode lasers. Typical emission wavelengths can be varied in the mid-infrared range 3.5–24 μm up to THz. Moreover, they are characterized by single-frequency operation, narrow linewidth (< 30 MHz), high powers (up to few watts), and continuous wave (cw) operation also at room temperature.

To date, QCLs have been used for measurements of different gases (NO, CO_2, NH_3 etc.) in industrial and vehicle exhaust emission with sensitivities in the range of the parts per million/trillion by volume (ppmv/pptv) (Kosterev & Tittel, 2002a; Elia et al., 2011).

In the following sections, innovative quantum cascade laser-based optical sensors for NO detection, in automotive exhaust applications, will be reviewed.

2.2.1 Direct absorption spectroscopy

Direct absorption spectroscopy based on quantum cascade laser has been extensively studied in the last years for in situ measurements of gas components in vehicle exhaust (Weber et al., 2002; McCulloch et al., 2005; Kasyutich et al., 2009; Hara at al., 2009).

QCL-based absorption analysers offer an improvement in performance, higher sensitivity and selectivity, even at low concentrations, with respect to the other spectroscopic schemes.

Kasyutich et al. (Kasyutich et al., 2009) measured NO in the direct exhaust gas of a gasoline engine, whose operating conditions can be varied in a controlled manner, with a sensitivity of 8 ppmv. They measured NO concentrations in real-time in the range 0 - 2000 ppmv with a temporal resolution (1 s) suitable to respond to changes in engine operating conditions.

Figure 1 reports the main components of the mid-IR spectrometer based on a thermoelectrically-cooled cw single mode quantum cascade laser used for NO measurements in the exhaust of a static internal combustion engine.

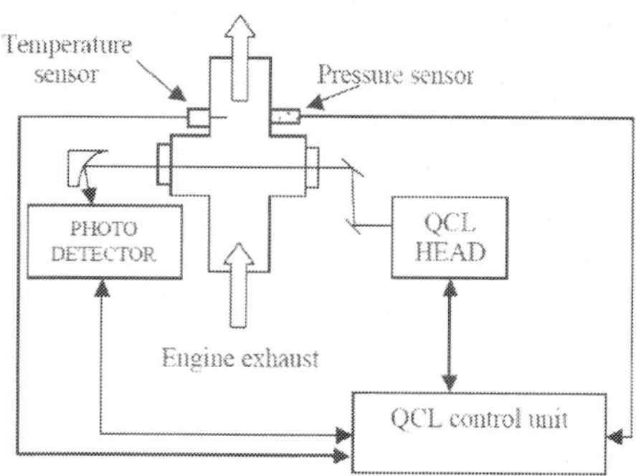

Fig. 1. Experimental setup for NO measurements on engine exhaust (reproduced with permission from Kasyutich et al., 2009. IOP Publishing).

The radiation emission of the single mode QCL was tuned over the strong NO absorption doublets $R(6.5)$ at 1900.075 (free of interference from strong water absorption lines) by applying a modulation voltage waveform. Concentrations were determined in real-time by a non-linear least-squares fitting routine by measuring the attenuation of the beam due to the light absorption by NO molecules. A minimum detectable optical depth of 0.0018 was estimated at signal to noise ratio (SNR) of 1 corresponding to a NO detection limit of 8 ppmv. The engine used in the tests was a Rover K series version. The mid-infrared QCL beam probed the 2-inch diameter stainless steel exhaust pipe 2.5 m away from the engine. The temperature and pressure of the exhaust gas were monitored just above the measurement point as reported in the figure. To allow optical access to the exhaust pipe, a cross-piece was inserted with two wedged sapphire windows mounted 23 cm apart.

In 2010 Horiba (Horiba, 2010) presented a new emission measurement system for NO, NO$_2$, N$_2$O and NH$_3$ gases (MEXA-1400QL-NX). Sample gas is fed into the gas cell and a laser pulse irradiates into the gas cell. The laser radiation emitted as continuous pulse is detected after a multiple reflection between two mirrors in the gas cell. From its inherent design and control, the wavelength of QCL radiation slightly varies with temperature therefore it is possible to scan the constant width of the wavelength in a particular region. Figure 2 shows the block diagram of the HORIBA QCL-based analyzer. Four laser elements corresponding to one measurement component respectively (NO, NO$_2$, N$_2$O and NH$_3$) are used in the device. The wavelengths of the respective laser elements are selected and controlled in order to have emission in a region where a spectrum peak falls with negligible interference from other environmental gases, such as CO, CO$_2$, H$_2$O. The analyzer design has two paths in a single sample cell; the short path with only few light reflections and the long path with multiple light reflections. The combination of two path lengths allows measuring both high concentration and low concentration gases providing a wide dynamic range measurement.

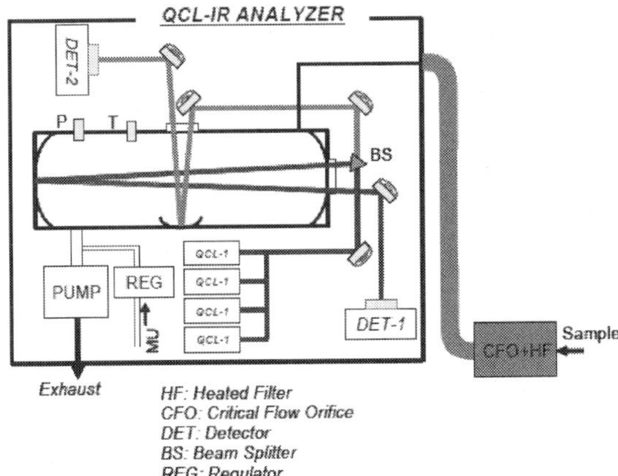

Fig. 2. Experimental setup of HORIBA QCL-based analyzer (reproduced with permission from HORIBA).

In comparison with FTIR systems, the results of measurements made using QCL technology are significantly more precise at low concentrations and offer a wider dynamic range of measurement. The finer resolution of the absorption spectrum makes QCL-based NO sensors less sensitive to incidental interference from other gases such as CO, CO_2, CH_4, H_2O and THC. In addition to the higher sensitivity and selectivity, these sensors also do not suffer from the interference of NO_x measurement by NH_3 typical of the chemoluminescence analysers.

2.2.2 Cavity ringdown spectroscopy

Cavity ringdown spectroscopy (CRDS), first demonstrated by O'Keefe and Deacon (O'Keefe & Deacon, 1988) in 1988, is one of the cavity enhanced spectroscopy methods which provides a much higher sensitivity than conventional long optical path length absorption spectroscopy due to its ability to achieve a long optical path in a compact sampling cell.

Sumizawa et al. reported the accurate and precise measurement of nitric oxide in automotive exhaust gas using a thermoelectrically cooled, pulsed quantum cascade laser as a light source and cavity ring-down spectroscopy (Sumizawa et al., 2010).

This technique, which can be performed with pulsed or continuous light sources, is based on the measurement of the decay time of an injected laser beam in a high-finesse optical cavity in the presence of an absorbing gas by measuring the time dependence of the light leaking out of the cavity. In particular, in the case of pulsed lasers, a short laser pulse is injected into a high finesse optical cavity to produce a sequence of pulses leaking out through the end mirror from consecutive traversals of the cavity by the pulse. Typically, the laser pulse is short and has a small coherence length compared to a relatively large physical cavity length. Under these conditions interference effects are avoided and the intensity of the cavity pulses decays exponentially with a time constant (ringdown time) defined by:

$$\tau = \frac{l}{c}\frac{1}{\alpha l + (1-R)}$$

Where l is the cavity length, c is the speed of light, R is the reflectivity of the cavity mirrors ($R\approx1$) and α is the absorption coefficient of the sample filling the cavity. Thus, the absorption coefficient can be determined by measuring the decay rate without (τ_{empty}) and with (τ) the absorbing gas present with the following equation:

$$\alpha = \frac{1}{c}\left(\frac{1}{\tau} - \frac{1}{\tau_{empty}}\right)$$

In Figure 3, the schematic of the CRDS-based NO sensor developed by Sumizawa and co-workers is reported. To detect fundamental vibrational transitions of NO, they used a pulsed QCL operated near 5.26 µm. The output of the QCL is focused with a lens (L$_1$) on the center of to the high finesse optical cell. At both ends of the stainless steel cell (500 mm long) two high-reflectivity ZnSe mirrors with a 25.4 mm diameter and a 1 m radius of curvature are mounted. A lens (L$_2$) was placed after the cell to collect the transmitted light on an amplified liquid nitrogen-cooled InSb detector. The ringdown decay curves were measured and then stored on a computer memory.

The vehicle used for real time NO monitoring in exhaust gas was a light duty truck with a 4.8 L diesel engine equipped with a common rail injection system and a diesel oxidation catalyst. The analyzed sample gas was made by diluting the exhaust gas with ambient air filtered by HEPA and charcoal filters using a constant volume sampler. The diluted exhaust gas was introduced into the cell at a constant flow; a membrane gas dryer was used to avoid interference by water. An HEPA filter was introduced to eliminate particles > 0.3 µm in diameter which would lead to arbitrary loss of light in the cell.

The sensor demonstrated a minimum detection limit of NO $\approx$50 ppbv in a 20 s averaging time for a signal-to-noise ratio of 2.

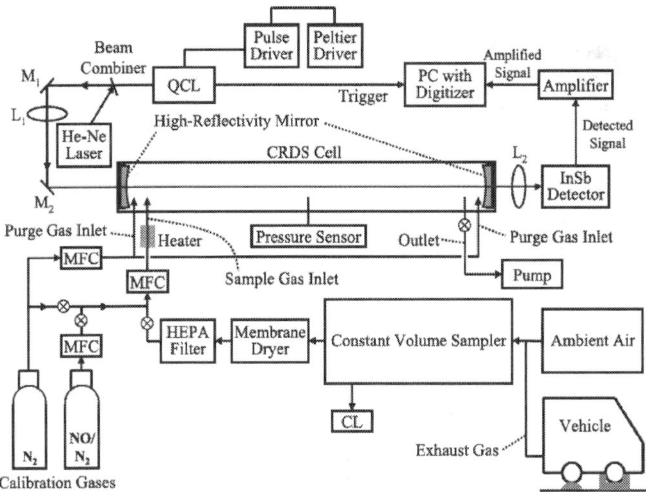

Fig. 3. Experimental setup of CRDS-based exhaust-gas NO sensor (reproduced with kind permission from Sumizawa et al., 2010. Springer Science+Business Media).

The NO exhaust measurement obtained from the CRDS-based NO sensor in a vehicle test run under the JE05 cold start cycle was in agreement with the simultaneous results from a conventional chemiluminescence NO_x sensor. Stable measurement in diluted exhaust gas sample with a concentration from sub-ppmv to 100 ppmv for more than 30 minutes and with a time resolution of 1 s was demonstrated.

2.2.3 Quartz enhanced photoacoustic spectroscopy

An effective method for sensitive trace gas detection in mechatronic applications is photoacoustic spectroscopy coupled with QCLs. PAS is an indirect technique in which the effect on the absorbing medium and not the direct light attenuation is detected. It is based on the photoacoustic effect, i.e. the generation of a pressure wave resulting from the absorption of modulated light of appropriate wavelength by gas molecules. The amplitude of this wave is directly proportional to the gas concentration and can be detected via a resonant transducer. Traditionally, the pressure wave is detected via one or more microphones (Elia et al., 2005; Di Franco et al., 2009; Elia et al. 2009). In 2002 (Kosterev et al., 2002b), an innovative method, the Quartz Enhanced Photoacoustic Spectroscopy (QEPAS), has been proposed by Kosterev et al.

The key issue of QEPAS is the detection of optically generated pressure wave via a rugged sharply resonant piezoelectric transducer, a quartz tuning fork (QTF), with a resonant frequency close to 32,768 (*i.e.*, 2^{15}) Hz (Kosterev et al., 2002b; Kosterev et al., 2005) The mode at this frequency corresponds to a symmetric vibration. A mechanical deformation of the QTF prongs caused electrical charges on its electrodes. The resulting system exhibits unique properties such as an extremely high quality factor (Q-factor) of > 10,000, small size, immunity to environmental acoustic noise and a large dynamic range.

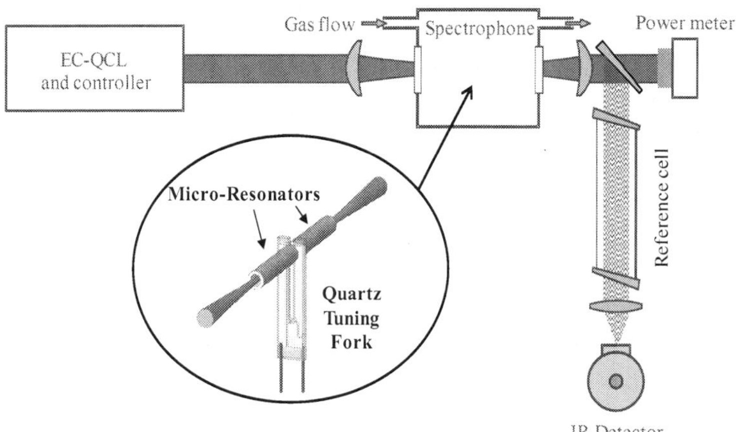

Fig. 4. Schematic of EC-QCL based QEPAS sensor.

In figure 4 typical QEPAS-based sensor configuration is reported. The module to detect laser-induced pressure wave is a spectrophone which consists of a QTF and a microresonator. It is made of a pair of thin tubes and increases the effective interaction length between the radiation-generated pressure wave and the QTF. The tubes are aligned perpendicular to the QTF plane. The distance between the free ends of the tubes is equal to half wavelength of sound in air at 32.75 kHz, thus satisfying the resonant condition. Experiments have shown that the microresonator yields a signal gain of 10 up to 20.

Spagnolo and co-workers reported the development and performances evaluation of a QEPAS based NO sensor, utilizing a cw, thermoelectrically cooled, external cavity quantum cascade laser (EC-QCL) as a light source operating at 5.26 μm (Spagnolo et al., 2010). The EC-QCL allows to access the strong and quasi interference-free NO absorption doublet R(6.5) at 1900.075 cm^{-1}. They performed a wavelength modulation technique by sinusoidally modulating the injection current of the laser at half of the QTF resonance frequency ($f=f_0/2\sim16.20$ kHz), while slowly scanning the laser wavelength. The corresponding photoacoustic spectra were obtained by demodulating the detected signal at the frequency f_0 using a lock-in-amplifier.

The NO detection in automotive exhaust assumes the presence of water vapor in the gas sample. Therefore, it is important to study the H$_2$O influence on the NO sensor performance. The V-T (vibration-to-translation) energy transfer time τ_{VT} for NO is dependent on the presence of other molecules and intermolecular interactions. The QEPAS measurements that are performed at a detection frequency 32 kHz are more sensitive to the vibrational relaxation rate compared to the conventional PAS which is commonly performed at < 4 kHz frequency f. In case of slow V-T relaxation with respect to the modulation frequency ($\omega\tau_{VT}\gg1$, where $\omega=2\pi f$), the translational gas temperature cannot follow fast changes of the laser induced molecular vibrational excitation. Thus, the generated photoacoustic wave is weaker than it would be in case of instantaneous V-T energy equilibration. Due to the high energy of first vibrational state of NO, the V-T energy transfer is slow, e.g. in dry N$_2$ the relaxation time is $\tau_{VT} = 0.3$ ms and so $\omega\tau_{VT}\gg1$. The addition of H$_2$O vapor enhances the V-T energy transfer rate and, thus, the detected QEPAS signal amplitude. Once the $\omega\tau_{VT}<1$ condition is satisfied, the amplitude of the PA signal is not affected by changes of τ_{VT}.

In this signal saturation condition, authors demonstrated a NO concentration resulting in a noise-equivalent signal of 15 ppbv. The higher sensor performances, the compactness, and the role of water (generally intereferent specie for other sensors) make QEPAS a promise in commercial sensors for automotive applications.

3. Electronic sensors

Electronic sensors are commonly produced by fabricating metal- or metal oxide-based nanostructured materials, which may provide long-term, reproducible and selective gas sensing performance. In fact, it is well-known that absorption or desorption of gas molecules on the surface of a metal oxide changes the conductivity (or resistivity) of the material, a phenomenon – first revealed by (Seiyama et al., 1962) using zinc oxide (ZnO) thin film layers. Since then, electronic (semiconductor) sensing has come a long way thanks to a huge flux of research in this field resulting into the achievement of sensitivities of electronic sensors to the order of parts per billion (ppb) toward various gases such as NO$_x$ (Gurlo et al., 1998; Guo et al., 2006; Kida et al., 2009). In addition, advances in fabrication technology enabled the production of low-cost sensors with improved sensitivity and reliability compared to those formed using conventional methods (Williams, 1999). Subsequent paragraphs give a brief introduction to the mechanism of electronic detection, and structure of the sensing elements and their types.

3.1 Principle of electronic sensors

According to band theory (Hoch, 1992), within a crystal lattice there exists a valence band and a conduction band, separated from each other as a function of energy (band gap), particularly the Fermi level that is defined as the highest available electron energy levels at a $T = 0$ K. Generally semiconductors have a sufficiently large energy gap i.e. in the range of 0.5-5.0 eV. Hence at energies below the Fermi level, conduction is not observed; while above it, electrons start occupying the conduction band, consequently enhancing the conductivity of a semiconductor. Over the years, band theory of solids attracted intense research with reference to semiconductor gas sensors (Yamazoe et al., 1979; Barsan et al., 1999). When gases like NO_x interact with the surface of an active layer i.e. generally through surface-adsorbed oxygen ions; it results in a change in the concentration of charge carriers of the active material. Such a change in charge carrier concentration transforms the conductivity (or resistivity) of the active layer. In an n-type semiconductor, majority of charge carriers are electrons, while a p-type semiconductor conducts with positive holes being the mainstream charge carriers. And being strongly oxidizing gases, the oxides of nitrogen (NO_x) serve to deplete the sensing layer of charge carrying electrons, thus resulting in a decrease in conductivity of the n-type semiconductor. Conversely, in a p-type semiconductor the opposite effects are observed with the sensing material i.e. showing an increase in the conductivity.

Wei and coworkers (Wei et al., 2004) postulated the sensing mechanism of different types of tin oxide (SnO_x) and single-walled carbon nanotube (SWCNT) composites. High sensitivity for the said nanocomposites was attributed to the expansion of depletion region on the surface of the SnO_2 particles and the p-n junction between n-type SnO_2 and p-type SWCNTs, when target gas molecules are adsorbed on the surface. However, in a recent work (Hoa et al., 2009a) on similar nanocomposite system, authors propose that owing to the different morphology, the characteristic response of the composite originates from the SnO_x nano-beads aligned together on the surface of SWCNTs. When these nano-beads are exposed to air/NO_x, the adsorbed O_2/NO_x molecules extract electrons out of the SnO_x beads, leading to the formation of an electron depletion layer, as shown in Figure 5.

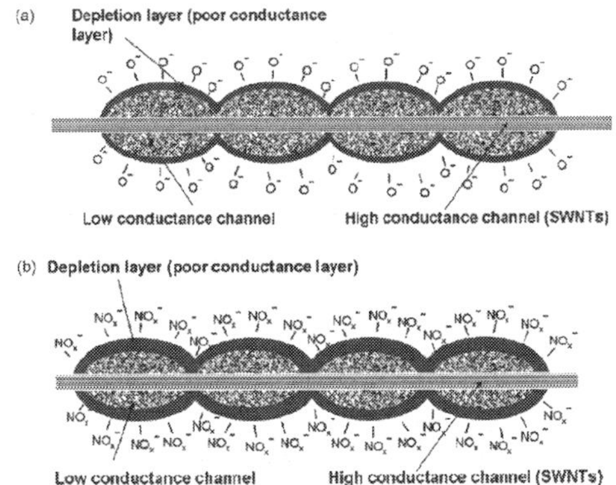

Fig. 5. The sensing mechanism model of tin oxide and SWCNTs nanocomposites based gas sensor. The surface state of nanowires structured tin oxide and SWCNTs nanocomposites in (a) air, and (b) NOx, respectively (reproduced with permission from Hoa et al., 2009a. Elsevier).

The adsorption of NO$_x$ on the surface of SnO$_x$ can be simplified as;

$$NO_x + e^- \rightarrow NO_x^-$$

Particularly, when NO$_x$ gas molecules adsorb on the surface of active layer, they capture electrons out of an n-type SnO$_x$ material (just like O$_2$ does), thus forming a depletion region. NO$_x$ adsorption, however, increases the depth of depletion region owing to the higher electron affinity (2.28 eV) of NO$_x$ as compared to the pre-adsorbed O$_2$ (0.43 eV) (Broqvist et al., 2004). Consequently, the resistance of the nanocomposite material increases and is measured as the electrical sensor response.

3.2 Device structure and types

Electronic sensors are usually based on Metal Insulator Semiconductor Field Effect Transistors (MISFET), and utilize metal and/or metal oxide nanostructured material as the catalytically active sensing layers. The field effect devices are sub-divided into different categories; transistors, Schottky diodes or capacitors, with transistors being the preferred choice for commercial applications (Mandelis and Christofides, 1993). Figure 6 presents an overview of the different types of devices.

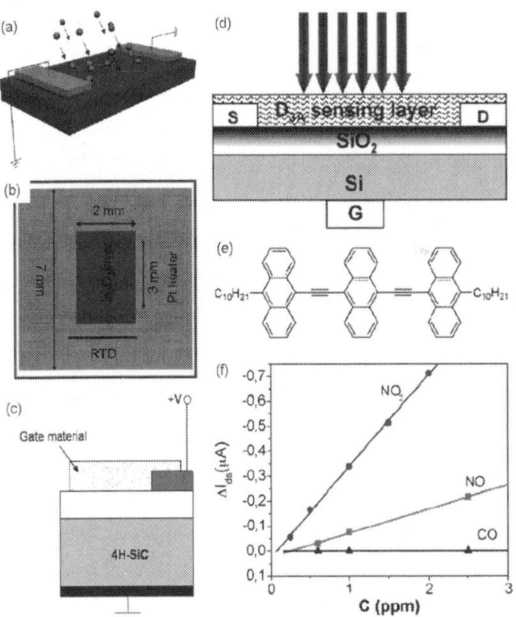

Fig. 6. The schematic diagrams of (a) graphene-based NO$_2$ gas sensor; (b) In$_2$O$_3$-based NOx sensor with inter-digitated electrodes; and (c) MIS field effect capacitive sensor. (d) A bottom gate organic thin film transistor (OTFT); (e) a novel organic semiconductor (D3A) used as the OTFT's active layer; and (f) its selective response to various gases at room temperature (adapted with permission from Ko et al., 2011; Kannan et al., 2010a; Marinelli et al., 2009. Elsevier).

A schematic diagram of the simplest electronic sensing device, a chemiresistor, is shown in Figure 6a. Ko et al (Ko et al., 2011) employed a graphene-based NO_2 gas sensor to study absorption/desorption of gas molecules on the surface of graphene. The device is fabricated by depositing graphene layers via standard scotch tape method (Novoselov et al., 2004) on a pre-defined SiO_2/Si substrate. The electron-beam lithography is used to form two metal contacts on to the graphene layer, followed by electron-beam evaporation of Pd/Au. The device is then used to obtain the current–voltage characteristics of graphene-based gas detectors at various concentrations of NO_2 gas.

Kannan and coworkers (Kannan et al., 2010a; 2010b) also fabricated NO_x sensors using a micro hotplate die, and a 0.3mm thick, 76.2mm Si wafer with 400nm thermally grown SiO_2 (Figure 6b). The resistance heater and inter-digitated Pt electrodes (IDE) are fabricated using a lift off process. They employed four distinct dies with varying IDE spacing. In_2O_3 based active films of controlled thickness are RF sputter deposited. These devices are subsequently tested for NO_x response discussed later in this chapter.

Figure 6c shows a typical field effect MIS (Metal–Insulator–Semiconductor) capacitor with a gate electrode. The capacitive sensor consists of p-doped Si as semiconductor, with a thermally grown oxide as insulator. The ohmic backside contact comprised of evaporated, annealed Al; while Cr/Au bonding pads are evaporated on top. The device is covered by a layer of catalytically active Au-NPs, and then mounted on a 16-pin holder along with a ceramic heater and a Pt-100 element to perform gas sensing measurements (Ieva et al., 2008; Cioffi et al., 2011).

In Figure 6d the structure of an organic thin film transistor (OTFT) is reported (Marinelli et al., 2009). It is a bottom gate OTFT with the organic semiconductor acting both as transistor channel and as sensing layer. Figure 6e shows the chemical structure of the 9, 10-bis[(10-decylanthracen-9-yl)ethynyl]anthracene molecule (D3A). The D3A is deposited as sensing layer via spin coating onto a SiO_2 (100nm)/n-doped Si substrate, where Au/Ti source (S) and drain (D) pads were photo-lithographically patterned. Authors reported that when the D3A OTFT was exposed to different gases like NO_2, NO and CO, concentrations as low as 250 ppb of NO_2 could be detected. The response–concentration regression lines for NO, CO and NO_2 are reported in Figure 6f, which shows that D3A OTFT sensor has very low cross sensitivities toward interfering gases and that linear dependency of sensor response on concentration is observed.

3.3 Active nanomaterials and sensor performance

A few exemplary electronic sensors, for instance, those based on graphene, indium oxide, gold nanoparticle, and organic semiconductor, have been discussed so far. To date, several other NO_x sensors have been proposed and examples of such devices include: semiconducting metal oxide sensors (GuO et al., 2008; Hwang et al., 2008; Wang et al., 2008; Hoa et al., 2009a; Navale et al., 2009, Qin et al., 2010; Firoz et al., 2010) and resistive sensors based on metal-phthalocyanines (Oprea et al., 2007; Shu et al., 2010), conjugated systems (Naso et al., 2003; Nomani et al., 2010) as well as carbon nanotubes (Sayago et al., 2008; Ueda et al., 2008) and nanocomposites (Balazsi et al., 2008; Kong et al., 2008; Hoa et al., 2009b; 2009c; Leghrib et al., 2010). The scientific research community is currently focusing on the development and investigation of novel materials, which are sensitive toward NO_x gases as well as appropriate and well-suited for the solid-state gas sensors. The most promising semiconductor materials for the fabrication of NO_x sensors are noble metals such as Gold

(Au) and metal oxides such as SnO$_2$, WO$_3$, In$_2$O$_3$, and ZnO nanostructures; since they afford high surface area for active layer-gas interaction, and satisfactory selectivity.

To adsorb as much of the target gas as possible on the surface, it is desirable that these active layers have a large surface area so as to give a stronger and easily measurable electrical signal e.g. to trace amounts of NO$_x$; and this has been accomplished by manipulating active materials into nano-regime. Manipulating and controlling sensing events at the molecular scale simultaneously avoids several problems associated with traditional sensor technologies. Incidentally, nanotechnology offers unique advantages to the sensor industry in terms of sensitivity, and rapid response and recovery kinetics due to larger surface-to-bulk ratio. An important feature of these nanomaterials, in addition, is the possibility to tailor the properties by varying the size and morphology of nanostructures, which in turn affects the electrical properties of materials.

Ahn et al (Ahn et al., 2009) fabricated ZnO nanowires on-chip via selective growth of nanowires on patterned gold catalysts thus forming nanowire air bridges (nano-bridges) between two Pt electrodes, as shown in Figure 7a. These nanowires were prepared by the carbo-thermal reduction process using a mixture of ZnO and graphite powders. Figures 7b and 7d shows side- and top-view scanning electron microscope (SEM) images of well-prepared ZnO nanowires grown on patterned electrodes. Researchers found that ZnO nanowires grow only on the patterned electrodes and many nanowire/nanowire junctions exist there, which act as electrical conducting path for electrons, as shown in Figure 7c.

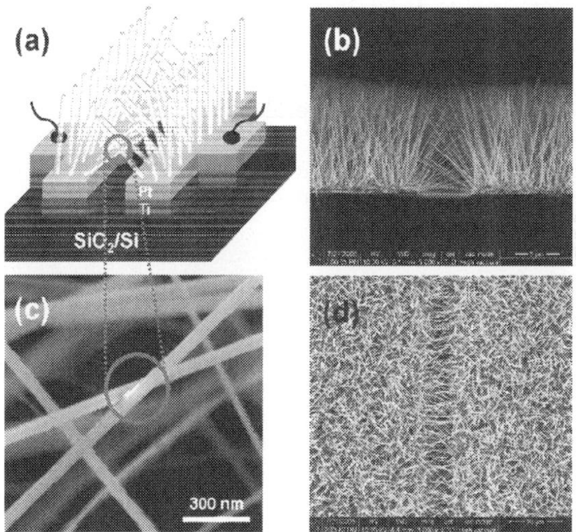

Fig. 7. (a) The schematic illustration of ZnO-nanowire air bridges over the SiO2/Si substrate. (b) Side-view and, (d) top-view SEM images clearly show selective growth of ZnO nanowires on Ti/Pt electrode. (c) The junction between ZnO nanowires grown on both electrodes (reproduced with permission from Ahn et al., 2009. Elsevier).

ZnO-based nanomaterials have rather good stability and sensing characteristics toward NO$_x$ gases combined with sufficient selectivity; that is why they have been studied widely in the past two years. Carotta et al (Carotta et al., 2009), for instance, compared ZnO nanoparticle-

and nanotetrapod-based thick film NO_x sensors. Oh and coworkers (Oh et al., 2009) fabricated a high-performance NO_2 gas sensor based on vertically aligned ZnO nanorod arrays grown via ultrasonic irradiation. Shouli et al (Shouli et al., 2010) studied different morphologies of ZnO nanorods and their sensing properties towards NO_2. These nanorods were grown via hydrothermal and sol-gel processes using different surfactants.

Zhang et al (Zhang et al., 2009) produced SnO_2 hollow spheres mediated by carbon microspheres. Authors report that carbon microspheres derived from hydrothermal conditions are hydrophilic with plenty of –OH and CO groups on the surface, which enable them to bind metal cations. Such carbon microspheres loaded metal cations give rise to hollow metal oxide spheres after calcination at high temperatures. Transmission electron microscopy (TEM) images of SnO_2 hollow spheres calcined at 450 °C are shown in Figure 8a. Researchers demonstrated that the hollow spheres have a rough morphology with a diameter in the range of 500–700 nm. Shell details are clearly visible in Figures 8b-8d, which reveal that the porous shells are formed with a thickness of about 25nm. Figure 8e shows the diffraction rings in the selected-area electron diffraction (SAED) pattern verifying the polycrystalline structure of SnO_2 hollow spheres. These hollow sphere NO_2 gas sensors present excellent selectivity and relatively swift response kinetics.

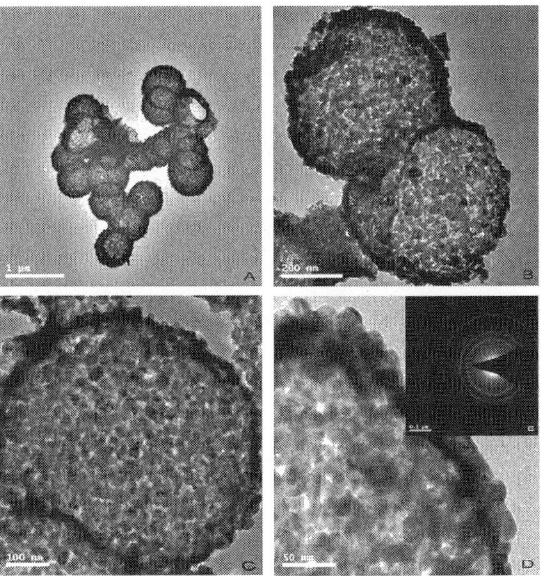

Fig. 8. (a-d) TEM images and (e) SAED pattern of SnO_2 hollow spheres calcined at 450°C (reproduced with permission from Zhang et al., 2009. Elsevier).

Zhang and coworkers (Zhang et al., 2010) fabricated atmospheric plasma-sprayed WO_3 coatings for sub-ppm (i.e. 0-450 ppb) level NO_2 detection. Park et al (Park et al., 2010) employed SnO_2-ZnO hybrid nanofibers as active layers for NO_2 sensing via combining electro-spinning and pulsed laser deposition methods. These nanofibers exhibited high response to NO_2 gas concentrations as low as 400 ppb at 200°C. In_2O_3-ZnO composite films

(Lin et al., 2010) were also fabricated to investigate NOx gas sensing characteristics, and it was found that composite films with In/Zn ratio of 0.67 reached detection limits of 12 ppb at 150°C.

The sensor responses of the gas sensors with channels composed of the as-pasted and the heat-treated ZnO nanoparticles are plotted in Figure 9a and 9b, respectively. Jun et al (Jun et al., 2009) reported that controlled heat-treatment of the ZnO nanoparticles at 400 °C led to their necking and coarsening, which resulted in a decrease in the number of particle junctions (junction potential barrier), thus reducing resistance in the presence of ambient air. However necking of the particles had an opposite effect when interacting with NO$_x$, as necked particles of small sizes become fully depleted due to removal of electrons, hence increasing the sensor response.

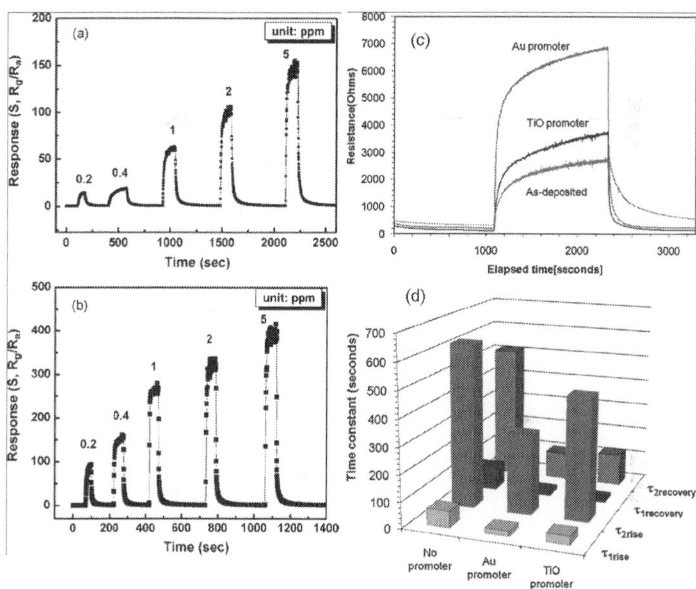

Fig. 9. Responses of the (a) as-pasted and (b) heat-treated ZnO NPs as a function of the injected NO$_2$ gas concentration at 200 °C. (c) Response of the as-deposited, TiO$_x$ promoter and Au promoter In$_2$O$_3$ films (d) Time constants for the rise and recovery from exposure to 25ppm NO$_x$ for as-deposited, TiO$_x$ promoter and Au promoter in ambient N$_2$ at 500 °C. (adapted with permission from Jun et al., 2010; and Kannan et al., 2010a. Elsevier).

Kannan et al (Kannan et al., 2010a; 2010b) fabricated chemiresistors with inter-digitated electrode (Figure 2b) with RF sputtered In$_2$O$_3$ thin film (150 nm) either with or without promoter layers, for instance, Au or TiO$_x$. Promoter layers act as additives on a semiconductor support (In$_2$O$_3$), and help improve the sensing characteristics. Figure 9c and 9d present the sensor response of 150nm thick In$_2$O$_3$ films as a function of the promoter layers to 25 ppm NO$_x$ in N$_2$ carrier gas operating at 500°C. In$_2$O$_3$ film with Au-promoter layer shows the faster and highest sensor response that is largely attributed to the spillover mechanism i.e. NO$_x$ strongly adsorbs on the Au surface and spills over to the In$_2$O$_3$ support.

Metallic nanoparticles such as gold (Au) exhibit high sensitivities toward NOx gases (Hanwell et al., 2006; Ieva et al., 2008). Cioffi and coworkers (Cioffi et al., 2011) synthesized core-shell gold nanostructure electrochemically, according to the so-called Sacrificial Anode Electrolysis (SAE). This kind of synthesis was carried out in the presence of quaternary ammonium salts dissolved in THF/acetonitrile mixture (ratio 3:1), which act both as the supporting electrolyte and as the stabilizer by forming a shell and thus giving rise to stable Au-colloidal solution. These nanoparticles were employed as gate material in field effect capacitive devices, shown in Figure 6c. TEM images of Au-NPs stabilized by different quaternary ammonium species are reported in Figure 10a-10c. It was found that Au nanoparticle sensor was able to detect 50-400 ppm NO_2, shown in Figure 10d. Authors suggest that voltage increase upon exposure to NO_2 instigates from the charge donating behavior of nitrogen oxides leading to an increase in the charge density at the Au nanoparticle film-insulator interface and thereby increasing the charge carriers in the semiconductor layer.

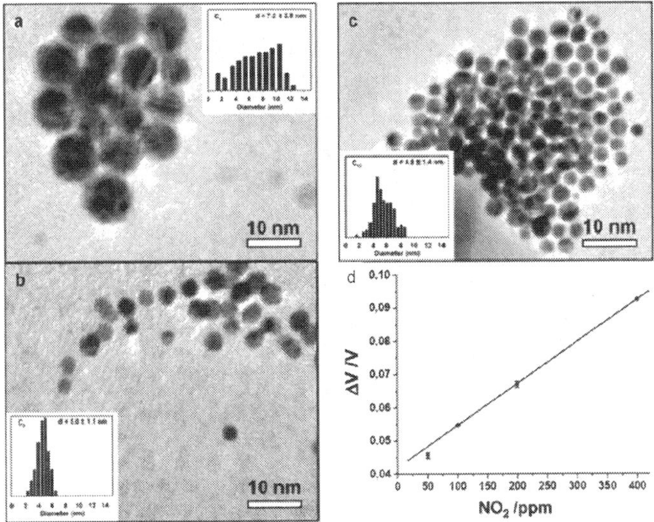

Fig. 10. TEM images and dimensional dispersion histograms (insets) of Au nanoparticles electro-synthesized in presence of (a) tetrabutyl ammonium chloride (TBOC), (b) tetraoctyl ammonium chloride (TOAC), and (c) tetradodecyl ammonium chloride (TDoAC). (d) Sensor response at operative temperature of 150°C for several NO_2 concentration levels in presence of an active gate layer of Au nanoparticles/TOAC (adapted with permission from Cioffi et al., 2011).

The electronic sensors based on core-shell Au-nanoparticles enable selective detection of NO_x gases; and the sensitivity of such systems is influenced by the particle size. These sensors show negligible sensitivity toward interfering gases such as H_2, CO, NH_3, and C_3H_6. Although further improvement and optimization of these systems is necessary, their favored characteristics could lead them becoming ever more important tools for real-time monitoring of NO_x gases.

4. Conclusion

Strict emission regulations and deeper environmental awareness have led to intense research into emissions reduction by engine producers and research organizations. Over the years, a huge number of optical and electronic technologies have been developed and applied to produce sensors with remarkable sensitivity toward nitric oxide and efficient response (and recovery) kinetics. In this chapter, a brief overview on the recent achievements and ground-breaking results obtained in the past few years, namely 2008-2011, in the field of both optical and electronic NO sensors have been reported.

Engine exhaust emissions have been successfully analyzed by FTIR and laser based spectroscopic techniques. Their advantages over more traditional measurement techniques include direct sampling of raw exhaust, measurement of many compounds with one analyzer and highly-stable calibrations. The FTIR technique is a validated methodology by several regulatory and standardization agencies for extractive gas-sampling analysis. Many highly reactive species in the exhaust can be simultaneously measured by FTIR even below part-per-million levels, replacing several discrete analyzers which may require complicated calibration procedures.

Among the absorption techniques, direct absorption spectroscopy, cavity enhancement approaches and photoacoustic spectroscopy coupled with laser sources have been demonstrated to give sensitive and selective sensors for on-line and in-situ applications, with high sensitivity, compact set-up, fast time response and portability. Infrared tunable semiconductor lasers represent the ideal radiation sources for gas sensing thanks to their excellent spectroscopic and technical properties, i.e., narrow linewidth, tunability, reliability and room-temperature operation.

QEPAS represents an innovative laser based spectroscopic technique promising in commercial sensor for automotive applications thanks to good performances in terms of selectivity and sensitivity, compactness and ease of operation.

Metal Insulator Semiconductor field effect devices used as gas sensors can be of different types: transistors, Schottky diodes or capacitors, with transistors being preferred for commercial devices. The gas sensing principle for field effect sensors is based on molecules adsorbing and dissociating on a catalytically active gate material on the sensor. These interactions create a change in the electric charges on the semiconductor surface, which in turn results in a shift in the sensor output voltage. The interactions of the gas molecules with the gate material depend on the operating temperature and the morphology and chemical characteristics of the gate material.

The use of nanostructured films as gate material has the potential to give sensors with increased sensitivity, and faster response and recovery time, due to the larger surface area available for interaction with the gas molecules, as compared to conventional thin film sensing layers. The optimum sensor performance has not been fully realized due to limited understanding of the catalytic properties of the active layer, the active layer-gas interactions, the effects of processing techniques and experimental conditions; which do influence the microstructure, morphology, and electrical properties of the nanomaterials. Improvements have been seen for selectivity, but operation at high temperature is still an open issue.

5. Acknowledgment

Financial support from the Apulian District on Mechatronics (MEDIS) is gratefully acknowledged.

6. References

Adachi, M. (2000). Emission measurement techniques for advanced powertrains. *Measurement Science and Technology*. Vol. 11, No. 10, (January 2000), pp. R113–R129, ISSN 0957-0233

Ahn, M. -W.; Park, K. -S.; Heo, J. -H.; Kim, D. -W.; Choi, K. J. & Park, J. -G. (2009). On-chip fabrication of ZnO-nanowire gas sensor with high gas sensitivity. *Sensors and Actuators B*, Vol. 138, No. 1, (April 2009), pp. 168-173, ISSN 0925-4005

Alving, K.; Weitzberg, E. & Lundberg, JM. (1993). Increased amount of nitric oxide in exhaled air of asthmatics. *European Respiratory Journal* Vol. 6, No. 9, (October 1993) pp. 1368–1370

Aroutiounian, V. M. (2007). Semiconductor NO$_x$ sensors. *International Scientific Journal for Alternative Energy and Ecology*, Vol. 4, No. 48, (April 2007), pp. 63-76, ISSN 1608-8298

Balazsi, C.; Sedlackova, K.; Llobet, E. & Ionescu, R. (2008). Novel hexagonal WO$_3$ nanopowder with metal decorated carbon nanotubes as NO$_2$ gas sensors. *Sensors and Actuators B*, Vol. 133, No. 1, (July 2008), pp. 151-155, ISSN 0925-4005

Barsan, N.; Schweizer-Berberich, M. & Göpel, W. (1999). Fundamental and practical aspects in the design of nanoscaled SnO2 gas sensors: a status report. *Fresenius Journal of Analytical Chemistry*, Vol. 365, No. 4, (October 1999), pp. 287-304, ISSN 0937-0633

Broqvist, P.; Gronbeck, H.; Fridell, E. & Panas, I. (2004). NO$_x$ storage on BaO: theory and experiment. *Catalysis Today*, Vol. 96, No. 3, (October 2004), pp. 71-78, ISSN 0920-5861

Carotta, M. C.; Cervi, A.; di Natale, V.; Gherardi, S.; Giberti, A.; Guidi, V.; Puzzovio, D.; Vendemiati, B.; Martinelli, G.; Sacerdoti, M.; Calestani, D.; Zappettini, A.; Zha, M. & Zanotti, L. (2009). ZnO gas sensors: a comparison between nanoparticles and nanotetrapods-based thick films. *Sensors and Actuators B*, Vol. 137, No. 1, (March 2009), pp. 164-169, ISSN 0925-4005

Choi, K. J. & Jang, H. W. (2010). One-dimensional oxide nanostructures as gas-sensing materials: reviews and issues. *Sensors*, Vol. 10, No. 4, (April 2010), pp. 4083-4099, ISSN 1424-8220

Cioffi, N.; Colaianni, L.; Ieva, E.; Pilolli, R.; Ditaranto, N.; Angione, M. D.; Cotrone, S.; Buchholt, K.; Spetz, A. L.; Sabbatini, L. & Torsi, L. (2011). Electrosynthesis and characterization of gold nanoparticles for electronic capacitance sensing of pollutants. *Electrochimica Acta*, Vol. 56, No. 10, (April 2011), pp. 3713-3720, ISSN 0013-4686

Comini, E. & Sberveglieri, G. (2010). Metal oxide nanowires as chemical sensors. *Materials Today*, Vol. 13, No. 7-8, (Jul-Aug 2010), pp. 36-44, ISSN 1369-7021

Di Franco, C.; Elia, A.; Spagnolo, V.; Scamarcio, G.; Lugarà, P.M.; Ieva, E.; Cioffi, N.; Torsi, L.; Bruno, G.; Losurdo, M.; Garcia, M.A.; Wolter, S.D.; Brown, A. & Ricco, M. (2009). Optical and electronic NO$_x$ sensors for applications in mechatronics. *Sensors*, Vol. 9, No. 5, (May 2009), pp. 3337-3356, ISSN 3337-3356

Durbin, T. D.; Wilson, R. D.; Norbeck, J. M.; Wayne, J. M.; Huai, T. & Rhee, S. H. (2002). Estimates of the emission rates of ammonia from light-duty vehicles using standard chassis dynamometer test cycles. *Atmospheric Environment*, Vol. 36, No. 9, (January 2002), pp. 1475–1482, ISSN 1352-2310

Elia, A.; Lugarà, P.M. & Giancaspro, C. (2005). Photoacoustic detection of nitric oxide by use of a quantum-cascade laser. *Optics Letters*, Vol. 30, No. 9, (May 2005), pp. 988-990, ISSN 1424-8220

Elia, A.; Lugarà, P.M.; Di Franco, C. & Spagnolo, V. (2009). Photoacoustic Techniques for Trace Gas Sensing Based on Semiconductor Laser Sources. *Sensors*, Vol. 9, No. 12, (December, 2009), pp. 9616-9628, ISSN 1424-8220

Elia, A.; Lugarà, P.M.; Di Franco, C. & Spagnolo, V. (2011). Quantum Cascade Laser Technology for the Ultrasensitive Detection of Low-Level Nitric Oxide, In: *Nitric Oxide - Methods and Protocols*, Helen O. O. McCarthy, Jonathan A. A. Coulter, pp. 115-133, Humana Press, ISBN: 978-1-61737-963-5

EPA: Method 320 (1998) Measurement of vapor phase organic and inorganic emissions by extractive Fourier transform infrared (FTIR) spectroscopy, in: *Federal Register, Environmental Protection Agency, 40 CFR Part 63, Appendix A to Part 63-Test Methods*, pp. 14 219–14 228

Faist, J.; Capasso, F.; Sivco, D. L.; Sirtori, C.; Hutchinson, A. L. & Cho, A. Y. (1994). Quantum cascade laser. *Science,* Vol. 264, No. 5158, (April 1994), pp. 553-556, ISSN 0036-8075.

Fine, G. F.; Cavanagh, L. M.; Afonja, A. & Binions, R. (2010). Metal oxide semi-conductor gas sensors in environmental monitoring. *Sensors*, Vol. 10, No. 6, (June 2010), pp. 5469-5502, ISSN 1424-8220

Firooz, A. A.; Hyodo, t.; Mahjoub, A. R.; Khodadi, A. A. & Shimizu, Y. (2010). Synthesis and gas sensing properties of nano- and meso-porous MoO$_3$-doped SnO$_2$. *Sensors and Actuators B*, Vol. 147, No. 2, (June 2010), pp. 554-560, ISSN 0925-4005

Guo, Y.; Zhang, X. W. & Han, G. R. (2006) Investigation of structure and properties of N-doped TiO$_2$ thin films grown by APCVD. *Material Science and Engineering B*, Vol. 135, No. 2, (November 2006), pp. 83-87, ISSN 0925-4005

Guo, Z.; Li, M. & Liu, J. (2008). Highly porous CdO nanowires: preparation based on hydroxyl- and carbonate- containing cadmium compound precursor nanowires, gas sensing and optical properties. *Nanotechnology*, Vol. 19, No. 24, (June 2008), pp. 245611, ISSN 0957-4484

Gurlo, A.; Bârsan, N.; Ivanovskaya, M.; Weimar, U. & Göpel, W. (1998). In$_2$O$_3$ and MoO$_3$-In$_2$O$_3$ thin film semiconductor sensors: interaction with NO$_2$ and O$_3$. *Sensors and Actuators B*, Vol. 47, No. 1-3 (April 1998), pp. 92-99, ISSN 0925-4005

Hanwell, M. D.; Heriot, S. Y.; Richardson, T. H.; Cowlam, N. & Ross, I. M. (2006). Gas and vapor sensing characteristics of Langmuir-Schaeffer thiol encapsulated gold nanoparticle thin films. *Colloids and Surfaces A*, Vol. 284-285, (August 2006), pp. 379-383, ISSN 0927-7757

Hara, K.; Nakatani, S.; Rahman, M.; Nakamura, H.; Tanaka Y. & Ukon, J. (2009). Development of Nitrogen Components Analyzer Utilizing Quantum Cascade Laser". *SAE Technical Paper 2009-01-2743*, doi:10.4271/2009-01-2743

Hoa, N. D.; Quy, N. V.; Cho, Y. & Kim, D. (2009b). Porous single wall carbon nanotube films formed by in situ arc-discharge deposition for gas sensors application. *Sensors and Actuators B*, Vol. 135, No. 2, (January 2009), pp. 656-663, ISSN 0925-4005

Hoa, N. D.; Quy, N. V. & Kim, D. (2009a). Nanowire structured SnO$_x$-SWNT composites: high performance sensor for NO$_x$ detection. *Sensors and Actuators B*, Vol. 142, No. 1, (October 2009), pp. 253-259, ISSN 0925-4005

Hoa, N. D.; Quy, N. V.; Tuan, M. A. & Hieu, N. V. (2009c). Facile synthesis of p-type semiconducting cupric oxide nanowires and their gas sensing properties. *Physica E*, Vol. 42, No. 2, (December 2009), pp. 146-149, ISSN 1386-9477

Hoch, P. K. (1992). The development of the band theory of solids, 1933-1960, In: *Out of the Crystal Maze – Chapters from the History of Solid State Physics*, Hoddeson, L.; Braun, E.; Teichmann, J.; Weart, S. (Eds.), pp. 182-235, Oxford University Press, ISBN 978-0-19-505329-6, New York, USA

Horiba (2010) Vehicle Test Systems. 04/18/2011 Available from: <http://www.horiba.com/it/automotive-test-systems>

Hwang, H. -S.; Song, J. -T.; Yeo, D. -H.; Shin, H. -S.; Hong, Y. -W.; Kim, J. -H. & Lim, D. G. (2008). Characteristics of a NO_x gas sensor based on a low temperature co-fired ceramic and Ga-doped ZnO. *Journal of the Korean Physical Society*, Vol. 53, No. 3, (September 2008), pp. 1384-1387, ISSN 0374-4884

Ieva, E.; Buchholt, K.; Colaianni, L.; Cioffi, N.; Sabbatini, L.; Capitani, G. C.; Spetz, A. L.; Kall, P. O. & Torsi, L. (2008). Au nanoparticles as gate material for NO_x field effect capacitive sensors. *Sensors Letters*, Vol. 6, No. 1, (February 2008), pp. 1-8, ISSN 1546-1971

Jun, J. H.; Yun, J.; Cho, K.; Hwang, I. -S.; Lee, J. -H. & Kim, S. (2009). Necked ZnO nanoparticle –based NO_2 sensors with high and fast response. *Sensors and Actuators B*, Vol. 140, No. 2, (July 2009), pp. 412-417, ISSN 0925-4005

Kannan, S.; Rieth, L. & Solzbacher, F. (2010a). NO_x sensitivity of In_2O_3 thin film layers with and without promoter layers at high temperatures. *Sensors and Actuators B*, Vol. 149, No. 1, (August 2010), pp. 8-19, ISSN 0925-4005

Kannan, S.; Steinebach, H.; Rieth, L. & Solzbacher, F. (2010b). Selectivity, stability and repeatability of In_2O_3 thin films towards NO_x at high temperatures (>500°C). *Sensors and Actuators B*, Vol. 148, No. 1, (June 2010), pp. 126-134, ISSN 0925-4005

Kasyutich, V.L., Holdsworth, R.J. & Martin, P.A. (2009). In situ vehicle engine exhaust measurements of nitric oxide with a thermoelectrically cooled, cw DFB quantum cascade laser. *Journal of Physics: Conference Series*, Vol. 157, No. 1, (January 2009), pp. 1-6, ISSN 1742-6596

Kida, T.; Nishiyama, A.; Yuasa, M.; Shimanoe, K. & Yamazoe, N. (2009). Highly sensitive NO_2 sensors using lamellar-structured WO_3 particles prepared by an acidification method. *Sensors and Actuators B*, Vol. 135, No. 2, (January 2009), pp. 568-574, ISSN 0925-4005

Ko, G.; Kim, H. -Y.; Ahn, J.; Park, Y. -M.; Lee, K. -Y. & Kim, J. (2010). Graphene-based nitrogen dioxide gas sensors. *Current Applied Physics*, Vol. 10, No. 4, (July 2010), pp. 1002-1004, ISSN 1567-1739

Kong, F.; Wang, Y.; Zhang, J.; Xia, H.; Zhu, B.; Wang, Y.; Wang, S. & Wu, S. (2008). The preparation and gas sensitivity study of polythiophene-SnO_2 composites. *Material Science and Engineering B*, Vol. 150, No. 1, (April 2008), pp. 6-11, ISSN 0925-4005

Kosterev A.A., & Tittel F. (2002a). Chemical sensors based on quantum cascade lasers. *IEEE Journal of Quantum Electronics*, Vol. 38, No. 6, (June 2002), pp. 582-591, ISSN: 0018-9197

Kosterev, A.A.; Bakhirkin, Y.A.; Curl, R.F. & Tittel, F.K. (2002b). Quartz-enhanced photoacoustic spectroscopy. *Optics Letters*, Vol. 27, No. 21, (November 2002), pp. 1902–1904, ISSN 0146-9592

Kosterev, A.A.; Tittel, F.K.; Serebryakov, D.; Malinovsky, A. & Morozov, A. (2005). Applications of quartz tuning fork in spectroscopic gas sensing. *Review of Scientific Instruments*, Vol. 76, (March 2005), pp. 0431051- 0431059, ISSN 0034-6748

Kupriyanov, L. Y. (1996). Semiconductor sensors in physico-chemical studies, In: *Handbook of Sensors and Actuators, Vol. 4*, Middelhoek, S. (Ed.), pp. 1-400, Elsevier Science B. V., ISBN 0-444-82261-5, Amsterdam, The Netherlands

Lee, J. -H. (2009). Gas sensors using hierarchical and hollow oxide nanostructures: overview. *Sensors and Actuators B*, Vol. 140, No. 1, (June 2009), pp. 319-336, ISSN 0925-4005

Leghirib, R.; Pavalko, R.; Felten, A.; Vasiliev, A.; Cane, C.; Gracia, I.; Pireaux, J. & Llobet, E. (2010). Gas sensors based on multi-wall carbon nanotubes decorated with tin oxide nanoclusters. *Sensors and Actuators B*, Vol. 145, No. 1, (March 2010), pp. 411-416, ISSN 0925-4005

Lin, C. Y.; Fang, Y. -Y.; Lin, C. -W.; Tunney, J. J. & Ho, K. -C. (2010). Fabrication of NO$_x$ gas sensors using In$_2$O$_3$-ZnO composite films. *Sensors and Actuators B*, Vol. 146, No. 1, (April 2010), pp. 28-34, ISSN 0925-4005

Lou, X. W.; Archer, L. A. & Yang, Z. (2008). Hollow micro-/nanostructures: synthesis and applications. *Advanced Materials*, Vol. 20, No. 21, (September 2008), pp. 3987-4019, ISSN 0935-9648

Mandelis, A. & Christofides, C. (1993). *Physics, Chemistry and Technology of Solid State Gas Sensor Devices*, John Wiley & Sons, Inc., ISBN 0-471-55885-0, New York, USA

Marinelli, F.; Dell'Aquila, A.; Torsi, L.; Tey, J.; Suranna, G. P.; Mastrorilli, P.; Romanazzi, G.; Nobile, C. F.; Mhaisalkar, S. G.; Cioffi, N. & Palmisano, F. (2009). An organic field effect transistor as a selective NO$_x$ sensor operated at room temperature. *Sensors and Actuators B*, Vol. 140, No. 2, (July 2009), pp. 445-450, ISSN 0925-4005

McCulloch, M.T.; Langford, N. & Duxbury, G. (2005). Real-time trace-level detection of carbon dioxide and ethylene in car exhaust gases. *Applied Optics*, Vol. 44, No. 14, (May 2005) pp. 2887-2894, ISSN 1559-128X

Naso, F.; Babudri, F.; Colangiuli, D.; Garinola, G. M.; Quaranta, F.; Rella, R.; Tafuro, R. & Valli, L. (2003). Thin film construction and characterization and gas sensing performances of a tailored phenylene-thienylene copolymer. *Journal of the American Chemical Society*, Vol. 125, No. 30, (July 2003), pp. 9055-9061, ISSN 0002-7863

Navale, S. C.; Ravi, V. & Mulla, I. S. (2009). Investigations on Ru-doped ZnO: strain calculations and gas sensing study. *Sensors and Actuators B*, Vol. 139, No. 2, (June 2009), pp. 466-470, ISSN 0925-4005

Nakamura H.; Haga I.; & Murakami K. (2006). Trend of Exhaust Emission Standards for Diesel-Powered Vehicles. *RTRI Report (Railway Technical Research Institute)*, Vol. 20, No. 7, pp. 53-56, ISSN:0914-2290

Nomani, M. W. K.; Shishir, R.; Qazi, M.; Diwan, D.; Shields, V. B.; Spencer, M. G.; Tompa, G. S.; Sbrockey, N. M. & Koley, G. (2010). Highly sensitive and selective detection of NO$_2$ using epitaxial graphene on 6H-SiC. *Sensors and Actuators B*, Vol. 150, No. 1, (September 2010), pp. 301-307, ISSN 0925-4005

Novoselov, K. S.; Geim, A. K.; Morozov, S. V.; Jiang, D.; Zhang, Y.; Dubonos, S. V.; Grigorieva, I. V. & Firsov, A. A. (2004). Electric field effect in atomically thin carbon films. *Science*, Vol. 306, No. 5696, (October 2004), 666-669, ISSN 0036-8075

Oh, E.; Choi, H. -Y.; Jung, S. -H.; Cho, S.; Kim, J. C.; Lee, K. -H.; Kang, S. -W.; Kim, J.; Yun, J. -Y. & Jeong, S. -H. (2009). High performance NO_2 gas sensor based on ZnO nanorod grown by ultrasonic irradiation. *Sensors and Actuators B*, Vol. 141, No. 1, (August 2009), pp. 239-243, ISSN 0925-4005

O'Keefe, A. & Deacon D.A. G. (1988). Cavity ring-down optical spectrometer for absorption measurements using pulsed laser sources. *Review of Scientific Instruments*, Vol. 59, (December 1988), pp. 2544-2551, ISSN 0034-6748

Oprea, A.; Frerichs, H. -P.; Wilbertz, C.; Lehmann, M. & Weimar, U. (2007). Hybrid gas sensor platform based on capacitive coupled field effect transistors: ammonia and nitrogen dioxide detection. *Sensors and Actuators B*, Vol. 127, No. 1, (October 2007), pp. 161-167, ISSN 0925-4005

Park, J. -A.; Moon, J.; Lee, S. -J.; Kim, S. H.; Chu, H. Y. & Zyung, T. (2010). SnO_2-ZnO hybrid nanofibers-based highly sensitive nitrogen dioxide sensors. *Sensors and Actuators B*, Vol. 145, No. 1, (March 2010), pp. 592-595, ISSN 0925-4005

Qin, Y.; Hu. M. & Zhang, J. (2010). Microstructure characterization and NO_2 sensing properties of tungsten oxide nanostructures. *Sensors and Actuators B*, Vol. 150, No. 1, (September 2010), pp. 339-345, ISSN 0925-4005

Reyes F.; Grutter M.; Jazcilevich A. & Gonz´alez-Oropeza R. (2006). Analysis of non-regulated vehicular emissions by extractive FTIR spectrometry: tests on a hybrid car in Mexico City. *Atmospheric Chemistry and Physics.*, Vol. 6, No. 4, (November 2006), pp. 5339-5346

Sayago, I.; Santos, H.; Horrillo, M. C.; Aleixandre, M.; Fernandez, M. J.; Derrado, E.; Tacchini, I.; Aroz, R.; Maser, W. K.; Benito, A. M.; Martinez, M. T.; Gutierrez, J. & Munoz, E. (2008). Carbon nanotube networks as gas sensors for NO_2 detection, *Talanta*, Vol. 77, No. 2, (December 2008), pp. 758-764, ISSN 0039-9140

Sberveglieri, G. (1992). *Gas Sensors: Principles, Operation and Developments*, Kluwer Academic Publishers, ISBN 0-7923-2004-2, Dordrecht, The Netherlands

Shen, G.; Chen, P. -C.; Ryu, K. & Zhou, C. (2009). Devices and chemical sensing applications of metal oxide nanowires. *Journal of Materials Chemistry*, Vol. 19, No. 7, (July 2009), pp. 828-839, ISSN 0959-9428

Shouli, B.; Liangyuan, C.; Dianqing, L.; Wensheng, Y.; Pengcheng, Y.; Zhiyong, L.; Aifan, C. & Liu, C. C. (2010). Different morphologiesof ZnO nanorods and their sensing property. *Sensors and Actuators B*, Vol. 146, No. 1, (April 2010), pp. 129-137, ISSN 0925-4005

Shu, J. H.; Wikle, H. C.; Chin, B. A. (2010). Passive chemiresistor sensor based on iron (II) phthalocyanine thin films for monitoring of nitrogen dioxide. *Sensors and Actuators B*, Vol. 148, No. 2, (July 2010), pp. 498-503, ISSN 0925-4005

Seiyama, T.; Kato, A.; Fujiishi, K.; Nagatani, M. (1962). A new detector for gaseous components using semiconductive thin films. *Analytical Chemistry*, Vol. 34, No. 11, (October 1962), pp. 1502-1503, ISSN 0003-2700

Seinfeld J.H. & S.N. Pandis (1998). Atmospheric Chemistry and Physics, In: *From Air Pollution to Climate Change*, John Wiley & Sons, ISBN: 9780471178163, New York (NY), USA

Spagnolo, V.; Kosterev, A.A.; Dong, L.; Lewicki, R. & Tittel, F.K. (2010). NO trace gas sensor based on quartz-enhanced photoacoustic spectroscopy and external cavity quantum cascade laser. *Applied Physics B*, Vol. 100, No. 1, (July 2010), pp. 125–130, ISSN 0946-2171

Sumizawa, H.; Yamada, H. & Tonokura, K. (2010). Real-time monitoring of nitric oxide in diesel exhaust gas by mid-infrared cavity ring-down spectroscopy. *Applied Physics B*, Vol. 100, No. 4, (July 2010), pp. 925–931, ISSN: 0946-2171

Sze, S. M. (1994). *Semiconductor Sensors*, John Wiley & Sons, Inc., ISBN 0-471-54609-7, New York, USA

Thermo Fisher Scientific Inc. (2007). The Use of FT-IR to Analyze NOx Gases in Automobile Exhaust, In: *Application Note: 50649*. 04/18/2011, Available from: https://www.thermo.com/eThermo/CMA/PDFs/Articles/articlesFile_7219.pdf

Tingvall, B.; Pettersson, E. & Agren, E. (2007). Vehicle test system - A pilot study, In: *Luleå University of Technology Department of Human Work Sciences Division of Sound and Vibration – Technical Report*, 04/18/2011, Available from: <http://epubl.ltu.se/1402-1536/2007/11/LTU-TR-0711-SE.pdf>

Tricoli, A.; Righettoni, M. & Teleki, A. (2010). Semiconductor gas sensors: dry synthesis and applications. *Angewandte Chemie International Edition*, Vol. 49, No. 46, (October 2010), pp. 7632-7659, ISSN 1433-7851

Ueda, T.; Bhuiyan, M. M. H.; Norimatsu, H.; Katsuki, S.; Ikegami, T. & Mitsugi, F. (2008). Development of carbon nanotube based gas sensors for NO$_x$ gas detection working at low temperature. *Physica E*, Vol. 40, No. 7, (May 2008), pp. 2272-2277, ISSN 1386-9477

Van Asselt H. & Biermann, F. (2007). European emissions trading and the international competitiveness of energy-intensive industries: a legal and political evaluation of possible supporting measures. *Energy Policy*, Vol. 35, No.1, (January 2007), pp. 497-506

Wang, C. Y.; Ali, M.; Kups, T.; Rohlig, C. –C.; Cimella, V.; Stauden, T. & Ambacher, O. (2008). NO$_x$ sensing properties of In$_2$O$_3$ nanoparticles prepared by metal organic chemical vapor deposition. *Sensors and Actuators B*, Vol. 130, No. 2, (March 2008), pp. 589-593, ISSN 0925-4005

Weber, W.H.; Remillard, J.T.; Chase, R.E.; Richert, J.F.; Capasso, F.; Gmachl, C.; Hutchinson, A.L.; Sivco, D.L.; Baillargeon, J.N. & Cho, A.Y. (2002). Using a Wavelength-Modulated Quantum Cascade Laser to Measure NO Concentrations in the Parts-per-Billion Range for Vehicle Emissions Certification. *Applied Spectroscopy*, Vol. 56, No. 6, (June 2002), pp. 706-714, ISSN 0003-7028

Wei, B. Y.; Hsu, M. C.; Su, P. G.; Lin, H. M.; Wu, R. J. & Lai, H. J. (2004). A novel SnO$_2$ gas sensors doped with carbon nanotubes operating at room temperature. *Sensors and Actuators B*, Vol. 101, No. 1-2, (June 2004), pp. 81-89, ISSN 0925-4005

Williams, D. E. (1999). Semiconducting oxides as gas-sensitive resistors. *Sensors and Actuators B*, Vol. 57, No. 1-3, (September 1999), pp. 1-19, ISSN 0925-4005

Yamazoe, N.; Fuchigami, J.; Kishikawa, M. & Seiyama, T. (1979). Interactions of tin oxide surface with O_2, H_2O and H_2. *Surface Science*, Vol. 86, (July 1979), pp. 335-344, ISSN 0039-6025

Zhang, C.; Debliquy, M.; Boudiba, A.; Liao, H. & Coddet, C. (2010). Sensing properties of atmospheric plasma sprayed WO_3 coating for sub ppm NO_2 detection. *Sensors and Actuators B*, Vol. 144, No. 1, (January 2010), pp. 282-288, ISSN 0925-4005

Zhang, J.; Wang, S.; Wang, Y.; Wang, Y.; Zhu, B.; Xia, H.; Guo, X.; Zhang, S.; Huang, W. & Wu, S. (2009). NO_2 sensing performance of SnO_2 hollow-sphere sensor. *Sensors and Actuators B*, Vol. 135, No. 2, (January 2009), pp. 610-617, ISSN 0925-4005

13

Transdisciplinary Approach of the Mechatronics in the Knowledge Based Society

Ioan G. Pop[1] and Vistrian Măties[2]
[1]Emanuel University Oradea
[2]Technical University of Cluj Napoca
Romania

*"The intuitive mind is a sacred gift and the rational mind is a faithful servant.
We have created a society that honor the servant and has forgotten the gift "*
(A. Einstein)

1. Introduction

Mechatronics and transdisciplinarity came into the light in the 1970's as multiple integrative possibilities to understand the way to achieve, transfer and incorporate knowledge in the context of the new informational society, the third wave of evolutionary process towards the informergical, knowledge based society, by transdisciplinary mechatronical revolution (Masuda, 1980; Toffler, 1983; Peters & Van Brussel, 1989; Kajitani, 1992; Klein, 2002; Pop & Vereş, 2010).

When the word *"mecha-tronics"* was invented, most people have had no idea about what it could be (Mori, 1969). Mechatronics has been associated with many different topics including manufacturing, motion controle, robotics, intelligent controle, system integration, vibration and noise controle, automotive systems, modeling and design, actuators and sensors, as well, as microdevices, as electromechanical systems, or controle and automation engineering (Kajitani, 1992; Erdener, 2003; Bolton, 2006). The term mechatronics is represented as a combination of words, *mecha*nisms and elec*tronics*, other some combinations being created before, as *"minertia,"* a name for a servomotor line that used *"minimum inertia"* to develop super-fast ability for machine to start and stop. Another term, *"mochintrol"*, a short name for *"motor, machine and control"*, represents electrical actuators able to controle freely mechanical components (Ashley, 1997). The *mechatronics* is the most used, most representative term, and finally accepted to define this new engineering field of knowledge, which began to gain popularity until the middle of 1980's (Auslander, 1996), the most commonly used one emphasizes synergy ([1]), *"*mechatronics is the synergistic integration of mechanical engineering with electronics and intelligent computer control in the design and manufacture of industrial products and processes"(Harashima et al, 1996). During the 1970's, mechatronics focused on *servotechnology*, in which simple implementation aided technologies related to control methods such as automatic door openers and auto-focus cameras (Bolton, 2006). In the 1980's, mechatronics was used to focus on *information technology* whereby microprocessors were embedded into mechanical systems to improve performance (Kyura & Oho, 1996; Gomes et al, 2003). Finally, in the 1990's, mechatronics centered on *communication technology* to connect products into large networks, including the production of the intelligent systems, technologies and products (Auslander, 1996; Isermann, 2000). Mechatronics is increasingly focused on the development of systems that synergize wide range of technologies and techniques, such as intelligent and precise mechanisms, smart sensors, to enhance information feedback computation power and information processing capabilities motion devices (Siegwart, 2001; Bolton, 2006). Mechatronics has been increasingly accepted as a methodology and as a new way of thinking in its own parameters. Mechatronical thinking, methodologies, and practices were applied to develop products with incorporated intelligence with multiple functionalities and enhanced by people as inform-actional agents (2) (Auslander, 1996; Giurgiutiu et al., 2002; Pons, 2005; Bolton, 2006; Habib, 2007; Pop & Mătieş, 2008a).

The meaning of the word *mechatronics* is somewhat broader than the traditional term electromechanics, being at a glance only an ambiguous, amorphous, heterogeneous, and continually evolving concept with a lot of definitions, many of which with a broad or a narrow significance, mechatronics being considered as "an engineering design philosophy applied with the synergy of disciplines to produce smart, flexible and multifunctional products, processes and systems" (Kaynak, 1996; Erdener, 2003; Habib, 2007). Another definition consider that "mechatronics is a unifying interdisciplinary paradigm that is capable of fulfilling such challenges, which make possible the generation of simpler, economical, robust, reliable and versatile intelligent products and systems" (Habib, 200). There is a significant design trend that has a marked influence on the product-development process, in manufactured goods, the nature of mechanical engineering education and quite probably in engineering management (Kerzner, 2003). Today, as the need for mechatronics continues to expand, the term which defines this new integrative field of knowledge becomes more and more common, two things contributing to its growth, the shrinking global market and the need for reliable and cost-effective products (Kerzner, 2003; Arnold, 2008). To be competitive, companies must develop new technologies to design and manufacture their products, as a rapid reaction to change, for competitive product properties and shortened product cycles (Arnold, 2008; Montaud, 2008). While mechatronics still involves the merging of mechanics and electronics, it also includes software and information technology, melding new technologies to the existing, combining them to solve problems, creating products or even developing new ways to obtain things by integrating different technologies to solve efficiently the emerging problems (Bolton, 2006). If in the past engineers tried to use their own lines of study to solve a problem, now they need to use the thought processes of many different outlooks to enhance their research with the use of more efficient tools in a transdisciplinary framework (Arnold, 2008; Nicolescu, 1996; 2006; Pop & Mătieş, 2010). During the time and with technological advancements mechatronics has become a familiar term in the field of engineering worldwide, but although the foundations for mechatronics were set, its full potential is yet only partially expressed, mechatronics being considered an open system of the knowledge achievement (Nicolescu, 1998; Berian, 2010). About the future of mechatronics, the transdisciplinary approach opens new perspectives on its development, incorporating more and more ideas which will be accounted to improve the way to do things and to live in the new context of ever-changing needs and willings of a complex and complicated world, when   innovations and technologies have to be improved and developed with the rapidly changing times (Mieg, 1996; Nicolescu, 1996 ; Jack & Sterian, 2002; Pop & Maties, 2009). In the next years mechatronics will increasingly oriented on safety, reliability and affordability, with efficiency, productivity, accountability and controle, with a very important role in the biotechnology, as well as in computerized world and parts of industry-based manufacturing, incorporating the computer as a part of the machine that builds a product (Jack & Sterian, 2002). Mechatronics gives to the engineer a new perspective with greater possibility to achieve and to use knowledge, so that concepts can be developed more efficiently, the communications with other engineering disciplines being improved, the major goals in the field of mechatronics being oriented to the client and market satisfaction, as well (Harashima, 2005; Montand, 2008; Arnold, 2008).

The most important thing is to know what mechatronics is, what isn't and how does it work, mechatronics being not a simple discipline (3), a new postmodern utopia, working through the new transdisciplinary transthematic educational paradigm by its exemplifying selection (what), interactive communication (how), and functional contextual legitimation (why) aspects (Grimheden, 2006; Berian, 2010). Mechatronics can be considered as a synergistic integrative system of Scientia, as a new educational transdisciplinary paradigm (mechatronical epistemology), of Techne, working as a reflexive language of the integrative design (the creative logic of the included middle) (4), and as Praxis, through a new socio-interactive system of thought, living and action (mechatronical ontology) (Mieg, 1996; Wikander et al, 2001; Nicolescu, 2002; Grimheden & Hanson, 2001; Bridwell et al, 2006; Pop, 2009). At the same time mechatronics cannot be considered as a simple working methodology (Auslander, 1996; Giurgiutiu et al., 2002), but it works with specific synergistic synthesis methodologies (Erdener, 2003; Ashley, 1997; Pop, 2009a). Mechatronics has not simply multi(pluri)disciplinary (Day, 1992; Giurgiuţiu, 2002), nor an inter(cross)disciplinary character (Arkin, et al, 1997; Siegwart, 2001; Grimheden, 2006; Habib, 2008), but a transdisciplinary one (Ertas et al, 2000; Pop & Mătieş, 2008a; 2010; Pop, 2009; Berian 2010), generating new disciplines in a codisciplinary context (Pop & Mătieş, 2008; 2009) with flexible and contextual curricula (robotics, optomechatronics, biomechatronics, etc) (Hyungsuck, 2006; Cho, 2006; Mândru et al., 2008).

The proposed aim of the paper is to introduce a new transdisciplinary perspective on the mechatronical integration of knowledge in the context of the new framework, the knowledge based-society, considered as the informergic (informaction integrated in mattergy) society, based on advanced knowledge. Only the transdisciplinarity knowledge achievement can explain the way the creativity, with a synergistic signification (see 1), works as an intentional action through ideas, design, modelling, prototyping, simulation, incorporating informergically the inform-action in matt-ergy, to realize smart products, sustainable technologies and specific integrative methods to give solution to the emerging problems (5). Real experiences cannot be replaced by learning only with simulations, for this being necessary to use complementarily, the virtual tools as design, modelling, simulation and the real world representations as prototyping, building smart mechatronical products, technologies and systems. The transdisciplinary way of knowledge is the only way to realize the integration of the rational knowledge of things and relational understanding of the world (Nicolescu, 1996; Pop, 2009), so the mechatronical knowledge achievement can be fulfilled only through the transdisciplinarity, as an open system of the integrative knowledge (De Gruyter, 1998; Nicolescu, 2002; 2008; Berte, 2005; Berian, 2010). This new paradigm ([6]) of the knowledge achievement implies an intellectual convergence towards some comon principles articulated and distributed (defined, taught and trained), with a mastery of these by new practitioners, the mechatronicians (workers, technicians, engineers) ([7]). The paradigm shift requires a re-interpretation of prior theory, a re-evaluation of the prior fact, with a reconstruction applied to new situations and re-assessed in previous ones (Cleveland, 1993; Scott & Gibbons, 2001; Arnold, 2008). This new paradigm works in a new state of equilibrium until an another challenge comes to provide another paradigm transition. From this perspective mechatronics can be considered as a brand, searching the identity evolving through different stages, in an continually emerging crisis, considered as an evolutionary chain of levels of reality in the knowledge field, all the keywords presented showing important ingredients of the mechatronical system in a continuous and dynamic development of the market conditions as a direct result of generation of high technology products incorporating complex and increased number of functionalities. (Ramo & St Clair; 1998, Arnold, 2008; Mătieş et al, 2008).

2. How does really mechatronics work, disciplinary or transdisciplinary?

2.1 Transdisciplinary mechatronical knowledge system

Knowledge refers to the state of knowing, acquaintance with facts, truths, or principles from study or investigation. A discipline is a branch of knowledge, instruction, or learning which is held together by a shared epistemology, as assumptions about the nature of knowledge, by the barriers, methodologies as acceptable ways of generating or accumulating knowledge. The terms multi(pluri)disciplinary ([8]), inter(cross)disciplinary ([9]) and transdisciplinary ([10]) refer to "multiple disciplinary system", in the theory of knowledge, some disciplines being considered closer together, while other disciplines being deemed farther apart, with a very distinctive distance between disciplines (epistemological distance). On the basis of epistemological proximity, disciplines are often clustering into groups, or knowledge subsystems such as: natural sciences (physics, chemistry, biology), social sciences (psychology, sociology, economics), humanities (languages, music, visual arts), among others, some of them using quantitative methods, while other relying on qualitative methods. Disciplines that belong to the same knowledge subsystems are closer together, but those that belong to different subsystems are farther away from each other. The disciplinary level of knowledge is working at the thematic-curricular level in the predisciplinary, monodisciplinary or codisciplinary context ([11]), while the professional programs and reasearch groups generally operate on a multi(pluri)disciplinary model (methodological level), being more than disciplines, and in some cases may bridge across knowledge subsystems working at the synergistic level (structural-interdisciplinarity, functional-crossdisciplinarity and generative-transdisciplinarity) *as a multiple disciplinary thinking perspective* of the knowledge (Choi & Pak, 2008; Pop & Vereş, 2010). When are combined, disciplines more disparate or epistemologically different from one another are giving new insight for a complex problem or issue than disciplines that share similar epistemological assumptions, the differences between disciplines provide alternative methods and perspectives, making it possible to see all the facets of the reality in a complex context, leading to the cognitive process of emergence of new ideas and knowledge perspectives, the more disparate are the disciplines, the more different are the perspectives, with a greater chance of success in tackling the complex problems (Palmer, 1978; Arecchi, 2007). Knowledge is considered to be expressed in a large spectrum represented as a continuum at one side, where it is almost completely tacit, as semiconscious and unconscious knowledge held in people's heads and bodies, as hands-on knowledge (Nonaka & Takeuchi, 1994; Polanyi, 1997). At the other side of the spectrum, knowledge is almost completely explicit, accessible to people, other than the individuals originating it, represented as a line - or band - structured spectrum of the knowledge, as hands-in and hands-off knowledge. Explicit elements are objective, rational and created "then and there" (top-down level), while the tacit elements are subjective, experiential and created "here and now" (bottom-up level) (Leonard & Sensiper, 1998). It is interesting to study the way the knowledge can or cannot be quantified, captured, codified and stored as well, the predominant aspect in the management of tacit knowledge being to try to convert it in a form that can be handled using the "traditional" approach, through the transdisciplinary process of the knowledge integration: hands-on (passive knowledge), hands-in (passive-active knowledge) and hands-off (active knowledge). There is a difference between know-what (selection the message to be knowledge communicated), as an explicit, and know-how (the way the message is codified and transmitted), as implicit knowledge (Brown & Duguid, 1991; Pop, 2008), procedures being known as a codified form of know-how that guide people in how to perform a task. The organizational (communion-like) knowledge constitutes core-competency and it is more than "know-what", requiring the more elusive "know-how" - the particular ability to put know-what into practice" as know-how (Hildreth et al, 2000; Gomes et al, 2003). To develop knowledge through interaction with others in an environment where knowledge is created, nurtured and sustained the Communities of Practice (CoPs) provide for people an adequate environment (Wenger & Snyder, 2000; Hildreth & Kimble, 2004) ([12]), where transactive knowledge (the organisation's self - knowledge - knowing what you know) and resource knowledge (knowing who knows what) are focusing on the knowledge of the organisational environment ([13]) (Hildreth et al, 2000). In the knowledge based-society, the education and training build on option for transdisciplinarity, represent a necessity in the new context of education and a guarantee for future success, at the same time with a new attitude, an active participation, flexibility and adequation to the context, transforming any problem into an opportunity (Berte, 2003; Pop & Mătieş, 2010). Transdisciplinarity, as doing and being approach of knowledge achievement, is based on an active process that enables the actors of the educational training environment, as a teaching factory (Alptekin, 1996; Lamancusa et al, 1997; Berte, 2003; Quinsee & Hurst, 2005), to use successfully the information, to question, integrate, reconfigure, adapt or reject it (Nicolescu, 1996). The framework of transdisciplinary approach on education presupposes the formulation and affirmation of original opinions, the rational choice of an option, the problem solving, the responsible debate of ideas, the process of teaching-learning beyond matter boundaries, beyond even the traditional academic rules. The best space for the transdisciplinary approach of knowledge achievment is the University, where inquiry can roam freely, as the natural home of the synergistic integration (Castells, 2001), with its flexibility and adaptiveness in the knowledge economy, a space often deconstructed, if not completely under erasure, in a continuous possible reconfiguration in a combination of a high required degree of competence in the different disciplines (breadth approach), but with the necessity to have a deapth profile of the knowledge in research on own cognitive field ([14]) (Kaynak, 1997). Transdisciplinarity can also explain the sustainability concept, in education and in development of the achievment of integrative knowledge systems (Gibbons et al, 1994; Hmelo et al, 1995; Hildreth et al, 2000; Arnold, 2008; McGregor & Volckmann, 2010).

Because the knowledge resides in people, not in machines or documents at all, this very important aspect is determining the spiritual dimension of knowledge (Reason, 1998; Pop, 2009), because the contemporary man is considered as an agent involved in the knowledge process, through a balance between the rationality in the knowledge of things (by doing) and the relationship in order to understand the world (by being) (Nicolescu, 1996; 2008). The paradigm shift in the knowledge process is necessary to encourage and support necessary changes in education, identifying and acknowledging critically and creatively the major tendencies that have determined modifications of the education purposes leading to a reviewing of curriculum in a creative innovative context [15] (Langley et al, 1987; Boden, 1994).

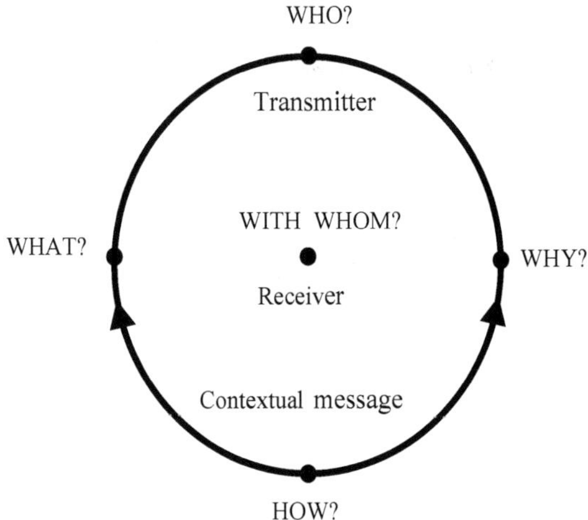

Fig. 1. The transdisciplinary contextual message model (Pop, 2008).

From a transdisciplinary point of view, disciplinary research concerns, at most, one and the same level of Reality, but in most cases, it only concerns fragments of one level of Reality, but transdisciplinarity concerns the dynamics engendered by the action of several levels of Reality at once, the discovery of these dynamics necessarily passes through disciplinary knowledge, being nourished by disciplinary research, and the codisciplinary research is clarified by transdisciplinary knowledge in a new fertile way (Nicolescu, 1996; Klein, 2002). In this sense, disciplinary (*deapth aproach*) and transdisciplinary research (*breadth approach*) are not antagonistic but they are working in the *"breadth through deapth"* complementarly paradigm, opening a new vision in the knowledge achieving process (Kaynak, 1996). In order to explain in an integrative way the process of knowledge achievement in the transdisciplinary context, was elaborated the *transdisciplinary contextual message model* [16] (fig.1), as a systemic perspective of the knowledge achieving process by communication, with functional structures, producing signs, signifying them and valuing the educational products of knowledge processes in an ethic-semiotic context, with the key synergistic significant questions: *who-with whom, what, how* and *why* (Pop, 2008). The questioning paradigm *"what, how* and *why"* of the mechatronics is a very important transdisciplinary approach for the emergence of the brand profile of the mechatronics itself (Harashima et al, 1996; Bradley, 1997; Buckley, 2000; Grimheden & Hanson, 2003; 2005).

2.2 The transdisciplinary knowledge search window

In the context of the necessity of a new kind of the mechatronical knowledge achievement are identified two well known problem solving strategies, namely bottom-up and top-down approaches in design literature as knowledge search window ([17]), as a new transdisciplinary approach, that of the included middle, with creativity in action and authenticity through participation (Lupasco, 1987 ; Nicolescu, 1996 ; Waks, 1997; De Bono, 2003; Pop & Mătieş, 2008a). The creativity in action is a very important way to facilitate the rational knowledge of things (by doing) through the adequateness and innovation for creativity and through competition and competence for action, to teach the disciples to improve their thinking to reflect on their creations and to find possibilities how to develope them in general patterns of lateral and vertical thinking, complementarly in their technological projects (Waks, 1997; De Bono, 2003; Pop & Mătieş, 2008a). A creative system must be able to detect the original ideas, to perform an efficient exploration with intelligent search strategy for admissible states and for moving from one state to another (Boden, 1994; Langley et al 1987; Savary, 2006). To be creative does mean to explore and possibly to transform the "conceptual space"([18]), the most important thing being "the identification, stimulation and evaluation of creativity" (De Vries, 1996; Doppelt & Schunn, 2008). Tradition and innovation are not opposed one to another, they are working together, the most creative individuals being considered those who explore a conceptual structure going beyond them, in a transdisciplinary way, the real giants being those determined individuals who manage to discern and articulate new structures which transgress the existing ones (Boden, 1994; Schäfer, 1996).

Today a mechatronical engineer has to understand and to work in the new synergistic relationship between precision engineering, control theory, computer technology and sensors and actuators technology. Achieving this objective requires a paradigm shift from the sequential to simultaneous engineering, in an integrative educational approach that seeks to develop systemic thinking learners and teachers as well. As an engineering field mechatronics is focused by training professionals to master the practical skills necessary for mechatronical systems design and maintenance, the new educational principles being focused on the creative concurrent design and development process. There are several intuitive touchstones for creative achievement, such as the complexity of the questions answered, its centrality or importance for the field explored. To learn the trade is to learn these structures, and to be creative is to produce new applications at the individual P-creativity level, or at the scientific community H-creativity level (Boden, 1994).

The knowledge search window is introduced as a methodological concept to explain the bottom-up/top-down mechanism of the teaching-learning process in the mechatronical educational paradigm from a transdisciplinary perspective (Pop & Mătieş; 2008a; Pop, 2009a). This methodology is working in achieving mechatronical knowledge process by learning, understanding and practicing mechatronical skills, being based on an active-reactive understanding-learning process, occurring either intentionally or spontaneously, enabling to control information, to question, integrate, reconfigure, adapt or reject it (Nicolescu, 1996; Berte, 2005). The teacher is considering as acting from a top-down perspective, while the disciples from a bottom-up perspective, the ranks of authority of the teacher and the disciples being alternatively in a symmetrical and complementary interaction state, depending of the synergistic context, in order to avoid potential conflicts, building bridges, avoiding the barriers, working and living together, as human beings in a permanent connection between them and with intelligent systems, technologies and products, as well (Nicolescu, 1996; Berte, 2005; Lute, 2006; Mătieş et al, 2008; Pop & Vereş, 2010). It is necessary to develop in each student a balance between these top-down and the bottom-up perspectives on mechatronical approach of knowledge, studying in depth the key areas of technology on which successfully mechatronical design are based and thus lays the foundation for the students to become true mechatronicians (workers, technicians, engineers) (Day, 1992; Pop, 2009a) in the vocational educational training systems (VETS), as a knowledge factory (Stiffler, 1992; Alptekin, 1996, Lamancusa et al, 1997; Rainey, 2002; Erbe & Bruns, 2003). To fulfil the demands for multi-skilled technicians and skilled workers, vocational educational training systems (VETS) together with industry are confronted with the need to develop theoretical sequences (top-down perspective) integrated with practical learning sequences (bottom-up perspective), in acquiring key competences and update the skills as a continuous all life learning process. There are considered three areas in order to achieve the proposed objectives by sustainable long term efforts: (1) *raising advanced knowledge level (as wisdom and skill achievement, as well)*, in order to avoid the risks of economic and social exclusion (the future labour markets in the knowledge-based society will demand higher skill levels from a shrinking work force); (2) *all the life learning (lifelong learning, lifewide learning and learning for life) strategies,* including all levels of education, the qualification frameworks and the validation of non-formal and informal learning, as well; (3) *the knowledge triangle, education, research and innovation,* which plays a key role in boosting jobs and growth, accelerating reform, promoting excellence in higher education and university-business partnerships, ensuring that all the fields of education and training are ready to play a full role in promoting creativity, innovation and development (Schäfer, 1996; Barak & Doppelt, 2000; De Bono, 2003; Derry & Fischer, 2005; Pop, 2009a; Pop & Mătieş, 2010).

As the bottom-up strategy produces solutions at physical, practical level, top-down design strategy looks for original ideas at functional level before investigating physical solution alternatives, being possible to explain what mechatronics is in a general engineering framework. The possibility to approach the mechatronical evolution from a top-down perspective as a living conceptual system, with a specific language and with strong educational skills in the knowledge based society is connected with the bottom-up perspective in the approach of reaching knowledge, the integration of new products, technologies and systems. This process is based on the mechatronical synergistic synthesis with complexity, increased performance to achieve skills in a transdisciplinary apprenticeship relation between the teacher and the disciples as transmitter and receiver of the contextual synergistic message. The key questions "what, how and why" in the mechatronical knowledge process, as a communicational interface between the teaching-learning fields of knowledge environment, „who with whom", are the fundamental pillars of the knowledge based society building (Gibbons et al, 1994; Harashima et al, 1996; Bradley, 1997; Buckley, 2000; Fuller, 2001; Klein, 2002; Pop, 2008a; Fricke, 2009).

Mechatronics can be considered as an educational paradigm, as a reflexive contextual language and as a socio-interactive way of being, as a lifestyle (thinking, living, acting), with a methodology to achieve an optimal design of intelligent products, to put in practice the ideas and techniques developed during the transdisciplinary process to raise synergy and provide a catalytic effect for finding new and simpler solutions to traditional complex problems (Berte, 2005; Everitt & Robertson, 2007; Nicolescu, 2008; Berian, 2010). The integrative design process is an top-down evaluation approach of the mechatronical knowledge perspective, being a very important component of the new transdisciplinary reflexive, creative integrated design language, as a new informergic transdisciplinary code (Pop, 2009). The design process could be finalized only by a team of specialists from different fields who must learn to communicate in a new transdisciplinary manner, each researcher working synergistically rather than sequentially, from his own field of research, with an obvious difference between the traditional, fragmented, sequential and the mechatronical integrative concurent design (Hewit & King, 1996; Lamancusa et al, 1997; 9 Ertas et al, 2000; Habib, 2008; Doppelt & Schunn, 2008). The principles of mechatronical education can be applied successfully to all teaching levels, creating the necessary teaching-learning environment, as a teaching factory, as a mobile mechatronical platform, or as another educational systems (Stiffler, 1992; Alptekin, 1996; Lamancusa et al, 1997; Arkin et al, 1997; Rainey, 2002; Erdener, 2003; Erbe & Bruns, 2003; Quinsee, 2005; Papoutsidakis et al, 2008; Măties et al, 2008). It is necessary to define curricular areas with the possibility to switch from a unilateral monodisciplinary thinking, based on a single discipline, to a flexible, global thinking, which assures an integrating approach to the educational process, as a synergistic generative transdisciplinary way (Berte, 2005; Rainey, 2002; Grimheden & Hanson, 2005).

The key aims of the mechatronical approach of knowledge are to promote relevant education and training, support the development of research programs and diffuse information relating to the application of techniques across all industrial fields (Kyura & Oho, 1996; Iserman, 2000; Minor & Meek, 2002). These advantages have been stimulated by factors including developments in microprocessor industry, new and improved sensors and actuators, advances in design and analysis methods, simulation tools and novel software techniques (Stiffler, 1992; Langley et al, 1987; Wikander et al, 2001; Shetty, 2002; Pons, 2005; Măties et al, 2005). Mechatronics is studied at a theoretical and practical level, as a balance between theory and practice, through the included middle approach of knowledge (Lupasco, 1987; Nicolescu, 2008), based on the physical understanding rather than on the mathematical formalism, in a mechatronical integration process of the physics as phenomenological, methodological and material sciences points of view emphasized through analysis and hardware implementation (Langley et al, 1987, De Vries, 1996, Wikander et al, 2001, Bolton, 2006). To evaluate concepts generated during the design process the mechatronicians must be skilled in the modelling, analysis, and control of dynamic systems and understand the key issues by computational explorations of the creative process (Jack & Sterian, 2002; Măties et al, 2005). The true mechatronical expert (engineer, worker, technician) has a genuine interest and ability across a wide range of technologies, and takes delight in working across disciplinary boundaries in a transdisciplinary way, to identify and use the particular blend of technologies which will provide the most appropriate solution to the emerging problems. Such an expert has to be a high communicator who has the knack of being able to motivate others about technologies and to promote alternative approaches (Rainey, 2002; Quinsee & Hurst, 2005). It is very important to develop a hierarchy of physical models for dynamic systems, from a real, natural model to a design model, and understand the appropriate use of this hierarchy of models, and its vertically structural system levels, to achieve the key elements of a measurement system and the basic performance specifications and digital motion sensors, the characteristics and models of various actuators, analogical and digital circuits and components, with semiconductor electronics (Comerford, 1994; Isermann, 2000; Bolton,

2006). At the same time the mechatronician has to be able to apply various control system design techniques, the digital implementation of controle and basic digital controle design techniques have to be learned and understood in order to be able to use a microcontroller as a mechatronical system component, to understand programming and interfacing issues, and to apply all these skills to the design of mechatronical systems and intelligent products (Yamazaki & Miyazawa, 1992; Minor & Meek, 2002; Mortensen & Hinds, 2002; Brazell et al, 2006; Bolton, 2006; Habib, 2008).

Transdisciplinarity as understanding (top-down approach), learning and practicing (bottom-up approach) is based on an active process, occurring either intentionally or spontaneously, that enables to control information, thus to question, integrate, reconfigure, adapt or reject it (Nicolescu, 1996; 2002). There are four pillars of the transdisciplinary knowledge: learning to know, learning to do, learning to be and learning to live with other people (Delors, 1996). To learn and to understand are the most two important issues of the transdisciplinary mechatronical knowledge in the integrative process through modelling and control in the design of mechatronical systems. To achieve knowledge in the transdisciplinary mechatronical context, it is necessary to reconfigure the framework of the way the four pillars of transdisciplinary knowledge are working, for this reason they are put together, in a new framework, learning as achieving information and knowledge, as an objective rational extrinsic logical issue, and understanding as an ethic-semantic issue, the subjective relational dimension of knowledge. „Learning to learn to know by doing" and „learning to understand to be by living together with other people" is the multiple transdisciplinary paradigm, working as guidelines to achieve both necessary integrative semiophysical skills in a synergistic communicational context, through the structural-functional semiophysical system, with its technical efficiency (knowing what and how we know), and ethic-semantic value of semiosycal products in an ethic authoritative context with its axiological coefficient (knowing how and why we live) (Pop & Veres, 2010; Pop, 2008). Every pillar of transdisciplinary knowledge can be integrated in this framework to explain the mechatronical perspective of achieving knowledge in the informational knowledge based society with a new transdisciplinary mechatronical epistemology, a new creative logic of the included middle and a new mechatronical ontology. Learning to know becomes a ring of the extrinsic active knowledge chain, with "what, how and why" epistemic questioning paradigm (Harashima et al, 1996; Bradley, 1997; Buckley, 2000; Pop, 2008a), related with the message (quantitative and qualitative aspects, know what), with the manner of the communicational process, code and channel (know how), and finally with the context (know why) (see fig.1). The ring "by doing", of the extrinsic active knowledge chain represents the "acquiring a profession necessarily passing through a phase of specialization in a challenging world, with changes induced by the computer revolution with excessive specialization risks, reconciling the exigency of competition with equal chance and opportunity for all (Nicolescu, 2002). Learning by doing could be, in the transdisciplinary approach of mechatronics, an apprenticeship in creativity (Siegwart, 2001), discovering what is new, bringing in actuality as innovation the creative potentialities, generating the conditions for the emergence of the authentic person, working at the top level of the creative potentialities (Boden, 1994; Waks, 1997). The intrinsic reactive approach of the mechatronical transdisciplinary knowledge, the learning to understand involves the spiritual dimensions of the knowledge process without which the knowledge couldn't be understandable (Nicolescu, 1996; Reason, 1998). The first step is "learning to be", a permanent communitarian apprenticeship in which teachers inform the disciples, as much as disciples inform the teachers, in a continuous teaching-learning process, so that the shaping of a person passes inevitably through a transpersonal dimension with fundamental tensions between the rational approach and the relational approach, discovering the harmony or disharmony between individuals and social life, testing the foundations of the personal believes in order to discover that which is found underneath, questioning in a scientific spirit being a precious guide for all the people (Nicolescu, 1996; 2002 ; Berte, 2003; 2005; Pop & Vereş, 2010). This can be done only by living together with other people in communion, supposing that the transgresive attitudes can, and must to be learned, allowing to a better understanding of own culture, to better defend the personal and collective identity with all its components. The transdisciplinary approach is based on the equilibrium between the outside (with its extrinsic active knowledge aspect) of the person and his inside (with its intrinsic reactive knowledge aspects) (Reason, 1998; Nicolescu, 2008; Pop & Mătieş, 2008a). So, transdisciplinary mechatronical knowledge, with its extrinsic active (learning to know by doing) and intrinsic reactive (understand to be by living with others) components can be presented in a new original manner. The rational knowledge process, by „learning to learn to know by doing" involves „creativity through adequateness and innovation (to know-what, how, why)", combined with „action through competence and performance (by doing-who, what, how and why)", as extrinsic active component, characterized by the efficiency of knowledge process. On the other hand there is the relational knowledge process, by "learning to understand to be by living with other people", which presupposes „authenticity through integrity and excellence (to be-who, how)", together with „participation through communion and apprenticeship (by living with-to whom)", as intrinsic reactive component, characterized by its axiological ethic-semantic parameter (Pop & Mătieş, 2008a; Pop & Vereş, 2010). It is very important to know the way mechatronics does work as a synergistic synthesis process of achieving knowledge, by integrating these two isues, rational (by doing) and relational (by being) as branches of the informergy, a transdisciplinary integration of the mattergy (matter and energy) with informaction (information and intentional action) (Pop & Vereş, 2010).

The existing models for educational mechatronics (Grimheden & Hanson, 2003; 2005) consider mechatronics as an engineering discipline, working only from an interdisciplinary perspective, different from the known disciplinary identity through three didactical oppositions: exemplifying - representative selection (what), interactive – active communication (how) and functional contextual – formal legitimation (why). The stages represented are going from the disciplinary identity (1), through multi(pluri)disciplinarity, with old courses (2); the cross(inter)disciplinarity new courses (3), followed by the curriculum stage with new programs (4); the organizational stage with new organizations (5) and finally, the so named thematic identity of the mechatronical education stage (6). As it is presented in other papers (Fuller, 2001; Mittelstrass, 2004; Habib, 2007; 2008) the inter(cross)disciplinary stage is considered as a final stage, for the knowledge attendable level, or there is a confusing or a missunderstanding about the difference between the inter(cross)disciplinarity and the transdisciplinarity (Jantsch, 1972; Fuller, 2001; Mittelstrass, 2004). It is very clear, that the knowledge process can not be closed (De Gruyter, 1998; Nicolescu, 1998; 2008), mechatronics as a transdisciplinary system of knowledge being an open system (Berian, 2010), the structurative integrating process modelling considering the existence of the three levels of the integration of the knowledge, thematic - curricular, methodological and synergistic, with different stages, predisciplinary, monodisciplinary and codisciplinary at first level, multi(pluri)disciplinary at the second level and, very important, at the third level there are three stages, structural synergy - interdisciplinarity, functional synergy

– crossdisciplinarity, and generative synergy – transdisciplinarity (Everitt & Robertson, 2007; Pop & Vereş, 2010).

Is proposed an integrative transdisciplinary model (Pop & Mătieş, 2008) which tries to demonstrate that mechatronics cannot be considered as multi(pluri)disciplinary, inter(cross)disciplinary, nor a new discipline, neither a simple methodology, but a transdisciplinary approach of the knowledge in the informergic society, as is sustained through the semiophysical communicational contextual message model, as well (Pop, 2008a). The transdisciplinary knowledge integrative mechatronical model presents five stages of the evolution of the knowledge process from monodisciplinary to transdisciplinary, through codisciplinary, (multi)pluridisciplinary and (inter)crossdisciplinary (fig. 2) (Pop & Mătieş, 2009). This model of the mechatronical knowledge is considered more integrative then others model known, through the educational paradigm by its transthematic, with representative selection, interactive communication, functional legitimacy aspects (mechatronical epistemology) (Grimheden & Hanson, 2003; 2005), as a reflexive way of communication through design, modelling (the creative logic of the included middle) (Lupasco, 1987; Nicolescu, 1996) and a socio-interactive system of thinking, living and acting (mechatronical ontology).

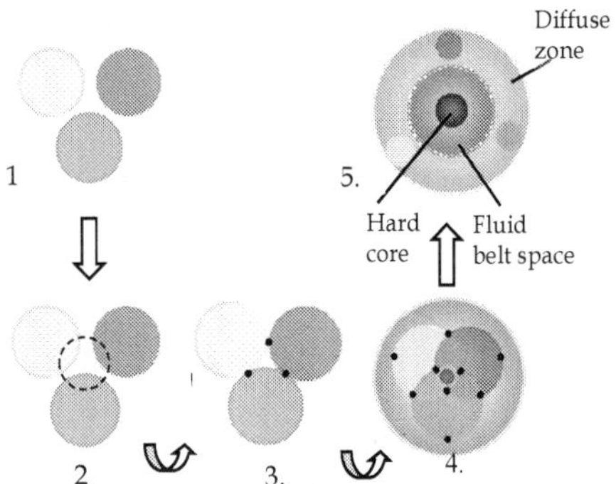

Fig. 2. The transdisciplinary knowledge integrative mechatronical model.

The three spheres represent different knowledge disciplinary fields of the mechanical engineering, electronic engineering and automation control engineering with computer engineeering systems) (Pop & Mătieş, 2008). Generally speeking, transdisciplinary knowledge integrative mechatronical model presented in fig. 2 could represent any synergistic context where two, three, or more disciplinary fields are working together in a generative - synergistic way, such as it is the semiophysics (phenomenological physics, semiotics, ethics) (Pop & Vereş, 2010), optomechatronics (optoelectronics, mechanics, informatics) (Cho, 2006), biomechatronics (biomechanics, electronics, informatics) (Mândru et al, 2008). Others such examples, the synergistic synthesis of Scientia (Mechatronical Education, as a new educational transdisciplinary paradigm, the mechatronical epistemology), of Techne (Mechatronical Technology, working as a reflexive way of the integrative design, the creative logic of the included middle), and of Praxis (Mechatronical Economy of the intelligent products, through the mattergic embedded informaction, with a new socio-interactive system of thought, living and action, the mechatronical ontology), are working in the same way. The common significant generative-sinergistic space of the knowledge resulted here is the Meta-Mecatronics (Phylosofia Technologica Systemica), considered as an open transdisciplinary integrative system of the informergic knowledge (Mieg, 1996; Wikander et al, 2001; Nicolescu, 2002; Grimheden & Hanson, 2001; Bridwell et al, 2006; Pop, 2009).

In the transdisciplinary knowledge integrative mechatronical model *the first stage* (1) represents the distinct *separation* between disciplines working with specific metodologies, thematic curricula and having net barriers, acting in a deapth competence approach in order to achieve knowledge. At the *second stage* (2), could be detected *a virtual connection* beween the knowledge fields with possible statistical transfer of the contents, methods, rules, definitions, preparing the next steps for the integrative process. *The multi(pluri)-disciplinary approach, cooperation through contact (3)*, is characterized by different kinds of contacts between disciplines with radial mutual interactive flows through each contact point. A degree of competence in others disciplines is required, so in the multi(pluri)disciplinary research groups the individuals are working on related questions from different disciplinary perspectives sharing their expertise between them, the *inter(cross) - disciplinary approach, as a combination by overlapping (4),* with common creative - innovative spaces, with transfer of methods and content. Circular flows determine the emergence of a new systemic configuration in a paradigmatic way, a reflexive communicational language and a socio-interactive reorganization of the contents and methods, this kind of informational flow being prevalent. Inter(cross) - disciplinarity is a generic term for a plurality of activities that perform a range of functions with regard to disciplines, new fields, programs and projects, representing the situation where the main effort is to create inter(cross)disciplinary courses, being created new curricula suitable to the inter(cross) - disciplinary thinking and to the different identity of the subject. The structural synergistic stage is the interdisciplinary way of the integration of knowledge, as an application epistemological degree, with emergence of new disciplines at an organizational stage, with flexible borders, methods and with different ideas, themes and courses. When these flexible walls are penetrated, being possible to overpass the barriers, is talking about the synergistic functional stage, the crossdisciplinarity. The radial informational flows assure the fullfilment of the closed regions, which are growing from the initial points of contact to space filled with the separated elements and they are combining with the circular flows which become prevalent. Consequently are appearing new structures and new functions, with synergistic significance, with a new perspective, emerging as a necessity to reconfigure the inquiry space of the teaching - learning environment. In many cases, inter(cross) - disciplinary work can propel forward discipline - based work, designing structures that overcome the tension between disciplinarity and inter(cross)disciplinarity as a challenging task with different strategies appropriate in different contexts. Inter(cross) - disciplinarity is not a simple call for opening or overpass the borders between disciplines, the inter(cross) - disciplinary borrowings being tolerated and even appreciated for the value added to solve problems in one's home discipline, rather, the persistent need for inter(cross)disciplinary solutions to disciplinary problems brings out the inherently conventional character of disciplines (Pop & Măties, 2009).

While inter(cross)disciplinarity may not respect disciplinary boundaries ([19]), it needs own boundaries to protect its free-ranging activities, the goal of the inter(cross) - disciplinary integration tends to be even a simple realignment of disciplinary boundaries, or a flexible translation of them, in an adaptiveness context, realizing good communication skills, without loosing any vital information in the pursuit of a common research project (Wikander et al, 2001). Even researchers are engaged in more transient or intermittent inter(cross)disciplinary collaborations they communicate for the purpose of specialized multi-disciplinary courses and may remain within their discipline-based units, being required a combination of strategies to foster, support and recognize the equally important contributions of both disciplinarity and inter(cross) - disciplinarity. Only the University is able to recognize the differential extent to which these kinds of initiatives have temporal contingencies or issues of sustainability (Jantsch, 1972; Kaynak, 1997; Wikander et al, 2001; Castells, 2001; Fuller, 2002; McGregor & Volcksmann, 2010). From this inter(cross)disciplinary point of view mechatronics could be considered only as yet another technological discipline, an evolutionary discipline with a curricula and specific organizational patterns and courses, while nothing is said about the next possible step, that of the transdisciplinary way of achieving knowledge in mechatronics, the evolution being considered as finished (Yamazaki & Miyazawa, 1992; Grimheden & Hanson, 2003; 2005; Habib, 2007; 2008). Consequently, the closed regions are growing from the initial points of contact to space filled with the separated elements (fulfilling the fields), so the last level of integrative, as *a transdisciplinary approach by synergistic generative synthesis (5)* emerges in a new transdisciplinary informational-functional structure, with ethic-semantic values, including the spiritual dimension (bridging the gaps). This system has a central hard synergistic core with flexible, deformable and penetrable boundaries, surrounded by a "fluid belt" through which are captured and modulated innovative ideas, new research themes and new courses in synergistic specific configurations and programs from the diffuse outer shell. The central zone is functioning as an integrative synergistic generative space, emerging a hierarchic-heterarchic rebuilding of the contents (could be just new transthematic disciplines as robotics, optomechatronics, biomechatronics, etc) (Cho, 2006; Hyungsuck, 2006; Mândru et al, 2008). The nodal points (inner, medium and outer) are considered as possible channels, knowledge search windows for explanation of the transdisciplinary mechatronical educational paradigm, through the specific creative innovative reflexive language of design, modeling prototyping, to create the socio-interactive way of understanding and practicing the mechatronics as a living, acting and thinking new lifestyle from the fourth wave perspective of knowledge, that of the informergic integration knowledge, functioning as a continuous synergistic integration of the knowledge as Science, Techne and Praxis (Pop & Vereş, 2010). If in the inter(cross)-disciplinary stage circular flows of knowledge are prevalent, in the transdisciplinary context there is a possible radial anisotropy of attractive-repulsive combining flows. In the transdisciplinary context is prevalent the radial anisotropy of attractive - repulsive combining flows, combined with inter(cross) - disciplinary circular flows of knowledge.

The presented transdisciplinary model for mechatronics give a better explanation then the existent models of the emergence of mechatronical epistemic teaching-learning paradigm, that of the synergistic identity of mechatronics, as a new transthematic generative discipline (20) (Pop & Mătieş, 2008; Berian, 2010). In this way it is possible to explain the appearing of the bridges between the different disciplines, as a step by step way through codisciplinary connection, multi(pluri)disciplinary combination, inter(cross)disciplinary overlap and transdisciplinary synergistic synthesis (De Conink, 1996; Klein, 2002; Mittelstrass, 2004; Choi & Pak, 2008). Due to the radial centripetal flows new mechatronical disciplines (optomechatronics, robotics, biomechatronics, etc) are emerging as satellites in the outer diffuse space, where the „codisciplinary outer nodal points"[21] are working as a resource spring generating mechatronical knowledge, expressed as a synergy between mechatronical transdisciplinary education, mechatronical design as a reflexive creative language and the mechatronical intelligent systems, technologies and products. The transdisciplinary perspective on mechatronical way of achieving knowledge gives the openness to a better understanding of the world from the mechatronical informergic (informaction as intentional action & information, and mattergy as energy incorporated in matter) integration process in the knowledge based society (Gitt, 1997; Pop & Vereş, 2010).

2.3 The mechatronician, a new synergistic job profile

Education and training with specific procedures and techniques are crucial in continuously economic and social changes, cultural differences and similarities concerning teaching/learning process and collaboration styles being present even not sufficiently integrated yet into curricula, courseware and teaching methods (Yamazaki & Miyazawa, 1992; Lyshevski, 2000). One of the transdisciplinary way to achieve knowledge in the mechatronical context is the Problem-Based Learning (PBL) method, working together with other active learning models, such as: *group work, guided design, work-based learning, learning by doing* and *case studies, learning by discovery,* being distinguished from these by solving a complex and realistic problem (Altshuller et al, 1989; Barret, 2005; Boud & Feletti, 1991; Fink, 2002; Grinko, 2008; Savary, 2006). The basic ideas behind Problem Based Learning are active learning, constant assessment, emphasis on meaning and not on facts, freedom and responsibility, access to resources. The work is organized in projects, small groups of disciples and guiders (instructor, teacher, mentor, facilitator) meeting together to discuss about a case, getting solutions in a creative-innovative framework, where the project aims working towards to solve a particular problem in a learning environment. The framework is characterized by a large responsibility of the disciples and teachers, as well, having a cognitive coaching role instead of a lecturing one, the disciples receiving from the guide an initial guiding plan-work (as a scenario), then they question the guide to get any additional information to solve the problem (Boud & Feletti, 1991; Barret, 2005; Savary, 2006). The computers have to be used as tools to provide alternative sources of learning material, interactive learning situations and simulation of systems that cannot be used in reality for reasons of cost, size or safety, including the Internet as the greatest source of information available for learning, as well as simulation tools with a number of benefits to education, available in industry. It is interesting to know how much of real experience can be replaced by learning with simulations, but is demonstrated that only the use of the computer simulations cannot replace all forms of applied training, in many branches of the science and technology-oriented programs hands-on activities in laboratories and workshops remaining an indispensable constituent of effective learning. Flexibility and adaptability should be characteristics most important to determine tertiary education ability of the institutions to contribute effectively to the capacity building needs of developing knowledge achievement skills and to react swiftly by establishing new programs, reconfiguring existing ones, to eliminate outdated courses without any administrative obstacles, in the context of systematic efforts to develop and implement a vision through strategic planning, by identifying both favorable and harmful trends in their immediate environment and linking them to a rigorous assessment of their internal strengths and weaknesses, so the institutions could better define their mission, market niche and medium-term development objectives and formulate concrete plans to achieve these objectives (Lyshevski, 2000; Pop, 2009a). To face effectively the challenges of economic development within a global marketplace, the new generation of engineering professionals has to be educated in a new framework, as a continuum educational program, to develop and strengthen the integrative skills in analysis, synthesis, and contextual understanding of problems and also, to expose them to the latest technologies in different engineering fields and the implications for sustainability of their use. The problem-based learning (PBL) approach, open-ended design problem solving by a multi(pluri)disciplinary team of disciples in a transdisciplinary context, simulation, modeling, prototyping, are integrated alltogether with the technology, economics, ecology and ethics, as four dimensions of the sustenability ([22]), considering them as parts of a synergistic - generative approach of knowledge integration (Grinko, 2008; Bras et al., 1995; Pop, 2008). Problem-based learning (PBL) is a contextualized approach to schooling, being centered to the disciples, where learning begins with a problem to be solved together, rather than mastering individually different contents of the research themes, courses, laboratory experiences (Grimheden & Hanson, 2003). PBL is based on the notion that learning occurs in problem-oriented situations is more likely to be available for later use in those contexts (Bras et al, 1995). PBL includes among its goals the developing of the scientific understanding through real-world cases; the reasoning strategies and the self-directed learning strategies. In PBL the focus is on what disciples learn, but more important becomes the way the knowledge could be applied, maintaining a balance between theory and practice (top-down in balance with bottom-up approach). The learning team (disciples) is evaluated by the teaching team (instructors), resulting a better coverage of specific problems, the results and experience of the research activity carried out by the teachers could be incorporated in the educational and training programs for disciples. Both, PBL and TT methods lead to more self-motivated and independent disciple, these learning methods preparing better the disciples (students, apprentices, pupils, adults, as well) to apply their learning to real-world situations (Mândru et al, 2008).

An alternative complementary method to PBL and TT is TRIZ (theory of inventive problem solving). The main point of this method is the observation that good ideas/solutions have the properties to resolve contradictions, to increase the "ideality" of the system and to use idle, easily available resources. To solve a technical problem has to find the contradiction in the definition of the problem, identifying it using available resources to arrive at the ideal final solution as closely as possible, choosing the good context, methods and the best possible way (Altshuller et al, 1989). All innovations emerge from the application of a very small number of inventive principles and strategies, technology evolution trends being highly predictable. The strongest solutions transform the unwanted or harmful elements of a system into useful resources, and also actively seek out and destroy the conflicts and trade-offs most design practices assume to be fundamental. TRIZ revolves around finding contradictions and using the collected knowledge and experience of decades is able to solve the problem. Universities and vocational training schools with their links to industry are under an increasing pressure placed on them to expose disciples to real working environments in education and training of multi-skilled technicians leading to a new type of job profile which contains a mix of electrical, mechanical and IT knowledge, a mechatronical one, to be trained for implementation and service using the education and training of engineers for design and manufacturing of mechatronical devices (Wikander, Torngren &

Hanson, 2001; Bruns, 2005; Mătieş et al, 2005). To get expertise as a vital and dynamic living treasure many enterprises rely on formal learning (off-the-job training), but the informal learning (on-the-job training) can be more close to the problems to be solved, being organized in a cooperative way, crossing the borders between different professions that are involved in a project (Jacobs & Jones, 1995). Experts work in projects (small groups of different professions) to solve problems, learn how to learn and think critically, learn how to understand, identifying the skills needed to meet the requirements emerged (bottom-up learning-teaching) and developing a personal theory of management, leadership or empowerment (top-down teaching-learning).

The design cycle for the intelligent products often take place in a competitive environment, where following the trends in technology itself and responding to innovative solutions from competitors create a challenging road for the engineering development process. Within this rapidly changing medium products, processes or systems need to be designed and developed satisfying both the customers and the developers. Web-based virtual laboratories, remote experience laboratories and access to digital libraries are some examples of the new learning enhancing opportunities to increase connectivity. In this context, tertiary institutions with virtual libraries can join together to established, inter-library loans of digitized documents on the Internet to form virtual communities of learning helping each other to apply and enrich available open education resources with significant challenges. In this way could be created a more active and interactive learning environment, called "instructional integration" with a clear vision to develop and create the new adequate technologies and the most effective way to integrate them in the design programs and delivery (Bridwell et al., 2006). Combining online and regular classroom courses gives to disciples more opportunities for human interaction, and developping the social aspects of learning through direct communication, debate, discussion in a synergistic communicational context (Pop, 2008). These requirements are applied also to the design and delivery of distance education programs which need to match learning objectives with appropriate technology support. The new types of distance education institutions and the new forms of e-learning and blended programs meet acceptable academic and professional standards, but a poor connectivity is a serious constraint in the use of the informational control technology related opportunities, with their limitations (Furman & Hayward, 2000). The use of simulation tools has a number of benefits in education, because the disciples are not strictly related with real world, and at the same time is able to explore a range of possible solutions, easily and quickly, with tools available in industry, with significantly less costs than the real world components and allows more participation and interaction than a limited demonstration. But, it is very clear that real experience can not be replaced by learning with simulations, being necessary to use complementarly, the virtual tools as design, modelation, simulation and real and the real world representations as prototyping, building smart mecahatronical products (Bridwell et al, 2006; Giurgiutiu et al., 2002). Only computer simulations cannot replace all forms of applied training, but in many branches of the science and technology-oriented programs hands-on activities in laboratories and workshops remain an indispensable constituent of effective learning. Flexibility and adaptability should be characteristics most important to determine tertiary education ability of the institutions to contribute effectively to the capacity building needs of developing knowledge achievement skills and to react swiftly by establishing new programs, reconfiguring existing ones, to eliminate outdated courses without any administrative obstacles, in the context of systematic efforts to develop and implement a vision through strategic planning, by identifying both favorable and harmful trends in their immediate environment and linking them to a rigorous assessment of their internal strengths and weaknesses. In the disciplinary educational system there is obvious the lack of flexibility and low level of adaptation to the changing conditions of the environment. A theoretical framework for this didactics requires more insight into how individual learning styles use individual learning methods, techniques and technologies, to outline paths to develop meaning and concepts from basic experiences with natural and technical phenomena, being important to analyze the transitions between concrete and abstract models of production systems and to specify abstract solution for an automation problem by a concrete demonstration (Schäfer, 1997; Bruns, 2005). To fulfill the demands for multi-skilled technicians and skilled workers vocational training schools together with industry are confronted with the need to develop theoretical integrated with practical learning sequences. Tasks and problem solving in mechatronics requires cognitive, operational knowledge and practical experience about building systems, diagnosis and maintenance techniques, a significant challenge being that these tasks are essentially characterized by the use of tele-medial systems, in a synergistic communicational networking system (Palmer, 1978; Grossberg, 1995; Arecchi, 2007; Baritz et al, 2010). To meet these requirements in education and training it has to elaborate concepts concerning pedagogical, technical and organizational aspects in a new significant synergistic way, that of the transdisciplinary educational paradigm (Pop, 2008; Pop & Maties, 2008) with *holistic-synergistic* problem solving or tasks distributed over time of training with increasing requirements to the learners, in a logical-creative framework, through included middle and lateral thinking (Lupasco, 1987; Waks, 1997; de Bono, 2003). Through this new didactical tansdisciplinary concept is avoided the disciplinary distribution of learning contents into separate classes for different separated disciplines, whereas the learners had been left alone to find out the connections between these contents. It has to be fulfilled every one of the four didactical principles in the transdisciplinary field of mechatronical training paradigm: synergistic transthematic identity, vertical exemplificative selection, interactive creative participation-communication and contextual functional legitimacy (Grimheden & Hanson, 2003; 2005; Berian 2010).

3. Conclusions

Mechatronics and transdisciplinarity are presented as multiple integrative possibilities to understand the way to achieve, transfer and incorporate knowledge in the context of the informergical knowledge based society. In order to know the way mechatronics does work in the transdisciplinary methodological approach it is very important to understand the new sinergistic-generative transdisciplinary model about the perspective of the integration from the thematic-curricular monodisciplinary level to the synergistic one, as structural, functional and generative stages, passing through methodological level.

The transdisciplinary knowledge search window, as a new methodology is working complementarily with the top-down and bottom-up levels of knowledge, integrating the rational knowledge of things expressed by „learning to learn to know by doing" with relational understanding of the world, working by „learning to understand to be by living together with other people". This multiple transdisciplinary paradigm ([23]), is integrating informergically (informaction integrated in mattergy) the creativity (adequateness and innovation) in action (competition and performance) and authenticity (character and competence) through participation (apprenticeship in communion).

Only the transdisciplinary knowledge achievement, as a new methodology, can explain the way the creativity, with a synergistic signification, works as an intentional action through ideas, design, modelling, prototyping, simulation, incorporating informergically the inform-action in matt-ergy, to realize smart products, sustainable technologies and specific integrative methods to give solution to the emerging problems. Real experiences cannot be replaced by learning only with simulations, for this being necessary to use complementarily, the virtual tools as design, modelling, simulation and the real world representations as prototyping, building smart mechatronical products, technologies and systems.

The proposed integrative model demonstrates that mechatronics cannot be considered as multi(pluri)disciplinary, inter(cross)disciplinary, nor a simple new discipline, neither a simple methodology, but a transdisciplinary approach of the mechatronical knowledge in the informergical society (informergy is information incorporated intelligently in mattergy), as is sustained through the semiophysical communicational contextual message model, with the "What-How-Why" questioning paradigm (24) of the mechatronics. The transdisciplinary knowledge integrative mechatronical model, with the five stages of the evolution of the knowledge process from monodisciplinarity to transdisciplinarity, through codisciplinarity, multi(pluri)disciplinarity and inter(cross)disciplinarity, is considered more integrative then the educational mechatronical model, integrating the transthematic aspect of the mecahatronics, with representative selection, interactive communication and functional legitimacy aspects (mechatronical epistemology), as a reflexive way of communication through design, modeling (the creative logic of the included middle) and a socio-interactive system of thinking, living and acting (mechatronical ontology).

The most important thing is to know what mechatronics is, what isn't and how does it work, mechatronics being not a simple discipline, but working through the new transdisciplinary transthematic educational paradigm by its exemplifying selection (what), interactive communication (how) and functional contextual legitimation (why) aspects.

Mechatronics can be considered as a synergistic integrative system of Scientia, as a new educational transdisciplinary paradigm (mechatronical epistemology), of Techne, working as a reflexive way of the integrative design (the creative logic of the included middle) and, as Praxis, through a new socio-interactive system of thought, living and action (mechatronical ontology)

About the future of integrative mechatronics, the transdisciplinary approach opens new perspectives on its development, incorporating more and more ideas which will be accounted to improve the way to do things and to live in the new context of ever-changing needs and willings of a complex and complicated world, when innovations and technologies have to be improved and developed with the rapidly changing times. The postepistemic economy will integrate in a synergistic-generative way the technical dimension with epistemic and with socioeconomical dimension, resulting the metamechatronics as a transdisciplinary engineering mecha-system (25).

4. Notes

[1]Synergy, synergistic signification is the transdisciplinary semiophysical process by which a system generates emergent properties resulting in the condition in which a system may be considered more then the sum of its parts (equal to the sum of its parts and their relationships) (synergy, $1 + 1 > 2$, more then everyone, and signification, $1 - 1 \neq 0$, otherwise then everyone) (Tähemaa, 2004; Bolton, 2006; Pop & Vereş, 2010).

[2]Agents are considered to be the ocupants of a knowledge system field (a semophysical system working through spatial participative sequence - space wise, temporal-connective sequence – time wise, actional – interactive sequence – act wise) (Pop, 1980);

[3]This a contextual adaptation of the apo-kataphatic approach of knowledge which does explain through the interparadigmatic dialogue the japanese roots of the mechatronics (Mushakoji, 1988).

[4]Principle of included middle (tertium quid) is the natural law by which triple is produced out of couple, rejecting the claim that the the mind (consciousness) and the body (object) are separated. Is proposed a change to the third classical linear logic axiom, submitting that a third term T does exist, being simultaneously A and non-A. Only considering this third term T, problem solvers would be able to integrate perspectives from different realities (economics with environmental), let alone integrate Subject (consciousness and perceptions) with Object (information) (Nicolescu, 2011).

[5] Smart mechatronical products, technologies and systems are considered sustainable if they are incorporating transdisciplinarily the informaction (information in action) in mattergy (matter and energy), with a high level of reciclable matter and low level of incorporated energy, in a modular configurational design, with a creative and responsible stewardship of resources in order to generate stakeholder value contributing to the well-being of current and future generations (Rzevski, 1995; Montaud, 2008).

[6]Paradigm is a set of fundamental beliefs, axioms, and assumptions that order and provide coherence to our perception of what is and how it works (a basic world view, also example cases and metaphors), refering to a thought pattern in any scientific discipline or other epistemological context, with theories, laws, generalizations and the experiments performed *(broadly, a philosophical or theoretical framework of any kind)* (Pop & Vereş, 2010);

[7]Mechatronician is a multi-skilled specialist, as engineer, technician, worker, envolved in the mechatronical design, creation and maintainance of smart products, technologies, systems (Rainey, 2002);

[8]The multi(pluri)disciplinary approache juxtaposes disciplinary/professional perspectives, adding breadth and available knowledge, information, and methods, speaking as separate voices; such activities involve researchers from various disciplines working essentially independently, each from own discipline specific perspective, to address a common problem; even multi(pluri)disciplinary teams do cross discipline boundaries; however, they remain limited to the framework of disciplinary research; Multidisciplinarity – a relationship between related disciplines occurring simultaneously without making explicit possible relationships or cooperation between them, working at methodological level of the integrative process of knowledge; Pluridisciplinarity – a relationship between various disciplines grouped in such a way as to enhance the cooperative relationships between them, working at the methodological level of the integrative process of knowledge (Pop & Mătieş, 2008);

[9]Inter(cross)disciplinarity is working on unity of knowledge differing from a complex, dynamic web or system of relations, but without producing a combination or synthesis which would go beyond disciplinary boundaries, for innovative solutions to knowledge questions, remaining in the disciplinary bounderies. Interdisciplinarity is a structural synergistic approach for a group of related disciplines having a set of common purposes and coordinated from a higher purposive level, that integrates separate disciplinary data, methods, tools, concepts, and theories in order to create a holistic view, or common understanding of complex issues, questions, or problem. Crossdisciplinarity is a functional

synergistic approach for various disciplines where the concepts or goals of one are imposed upon other disciplines, thereby creating a rigid control from one disciplinary goal (Habib, 2008, Pop & Mătieş, 2009, Fuller, 2001).

[10]Transdisciplinarity concerns with that is at once between the disciplines, across the different disciplines, and beyond all disciplines, connecting what is known (theory - what) to action (application - how), in order to accomplish specific goals in the context of human survival, sustainability and creativity (worldly problems and/or opportunities), creating new knowledge, new languages, new disciplines, new systems, new processes and new economic opportunities. Transdisciplinary approaches are comprehensive frameworks that transcend the narrow scope of disciplinary world views through an overarching synergistic generative sinthesis of knowledge, including cooperation within the scientific community with a permanent debate between research and the society at large, transgressing boundaries between scientific disciplines and between science and other societal fields, with deliberation about facts, practices and values, at the stages of conceptualization, design, analysis, and interpretation by integrated team approaches, realizing the coordination of disciplines and interdisciplines with a set of common goals towards a common system purpose (Jantsch, 1972; Nicolescu, 1996; Max Neef, 2005).Transdisciplinary methodology is working with three axioms, the ontological axiom (there are different levels of Reality of the Object and, correspondingly, different levels of Reality of the Subject); the logical axiom (the passage from one level of Reality to another is insured by the logic of the included middle) and the epistemological axiom (the structure of the totality of levels of Reality has a complex structure, every level being what it is because all the levels exist at the same time) (Nicolescu, 1996).

[11]Predisciplinarity stage is the first step of the lowest level, the thematic-curricular level of the integration knowledge process, the way a discipline is born; disciplinarity context is the classical mode of deapth approach of knowledge with own boundaries, methodologies, and specific content; codisciplinary context of the integration of knowledge is conecting, from a transdisciplinary point of view, the three levels, the thematic-curricular, the methodological level and the synergistic one (Pop & Mătieş, 2008).

[12]Communities of practice (CoPs), as knowledge achievement environments, are functioning as creative group of people who share an interest, a craft, and/or a profession, evolving naturally because of the common interst of the members in a particular domain or area, or it can be created specifically with the goal of gaining knowledge related to their field (Wenger & Snyder, 2000).

[13]Organisational educational environment is working with the principles of mechatronical education which can be applied successfully to all teaching levels, creating the necessary teaching-learning environment, as a teaching factory, as a mobile mechatronical platform, or as another specific educational systems (Nonaka & Takeuchi, 1994; Lamancusa et al, 1997; Doppelt & Schunn, 2008; Mătieş, 2009).

[14]Cognitive way of knowledge does explain the way stimuli (coming from the sensitive sensors, as a bottom up approach) and signals (at the brain level, as a top down approach) are working together in the ART (Adaptive Resonant Theory) (Grossberg, 1995);

[15]Creative innovative context is determined by the learning/teaching transdisciplinary environment, as teaching factory through all life learning aspects (lifewide learning, longlife learning and learning for life), that challenges perspective of the learners and facilitates the expansion of their worldview, promoting human fulfillment, enabling the learners to cope with uncertainty and complexity, empowering them to shape creatively change in order to

configurate the future through the synergistic design (Lamancusa et al, 1997;Alptekin, 2001; Erdener, 2003; Habib, 2008).

[16]Transdisciplinary semiophysical contextual message model is working with 7 questions: where (space wise sequence), when (time wise sequence), who, with whom, what, how and why (act wise sequence) (Bradley, 1997; Harashima et al, 1996; Buckley, 2000; Pop & Vereş, 2010).

[17]Knowledge search window is a methodological concept explaining the bottom-up/top-down mechanism of the teaching-learning process in the mechatronical educational paradigm using the included middle transdisciplinary perspective (Lupasco, 1987, Pop, 2009);

[18]Conceptual space presuposes to identify, to develop and to evaluate the creativity working in such a way to realise the equillibrium between tradition and innovation, the most creative individuals being considered those who explore a conceptual structure going beyond them in a transdisciplinary way, managing the reconfiguration of the new structures to achieve knowledge which transgress the barriers, bridging the gaps and filling the fields (Boden, 1994; Schafer, 1996;De Vries, 1996; Doppelt & Schunn, 2008).

[19]Boundaries are parametric conditions that are delimiting and defining a system, and set it apart from its environment;

[20]Mechatronics works as an opening new transthematic generative discipline, with a very transdisciplinary character, bridging the gaps between different disciplines, as a step by step way through codisciplinary connection, multi(pluri)disciplinary combination, inter(cross)disciplinary overlap, and transdisciplinary synergistic synthesis (Pop & Vereş, 2010);

[21]Codisciplinary outer nodal points are considered as resource springs generating mechatronical knowledge, expressed as a synergy between mechatronical transdisciplinary education, mechatronical design as a reflexive creative language and the mechatronical intelligent systems, technologies and products (Pop & Mătieş, 2008);

[22]Sustainability represents the creative and responsible stewardship of resources (human, natural and financial resources management) in order to generate stakeholder value while contributing to the well-being of current and future generations of all beings. Sustainable development is an individual, societal, or global process, which can be said to be sustainable (sociocultural, economical, educational, technological, and ecological as well) if it involves an adaptive strategy that ensures the evolutionary maintenance of an increasingly robust and supportive specific environment, such a process enhancing the possibility to generate a wellfaire state (Giovannini & Revéret, 1998);

[23]Multiple transdisciplinary paradigm represents the informergically integration (informaction integrated in mattergy) of the creativity (adequateness and innovation) in action (competition and performance) and authenticity (character and competence) through participation (apprenticeship in communion) (Pop & Mătieş, 2009).

[24]The "What-How-Why" questioning paradigm is a transdisciplinary knowledge integrative mechatronical model, integrating the transthematic aspect of the mecahatronics, with representative selection, interactive communication and functional legitimacy aspects (mechatronical epistemology), as a reflexive way of communication through design, modeling (the creative logic of the included middle) and a socio-interactive system of thinking, living and acting (mechatronical ontology) (Pop & Vereş, 2010).

[25]Meta-mechatronics is a transdisciplinary engineering mecha-system, resulting through synergistic synthesis of the Scientia (Educational Mechatronics), Techne (Technological

Mechatronics), and Praxis (Economical Mechatronics) at the top level of integration as informergical metamodel (Hug et al, 2009; Pop & Vereş, 2010).

5. References

Alptekin, S.E., (1996), Preparing the Leaders for Mechatronics Education, *FIE '96, 26th Annual Conference, Proceedings of the Frontiers in Education Conference*, Vol. 2, 6-9 Nov. 1996, pp. 975–979.

Altshuller, G., Zlotin, B., Zusman, A. & Filatov, V., (1989), *Search for New Ideas: from Insight to Technology*, Kartia Moldavenayska, Kishenev.

Arkin, R., Lee, K-M., McGinnis, L., & C. Zhou (1997),The Development of a Shared Interdisciplinary Intelligent Mechatronics Laboratory, *Journal of Engineering Education*. April, pp.113-118.

Arnold, S. (2008), *Thesis, Transforming systems engineering principles into integrated project team practice*, Cranfield University.

Arecchi, F.T., (2007), Physics of cognition: Complexity and creativity, *European. Phys. Journal, Special Topics*, Vol. 146, pp. 205–216.

Ashley, S., (ass. ed.), (1997), Getting a hold on mechatronics, , *Mechanical Engineering-CIME*, May, vol. 1. A.

Auslander, D.M. (1996), What is mechatronics? *IEEE/ASME Transactions on Mechatronics*, Vol.1.No. 1, pp. 5-9.

Barak, M., & Doppelt, Y. (1998). *Promoting creative thinking within technology education.* International Workshop for Scholars in Technology Education, WOCATE, George Washington University, Washington, DC.

Baritz M., Cotoros D., Repanovici A. & Rogozea L., (2010), Cognitive Virtual Learning Used for Improving the Long Term Handling Skills, *Proceedings of the 6th WSEAS/IASME International Conference on EDUCATIONAL TECHNOLOGIES (EDUTE'10), Tunisia,*.

Barret, T., ed. (2005), "Understanding Problem-Based Learning", in *Handbook of Enquiry & PBL.*

Berte, M. (2003), *Active learning and transdisciplinarity*, Cluj-Napoca: Promedia Plus Press.

Berte, M., (2005), Transdisciplinarity and Education: "The treasure within" -Towards a transdisciplinary evolution of education, *The 2nd World Congress of Transdisciplinarity, Brazil.*

Berian, S., (2010), *Thesis, Research about the transdisciplinary potential of the Mechatronics (Teză de doctorat, Cercetări privind potenţialul transdisciplinar al mecatronicii)*, Universitatea Tehnică Cluj-Napoca.

Bishop, H.B. & Ramasubramanian, M.K., (2002), *What is Mechatronics?* The Mechatronics Handbook, CRC Press.Boden, M., (1994), *What is Creativity?* In M. Boden (ed.), *Dimensions of Creativity*, Cambridge, Ma, and London: The MIT Press.

Bolton, W., (2006), *Mecatronics*, Pearson Education Limited, Essex, England.

Boud, D. & Feletti, G., (1991), *The Challenge of problem based learning.* New York: St. Martin's Press.

Brown, J. S. & Duguid, P., (1991), "Organizational Learning and Communities-of-Practice; Toward a Unified View of Working, Learning and Innovation", *Organization Science* 2, No. 1, pp. 40–57.

Bradley, D. A. (1997), "The what, why and how of mechatronics." *IEEE Journal of Engineering Science and Education*, Vol. 6(2), pp. 81-88.

Bradley, D., Seward, D., Dawson, D. & Burge, S., (2000), *Mechatronics and the design of intelligent machines and systems*. Stanley Thornes.

Bras, B., Hmelo, C., Mulholland, J., Realff, M., & Vanegas, J., (1995), "A problem based curriculum for sustainable technology", *The Second Congress on Computing in Civil Engineering*, Atlanta GA.

Brazell, J., Elliot, H. & Vanston, J. (2005). *Mechatronics: Integrating Technology, Knowledge & Skills*, Retrieved from http://system.tstc.edu/forecasting/techbriefs/mechatronics.asp.

Bridwell, W., Sánchez, J. N., Langley, P. & Billman, D., (2006), An interactive environment for the modeling and discovery of scientific knowledge, *International Journal of Human-Computer Studies*, 64, 11, pp. 1099-1114.

Brown, J. S. & Duguid, P., (1991), "Organizational Learning and Communities-of-Practice; Toward a Unified View of Working, Learning and Innovation," *Organization Science*, vol. 2 (1), pp. 40–57.

Bruns, F. W., (2005), Didactical Aspects of Mechatronics Education, *Proceedings of the 16th IFAC World Congress (International Symposium on Intelligent Components and Instruments for Control Applications)*, Prague.

Buckley, F. J., (2000), *Team Teaching: What, Why, and How?* London: Sage Publications, Inc.

Buur, J., (1990), *A Theoretical Approach to Mechatronics Design*, PhD Thesis, Institute for Engineering Design, Technical University of Denmark, Lyngby.

Castells,M., (2001), *Universities as dynamic systems of contradictory functions*. In J. Muller, N. Cloete and S. Badat (eds). *Challenge of globalisation*. South African Debates with Manuel Castells, pp. 206-224.

Cho, H., (2006), *Optomechatronics: Fusion of Optical and Mechatronic Engineering*, CRC Press, London.

Choi, B.C.K. & Pak, A.W.P., (2008), Multidisciplinarity, interdisciplinarity, and trans-disciplinarity in health research, services, education and policy: 3. Discipline, inter-discipline distance, and selection of discipline, CIM *Clin Invest Med, Vol 31, no 1, February*, E41-E48.

Cleveland, H., (1993), *Birth of a new world*, San Francisco (CA): Jossey- Bass Publishers. Comerford, R. Sr (Ed), (1994 Aug.). Mechatronics...what? *IEEE Spectrum*, pp. 46-49.

Craig, K. & Stolfi, F., (2000), *Teaching Control System Design Through Mechatronics: Academic and Industrial Perspectives*. Elsevier Science Ltd.

Day, C. B., (1992), "Mechatronics - meeting the technician's needs into the 21st century", *Conference on Mechatronics - The integration of Engineering Design*, Dundee.

De Bono, E., (2003), *Lateral Thinking (Gandirea laterală* - romanian translation), Bucuresti: Curtea Veche.

De Conink, P., (1996), De la disciplinarité à la transdisciplinarité, *Info-Stoper*, Vol. 4, No 1, Sherbrooke.

De Gruyter, W., (1998), *"Gödelian Aspects of Nature and Knowledge"*, in Gabriel Altmann and Walter A. Koch (ed.), Systems-New Paradigms for the Human Sciences, Berlin - New York, pp. 385-403.

Delors, J., (1996), *L'éducation, le trésor est caché dedans.* Rapport à l'UNESCO de la Commission internationale sur l'éducation pour le vingt et unième siècle, Twenty-first Century, UNESCO, (Romanian translation, Iasi: Polirom, 2000).

Derry, S.J. & Fischer, G., (2005), Toward a Model and Theory for Transdisciplinary Graduate Education, AERA Annual Meeting as part of *Symposium, "Sociotechnical Design for Lifelong Learning: A Crucial Role for Graduate Education"* April, University of Colorado-Boulder.

De Vries, M. J., (1996), Technology education: Beyond the "technology is applied science" paradigm. *Journal of Technology Education*, 8(1), 7-15.

Doppelt, Y. & Schunn, C. D., (2008), Identifying students' perceptions of the important classroom features affecting learning aspects of a design based learning environment? *Learning Environments Research, 11(3)*.

Erbe, H.H. & Bruns, F. W., (2003), Didactical Aspect Mechatronics Education, *International Conference on Microelectronical Systems Education*, 6.

Erdener, O. A., (2003), *Development of a Mechatronics Education Desk*, Thesis, The Middle East Technical University, Turkey.

Ertas, A., Tanik, M. M. & Maxwell, T. T., (2000), Transdisciplinary engineering education and research model, *Journal of Integrated Design and Process Science*, dec., Vol. 4 (4), pp. 1-11.

Everitt, D. & Robertson, A., (2007), Emergence and complexity: Some observations and reflections on transdisciplinary research involving performative contexts and new media, *International Journal of Performance Arts and Digital Media*, Vol. 3, No. 2 & 3, pp. 239-252.

Fink, F., (2002), Problem-Based Learning in Engineering Education. *World Transactions on Engineering and Technological Education*, vol. 1(1), pp. 29-32.

Fricke, M., (2009), The knowledge pyramid: a critique of the DIKW hierarchy, *Journal of Information Science*, 35(2).

Fuller, S. (2001), *"Strategies of Knowledge Integration"* in M.K. Tolba (ed.), Our FragileWorld: Challenges, Opportunities for Sustainable Development (EOLSS Publishers (for UNESCO), Oxford), pp.1215-1228.

Fuller, S. (2002), *The Changing Images of Unity and Disunity in the Philosophy of Science*, In I. Stamhuis, et al., eds. The Changing Image of the Sciences, pp. 173-196, Dordrecht: Kluwer.

Furman, B.J. & Hayward, G.P., (2000), Asynchronous hands-on experiments for mechatronics education, *Mechatronics 2000 - 7th Mechatronics Forum International Conference*, Georgia, USA.

Gibbons, M., Limoges, C., Nowotny, H., Schwartzman, S., Scott, P., & Trow, M., (1994), *The new production of knowledge: the dynamics of science and research in contemporary societies*. London: Sage.

Giovannini, B. & Revéret, J. P., (1998), *The Practice of Transdisciplinarity in Sustainable Development, Concept, Methodologies and Examples*, Paperback, University of Geneva: Departément de la matière condensée.

Gitt, W., (1997), *In the Beginning Was Information*, Christliche Literatur-Verbeitung e. V. Postfach, Bielefeld.

Giurgiutiu, V., Bayoumi, A.E., & Nall, G., (2002), Mechatronics and smart structures-emerging engineering disciplines of the third millennium, *Journal of Mechatronics*, Vol. 12(2), pp. 169–181.

Gomes, J.F., de Weerd-Nederhof, P.C., Pearson, A. & Cunha, M.P. (2003). 'Is more always better? An exploration of the differential effects of functional integration on performance in new product development.' *Technovation* 23(3), 185-192.

Grimheden, M., & Hanson, M., (2001), What is Mechatronics? Proposing a Didactical Approach to Mechatronics. *Proceedings of the 1st Baltic See Workshop on Education in Mechatronics*, Kiel.

Grimheden, M. & Hanson, M., (2005), Mechatronics - the Evolution of an Academic Discipline in Engineering, *Education Mechatronics*, vol. 15, pp.179-192.

Grinko, V., (2008), Some new aspects in education for specialists of Mechatronics, *The 7th France-Japan and 5th Europe-Asia Mechatronics Congress*, France, pp. 92 - 94.

Grossberg, S., (1995), ART (Adaptive Resonance Theory), *American Scientist*, Vol. 83, pp.439.

Habib, M.K., (2007), Mechatronics - A unifying interdisciplinary and intelligent engineering science paradigm, *Industrial Electronics Magazine, IEEE*, Vol.1, Issue2, pp. 12-24.

Habib, M.K., (2008), Interdisciplinary Mechatronics engineering and science: problem-solving, creative-thinking and concurrent design synergy, *International Journal of Mechatronics and Manufacturing Systems*, Vol. 1, No. 1, pp. 4 - 22.

Hakkarainen, K., Palonen, T., Paavola, S., & Lehtinen, E., (2004), *Communities of networked expertise: Professional and educational perspectives*. Pergamon Press.

Hanson, M., (1994), "Teaching Mechatronics at Tertiary Level", *Journal of Mechatronics*, Vol. 4, No.2, pp. 217-225.

Harashima, F., Tomizuka, M. & Fukuda, T., (1996), "Mechatronics -"What Is It, Why, and How?" An Editorial", *IEEE/ASME Transactions on Mechatronics*, Vol. 1 (1), pp. 1-4.

Harashima, F. (2005), 'Human adaptive mechatronics', *Proceedings of the 10th IEEE International Conference on Emerging Technologies and Factory Automation*, Italy.

Hewit, J.R. & King, T.G. (1996a) 'Mechatronics design for product enhancement', *IEEE/ASME Transactions on Mechatronics*, Vol. 1, No. 2, pp.111–119.

Hildreth, P., Kimble, C. & Wright, P., (2000), "Communities of Practice in the Distributed International Environment," *Journal of Knowledge Management* 4, No. 1, 27–38.

Hildreth, P., M. & Kimble, C., (2004), *Knowledge Networks: Innovation through Communities of Practice*, Idea Group Inc (IGI).

Hmelo, C.E., Shikano, T., Realff, M., Bras, B., Mullholland, J., & Vanegas, J.A., (1995), A problem-based course in sustainable technology. *Proceedings of the Frontiers in Education Conference*, vol. 2, 4a, pp. 31-35.

Hug, Ch., Front, A., Rieu, D. & Henderson-Sellers, B., (2009), A method to build information systems engineering process metamodels, *Journal of Systems and Software*, Vol. 82 (10), October, Publisher Elsevier Science Inc. New York, NY, USA.

Hyungsuck C., (2006), *Optomechatronics*, Taylor & Francis Group, LLC, U.S.A.

Isermann, R., (2000), Mechatronic systems: concepts and applications, *Transactions of the Institute of Measurement and Control*, Vol. 22, No. 1, pp. 29-55.

Jacobs, R. L. & Jones, M. J., (1995), Structured on-the-job training: Unleashing employee expertise in the workplace, *Human Resource Development Quaterly*, Vol. 6 (3), pp. 320-323.

Jack, H. & Sterian, A., (2002), *Control with Embedded Computers and Programmable Logic Controllers*. The Mechatronics Handbook, CRC Press.

Jantsch, E., (1972), *Towards Interdisciplinarity and Transdisciplinarity in Education and Innovation, Interdisciplinarity, Problems of Teaching and Research in Universities*, OECD, Paris.

Kajitani, M., (1992), What has brought Mecatronics into existence in Japan? *Procceding of the 1st France – Japan Congress on Mecatronics*, Besancon, France.

Kaynak, O., (1996), 'A new perspective on engineering education in mechatronics age', *Proceedings of FIE 1996 Conference, Mechatronics Education, vol.2, pp.970-974*.

Kerzner, H., (2003), *Project Management: A System Approach to Planning, Scheduling and Controling*, 8th Ed., Wiley.

Klein, J.T., (2002), *Unity of knowledge and transdisciplinarity. In Unity of knowledge (in transdisciplinarity research for sustainability* (Gertrude Hirsch Hadorn, Ed.) Paris: UNESCO Publishing-Eolss Publishers, Oxford, UK.

Korossy, K., (1999). Modeling knowledge as competence and performance. In D. Albert & J. Lukas (Eds.), *Knowledge Spaces: Theories, empirical research, applications*, pp. 103-132, Mahwah, NJ: Lawrence Erlbaum Associates.

Kyura, N. & Oho, H., (1996), Mechatronics - an industrial perspective, *IEEE/ASME Transactions on Mechatronics*, Vol.1 (1), pp. 10 – 15.

Lamancusa, J., Jorgensen, J. & Zayas-Castro, J., (1997), The Learning Factory-A New Approach to Integrating Design and Manufacturing into the Engineering Curriculum, *Journal of Engineering Education*, pp. 103-112.

Langley, P., Simon, H.A., Bradshaw, G.L. & Zytkow, J.M., (1987), Scientific discovery: computational explorations of the creative process, *Proceedings of the Thirtieth Annual Meeting of the Cognitive Science Society*, Washingon D.C.

Leonard, D. & Sensiper, S., (1998), "The role of tacit knowledge in group innovation" *California Management Review*, 40 (3), pp. 112-13.

Lupasco, S., (1987), Le principe d'antagonisme et la logique de l'énergie - Prolégomènes à une science de la contradiction, Hermann & Cie. *Colloque „Actualités scientifiques et industrielles"*, no. 1133, Paris, 1951; 2nd ed., Le Rocher, Monaco.

Lute, B. E., (2006), *Win/Win - The Art of Synergistic Communication*, Trafford Publishing.

Lyshevski, S.E., (2000), Mechatronics and New Directions in Engineering Education. *Mechatronics 7th Mechatronics Forum International Conference*, Georgia, USA.

Macedo, A., (2002), *Team Teaching: Who Should Really Be in Charge?*, Master of Arts Dissertation, University of Birmingham.

Masuda, Y., (1980), *The Information Society as Post-Industrial Society*, Transaction Publishers.

Max Neef, M.A., (2005), Foundations of transdisciplinarity, *Ecological Economics*, Vol. 53, pp. 5– 16.

Mătieş, V., Szabo, F. & Besoiu, S., (2005), Information links, flexibility and reconfigurability of the mechatronic systems. *The 9th IFToMM International Symposium on Theory of Machines and Mechanisms SYROM 2005*, September, 1 – 4, Bucharest, Romania.

Mătieş, V., Bălan, R. & Stan, S.D., (2008), Mechatronic Education in the Study of the Mechanisms, *The 7th France-Japan and 5th Europe-Asia Mechatronics Congress*, pp.145, May 21-23, France.

Mătieş, V., (ed.), (2009), Platforme mecatronice pentru educaţie şi cercetare (Mechatronical Platform for Education and Research), Editura Todesco, Cluj-Napoca.

Mandru, D., Lungu, I. & Tătar, O., (2008), An Approach to Problem-Based Learning and Team Teaching in Biomechatronics, *The 7th France-Japan and 5th Europe-Asia Mechatronics Congress*, pp. 133, May 21-23, France.

McGregor, S. & Volckmann, R., (2010), Transdisciplinarity in Higher Education, The Path of Arizona State University, *Integral Leadership Review*, vol.X. No.3.

McLuhan, M., (1962), *The Gutenberg Galaxy: The Making of Typographic Man*. Toronto: University of Toronto Press.

Mieg, H.A., (1996), Managing the interfaces between science, industry, and education. In UNESCO (ed.), *World Congress of Engineering Educators and Industry Leaders* Vol. I, UNESCO,Paris, 529-533.

Minor, M. A. & Meek, S. G., (2002), Integrated and Structured Project Environment in Mechatronics Education, *Proceedings of the American Society for Engineering Education Annual Conference & Exposition*.

Mittelstrass, J., (2004), "A new paradigm for relations between higher education and research", *The Europe of Knowledge 2020 "A Vision for University Based Research and Innovation"*, Liège 25 – 28, April.

Montaud, A., (2008), Le monde de mecatroniques, *Artema Mechatronics Infos*, pp. 1-2. Mori, T., (1969), Mecha-tronics, Yasakawa Internal Trademark Application Memo 21.131.01., July 12.

Mortensen, M. & Hinds, P.J., (2002), *Fuzzy teams: boundary disagreemant in distributed and collocated teams. Distributed works.* P.J. Hinds and S. Kiesler, Cambridge MA, MIT Press.

Mushakoji, K., (1988), Global Issues and Interparadigmatic Dialogue. Torino, Albert Meynier.

Nicolescu, B., (1996), *La Transdisciplinarité*, Rocher, Paris, (English translation: Watersign Press, Lexington, USA ; Ed. Polirom, Iaşi, 1999).

Nicolescu, B., (1998), *Gödelian Aspects of Nature and Knowledge*, in "Systems - New Paradigms for the Human Sciences", Walter de Gruyter, Berlin - New York, edited by Gabriel Altmann and Walter A. Koch.

Nicolescu, B., (2006), *Transdisciplinarity – Past, Present and Future*, în Haverkort B., Reijntjes C., Moving Worldviews - *Reshaping Sciences, Policies and Practices for Endogenous Sustainable Development*, COMPAS Editions, Holland, pp. 142-166.

Nicolescu, B., (2008), *Transdisciplinarity -Theory and Practice* (Ed.), Hampton Press, Cresskill, NJ, USA.

Nicolescu, B., (2010), Methodology of transdisciplinarity, *Transdisciplinary Journal of Engineering and Science*, 1(1), 19-38.

Nonaka, I. & Takeuchi, H., (1994), "A Dynamic Theory of Organizational Knowledge Creation", *Organizational Science* 5, No. 1, pp. 14–37. Onwubolu, G. C., (2005), *Mechatronics: principles and applications*, Technology & Engineering, Oxford, UK.

Palmer, S.E., (1978), *Fundamental aspects of cognitive representation,* in E. Rosch, & B.B. Lloyd (eds.), *Cognition and Categorization,* Erlbaum, Hillsdale, pp. 259–303.

Papoutsidakis, M.G., Andreou, I.K. & Chamilotoris, G.E., (2008), Challenging educational platforms in mechatronics: learning framework and practice, *Proceedings of the 5th WSEAS/IASME International Conference on Engineering Education*, pp. 121-126.

Peters, J. & Van Brussel, H., (1989), Mecatronics revolution and engineering education, *European Journal*, vol. 34, nr.1, pag. 5-8.

Polanyi, M., (1997), *"The Tacit Dimension,"Knowledge in Organizations*, L. Prusak, Editor, Butterworth-Heinemann, Woburn, MA.

Pons, J.L., (2005), *Emerging Actuator Technologies*, John Wiley & Sons, New York.

Pop, I. G., (2008), The semiophysical model of contextual synergistic communication in mechatronical knowledge, *The 11th Mechatronics Forum Biennial International Conference*, University of Limerick, Ireland.

Pop, I. G. & Mătieş, V., (2008), A transdiscpilinary approach to knowledge in mechatronical education, *The 7th France-Japan and 5th Europe-Asia Mechatronics Congress*, pp. 195-198, Grand Bornand, France.

Pop, I. G. & Mătieş, V., (2008a), A transidiciplinary approach of the mechatronical education in the context of knowledge based society, *"Problems of Education in the 21st Century"*; 8(8), pp. 90-96, Lithuania.

Pop, I. G., (2009), Creativity as a reflexive transdisciplinary language in the mechatronical knowledge, *"Problems of Education in the 21st Century"*, vol. 13, pg. 94-102, Lithuania.

Pop, I. G., (2009a), Considerations about the transdiciplinary knowledge search window in mechatronical education, *The 5th Balkan Region Conference on Engineering Education & 2nd International Conference on Engineering and Business Education*, pg. G159-165, Lucian Blaga University of Sibiu, Romania.

Pop, I. G. & Mătieş, V., (2009), Considerations about the mechatronical transdisciplinary knowledge paradigm; *The 5th IEEE International Conference on Mechatronics*, pp. 1-4, Málaga, Spain.

Pop, I. G. & Mătieş, V., (2010), Sustainainable strategies in mechatronical education as vocational training environment, *"Problems of Education in the 21st Century"*, vol. 19(19), pp. 94-102, Lithuania.

Pop, I. G. & Veres, O. L., (2010), *Elemente de Semiofizică Aplicată (Elementary Applied Semiophyics)*, Risoprint, Cluj-Napoca.

Quinsee, S. & Hurst, J., (2005). Bluring Bounderies? Support Students and Staff within an Online Learning Environment? *TOJDE (Turkish Online Journal of Distance Education)*, Vol.6, Nr.1, pp. 52-59.

Rainey, V.P., (2002), Beyond Technology – Renaissance Engineers, *IEEE Transactions on Education*, vol. 45, 1, pp. 4-5.

Ramo, S. & St Clair, R, (1998), *The system approach*, KNI, Anaheim, CA. Reason, P., (1998), Political, Epistemological, Ecological and Spiritual Dimensions of Participation. *Culture and Organization*, 4 (2), pp. 147-167.

Rzevski, G., (1995), Mechatronics - Designing Intelligent Machines, Vol. 1, Butterworth-Heinemann Ltd., England.

Savary, J., (2006), Overview of Problem-based Learning: Definitions and Distinctions. *The Interdisciplinary Journal of Problem-based Learning*, Vol. 1(1), pp. 9-20.

Schäfer, S., (1996), *"Making Up Discovery"*. In M. Boden (ed.), *Dimensions of Creativity*. Cambridge, Ma, and London: The MIT Press., pp.13-51.

Scott, P., & Gibbons, M., (2001), *Re-Thinking Science. Knowledge and the Public in an Age of Uncertainty* , Polity Press, Cambridge.

Shetty, D., (2002), *"Designing for Product Success"*, SME Publications.

Siegwart, R., (2001), Grasping the Interdisciplinarity of Mechatronics, *IEEE Robotics & Automation Magazine*, Vol. 8, No. 2, June, pp. 27-34.

Stiffler, A. K., (1992), *Design with Microprocessors for Mechanical Engineers*, McGraw-Hill.

Tahemaa, T., (2004), Trends in positive and negative synergy at mechatronic systems "Industrial engineering – innovation as competitive edge from SME" 29 –30th April 2004, Tallinn, Estonia.

Toffler, A., (1983), *The Third Wave*, Bucharest: Political Press (romanian translation).

Waks, S., (1997), Lateral thinking and technology education, *Journal of Science Education and Technology*, 6(4), pp. 245-255.

Wenger, E. & Snyder, B., (2000), "Communities of Practice: The Organizational Frontier," *Harvard Business Review* Vol. 78, No. 1, pp. 139–145.

Wikander, J., Torngren, M. & Hanson, M., (2001), The Science and Education of Mechatronics Engineering, *IEEE Robotics and Automation Magazine*, 8, 2, pp. 20-26.

Yamazaki, K., & Miyazawa, S., (1992), "A Development of Courseware for Mechatronics Education", *International Journal of Engineering Education*, Vol. 8, 1, pp. 61-70.